AMERICAN AGRICULTURE

AMERICAN

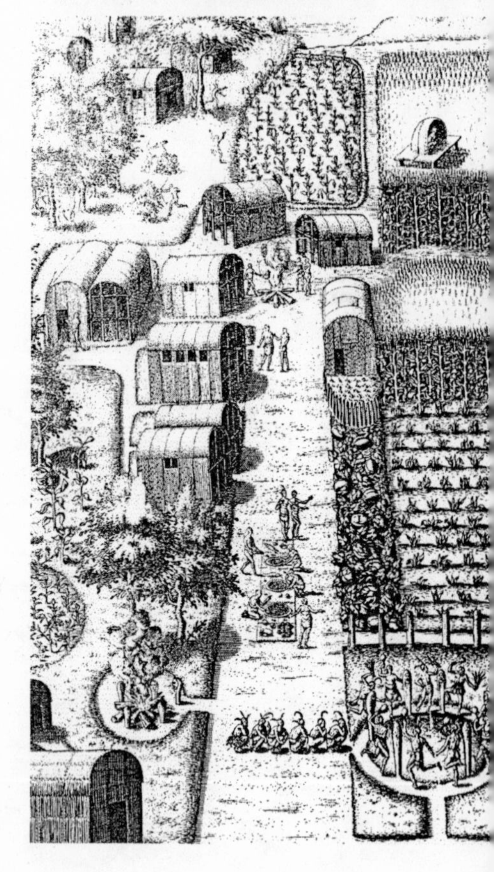

AGRICULTURE A Brief History

Revised Edition

by

R. DOUGLAS HURT

Purdue University Press
West Lafayette, Indiana

FOR **MARY ELLEN, ADLAI,** AND **AUSTIN**

R. Douglas Hurt is professor and director of the graduate program in Agricultural History and Rural Studies at Iowa State University. He is the author of ten books, including *The Dust Bowl, Indian Agriculture in America,* and *Agriculture and Slavery in Missouri's Little Dixie.*

♾ Printed on acid-free paper in the United States of America

Revised edition, 2002

Library of Congress Cataloging-in-Publication Data

Hurt, R. Douglas.
American agriculture : a brief history /
by R. Douglas Hurt.—rev. ed.
p. cm.
Includes bibliographical references and index.
ISBN 1-55753-281-8 (pbk. : alk. paper)
1. Agriculture—United States—History. I. Title.
S441.H918 2002
338.1'0973–dc20

CONTENTS

PREFACE

This book is intended for anyone who is coming to the history of American agriculture for the first time. It is also designed as a ready reference for the most important economic, social, and political developments in American agriculture. As such, it is an introduction and a guide that will provide the background necessary to understand the complexity of American agricultural history and keep the reader on course, directing the way to more advanced reading, research, and learning. Any book that attempts to survey more than four hundred years of American agricultural history must be sweeping in its approach but sufficiently specific to make the generalizations sound. Anything less would be superficial and useless. Although broad trends and specific developments are most easily recorded and manipulated by statistics to convey a great deal of information as briefly as possible, I have avoided emphasizing production statistics such as bushels and bales per acre, the number of livestock grazed or fed, the volume of exports, and various monetary compilations. Agricultural histories that use this technique are invariably dull. Moreover, the reader can easily lose the significance of the matter at hand by becoming overwhelmed with details, facing the proverbial problem of not being able to see the forest for the trees.

To avoid a tedious statistical approach, I have written a narrative. Not all production or related statistical matter has been omitted because it is often unavoidable or the specific point cannot be made without doing so. Instead, I have attempted to limit statistical discussions by stressing major topics in a chronological framework. These topics and corresponding economic, social, and political themes are carried through each chapter to give continuity to the study.

Because my primary intent has been to discuss the major developments in American agricultural history, I have not been able to trace every matter from its origin to its present condition, nor have I been able to document the text because of space limitations. Ideally, a synthesis of American agricultural history should extend through several volumes, but my task has been to discuss the essential developments of that history in one book. However, I have provided a list of suggested readings for each chapter. These essential studies will take the reader to the next level of study. Moreover, the Bibliographical Note will help the introductory student of American agricultural history to begin research on a variety of topics by consulting the appropriate aids.

It would be easy to write a laudatory history of American agriculture, and it would be basically accurate, as far as it went. But when considering the entire history of American agriculture, such an approach would be not only wrong but intellectually dishonest. Certainly there is much to praise about American agriculture during its long history. The production of American farmers remains unsurpassed by agriculturists in any other nation. Their risk taking, entrepreneurial spirit, and courage enabled the settlement of the greater part of the North American continent, and they helped build a strong nation to which other countries turn for guidance and aid—that is, leadership. Certainly the United States is respected regarding its agricultural achievements.

Yet America's agricultural history is not a story of continuous progress or moral certainty. Because it is a story of human actions over time and in the context of culture and place, it has been shaped by the best and the worst of human nature. Certainly inventive minds and skillful hands of men and women alike have played an instrumental role in American agricultural history, but not all have been as magnanimous as John and Mack Rust, who wanted to ensure that their cotton picker would not cause unemployment. Ideally, naively, and even nobly, they hoped to restrict its use with written controls designed for social ends. The history of American agriculture is also a story of daring fraud, insatiable greed, ruthless speculation, vicious racism, malicious corporate power, callous government policy, and incredible violence. Nor is it just a story of bountiful harvests that have conquered hunger and created a well-fed nation; it is also the history of poverty, hunger, and want. The history of American agriculture, then, is a story of achievement and failure as well as desperation and hope—that is, life. Much has been written about it, yet much remains to be done.

In any synthesis of this nature and scope the author must rely on the work of many historians, political scientists, economists, and sociologists who have gone before. Fortunately for the historian of American agriculture, the books and articles published since 1919, when the Agricultural History Society organized and essentially founded this subfield of American history, have been far-reaching and detailed as well as sound and lasting. Much of that research and publication has been devoted to economic and political affairs on local, regional, and national levels, although social history has become an important aspect of agricultural and rural history during the past twenty years. Unfortunately, not all locales, states, or regions have been studied by agricultural and rural historians to the extent that anyone who attempts a synthesis like this would wish (the agricultural and rural history of the twentieth-century West and Northeast are prime examples). Even so, enough grass-roots work has been completed to enable a historian to trace the main themes and developments of American agricultural history.

The writing of a synthesis, even a brief history, is filled with dangers because it must be broadly if not impressionistically written. With that recognition, I have taken a regional approach throughout while addressing major topics such as technological change, agricultural organizations, and rural life. This study is designed so that a reader can pursue only the topic or period of interest without reading the entire book. Or the book can be studied as a whole to gain an overview of the main developments in American agricultural history from approximately 5000 B.C. to the late twentieth century.

Although historians often debate whether farmers were subsistence or commercially oriented before the twentieth century, particularly in frontier or newly settled areas, agriculturists were always interested in both. Until the early twentieth century, many farmers sought to provide the subsistence needs of their families and to be as self-sufficient as possible. Yet commercial gain, which requires farming for profit, has dominated the thoughts, goals, and endeavors of most American farmers throughout the nation's history. When choices had to be made, farming for profit superseded agriculture as a way of life. Home manufacturing and production for family consumption often made good business sense, but not always. Although participation in a market economy might be selective at first, it eventually became essential for financial gain.

Commercial agriculture, not subsistence farming or self-sufficiency, enabled farmers to make the most productive use of their lands, encouraged technological and scientific change, and enabled financial gain. It also caused oppression of some farmers and agricultural workers, abuse of the land, and government regulation. Overall, then, the commercial bent of farm men and women shaped the economic, political, technological, scientific, and social history of American agriculture from the colonial period to the twenty-first century.

ACKNOWLEDGMENTS

Like most histories, this book has been in the making for a long time. Although I decided to write this study while serving as a Smithsonian postdoctoral fellow in the history of science and technology in 1976, it is always easier to get a good idea than to execute it. Between the beginning and end, however, I have happily incurred a number of debts, and I would be not only remiss but also ungrateful if I did not acknowledge them. Homer E. Socolofsky and John T. Schlebecker always proved supportive while reminding me of necessary cautions. Terry Sharrer at the Smithsonian Institution, Douglas Helms at the Soil Conservation Service, Douglas Bowers and Ann Effland in the History Section of the U.S. Department of Agriculture, and my colleague Alan I Marcus provided essential information. Michael W. Woolverton at the Thunderbird American Graduate School of International Management helped with a crucial map. James W. Whitaker, Thomas R. Wessel, and David B. Danbom read a preliminary draft of the manuscript. I am grateful for their critiques and the help of the anonymous readers. Wayne Petersen, Susan Congdon, Mary Jane Thune, and Kathy Patton in the Interlibrary Loan Department and Betty Erickson in the Special Collections of Parks Library at Iowa State University provided expert research help.

In writing this brief history of American agriculture, I have not only drawn on the essential works of my colleagues, both past and present, but relied heavily on my own work, and I am indebted to a number of publishers who have permitted me to use portions of previously published studies. Robin Higham generously granted permission to use portions of my work published in *American Farm Tools: From Hand Power to Steam Power* and *Agricultural Technology in the Twentieth Century.* He also provided a variety of research materials. Chelsea House Publishers enabled me to use portions of my *The Department of Agriculture;* selections from this study appear in revised form in the chapters on the antebellum and Civil War eras. In addition, Fred Woodard allowed me to draw upon my *Indian Agriculture in America: Prehistory to the Present* for Chapter 1. Christopher S. Duckworth permitted me to use portions of my work published in *Timeline.* The studies that I have drawn from are "REA: A New Deal for Farmers"; "Farmers at the Barricades"; "The Singing Plow"; and "From Superstition to Science: Veterinary Medicine." Robert Daugherty granted permission to use portions of my article "Ohio Agriculture Since World War II," which appeared in *Ohio History.* Iowa State University Press permitted me to draw upon my chapter "Northern Agriculture after the Civil War, 1865–1900," which appeared in *Agriculture and National Development: Views on the Nineteenth Century,* edited by Lou Ferleger.

A number of curators and archivists also provided essential aid. I am grateful for the help of Margery S. Long at the Walter P. Reuther Library, Wayne State University; Carolyn S. Parsons and Mary Sine, Virginia State Library and Archives; Carma Muir Berglund, Labor Archives and Research Center, San Francisco State University; Gail Miller, Georgia Department of Archives and History; Joy Werlink, Washington State Historical Society; LaVera Rose, South Dakota State Historical Society; Tani Graham and Vernon Will, Ohio Historical Society; John R. Lovett, Western History Collections, University of Oklahoma; Nancy Sherbert, Kansas State Historical Society; Christine Bobbish, Connecticut Historical Society; Genevieve Troka, California State Archives; John White, Southern Historical Collection, The University of North Carolina; Kathleen A. Correia, California State Library; Rebecca Kohl, Montana Historical Society; and Vicki Eller, Deere & Company Archives.

AMERICAN AGRICULTURE

1 • THE NATIVE AMERICAN EXPERIENCE

AMERICAN INDIANS BEGAN FARMING IN THE REGION OF THE CONTINENTAL United States perhaps as early as 5000 B.C. Indeed, Native Americans were the first farmers in the Western Hemisphere and on the North American continent. The most important Indian agricultural practices diffused north from Mexico into the American Southwest, Great Plains, South, and North, although local or indigenous agriculture also developed in the Midwest and Upper South. Indigenous agricultural practices included domesticating squash, sumpweed or marsh elder, sunflower, and chenopod during the second millennium B.C.,—all of which served as the most important cultigens until surpassed by maize or corn after A.D. 800.

By A.D. 1000, Indian farmers had developed a productive agricultural system based on corn, beans, and squash, along with a host of supplemental crops, that provided a dependable food supply and

BEFORE CONTACT with European civilization, the Native American cultures of the present eastern United States raised a variety of crops. This engraving by Theodore DeBry, after a painting by John White, shows a village in North Carolina that depended on agriculture. Note the plantings of corn, squash, sunflowers, and tobacco. *Library of Congress.*

reinforced the stability of villages. These farmers, who were primarily women east of the Mississippi River, became skilled plant breeders. They successfully adapted their crops to meet the environmental conditions of their region, developing corn varieties that would grow in the cool, moist region along the present Canadian border and in the hot, dry Southwest. They learned to clear land in the best manner their technology allowed and to store their crops for future use. They discovered the narcotic effect of tobacco and raised it for ceremonial purposes. In the Southwest, some cultural groups built extensive irrigation systems. This work took thousands of years to develop, but no one surpassed their achievements until the advent of modern agricultural science and the creation of the state agricultural experiment stations during the late nineteenth century.

THE NORTH

Before contact with white civilization, the Indians in the present northern United States cultivated the river valleys and flood plains where the soft alluvial soil could be worked relatively easily with bone and wooden hoes and digging sticks. The women often had the task of clearing brush and grubbing roots from the land. Indian women raised the traditional crops—corn, squash, and beans, the "three sisters."

Corn, however, was the most important crop, and they raised many varieties of flint, dent (flour), sweet, and popcorn—all selected according to the requirements of local growing conditions. The Huron, for example, developed a flint corn that matured in 100 days and a flour corn that ripened in 130 days to take advantage of the short, cool growing season along the Canadian border. When the corn dried, the women shelled the ears and stored the kernels in baskets. This most important food plant is a native of Mesoamerica; its cultivation spread northward after 3400 B.C. Corn growing may have moved northward along the east coast of Mexico into Texas and the Mississippi and Ohio valleys. Or the Indians may have introduced it from the American Southwest.

Contrary to popular belief, however, the Indians did not fertilize their corn with fish or teach the European immigrants to do so. Squanto apparently showed the Pilgrims how to use fish for fertilizer in 1621, but no evidence proves the Indians customarily followed this practice. Instead, Indian farmers abandoned exhausted crop lands and cleared new areas for cultivation. Squanto knew about fertilization, but he probably gained that knowledge while a captive in Europe. The Indians preferred to rotate their fields instead of fertilizing to maintain crop productivity. Some Indian farmers, such as the Huron, prolonged soil fertility by burning the brush in the fields before planting time. By using this technique, they added magnesium, calcium, potash, and phosphorus to the soil. Burning also reduced soil acidity, thereby promoting bacterial activity and formation of nitrogen. Native American farmers also planted beans in the hills where they seeded corn, and the bean crop also provided nitrogen to the soil. Frequent weedings prevented the growth of grasses that could not be chopped away with stone, bone, or shell hoes.

Among the Mohawk, at planting time in the spring, the women selected the seeds from the best ears of the previous harvest and planted the kernels in hills. Then the women erected watchtowers from which they and their older children guarded the crop from marauding birds. When the corn sprouted, the women cultivated the rows and at harvesttime pulled the ears from the stalks or uprooted the corn for removal of the ears later. They stripped the husks with the aid of a wooden peg. The Mohawk stored shelled corn in bark containers or underground pits, using the husks to make mats, baskets, and moccasins and to wrap bread for baking or boiling.

Among the Indians of the Midwest, the women gathered the corn in late summer and husked the ears, which they spread on mats outdoors to dry. When the corn dried, they threshed the kernels from the ears by beating them with sticks. The Indians of the Midwest, like those of the Northeast, traditionally stored corn in baskets that they placed in underground pits or caches. Those storage pits were frequently 8 feet in diameter and 6 feet deep. Each pit had an opening 2 or 3 feet wide, the top covered with grass and earth when the cache was full. Women had the task of placing the corn in storage. They also hollowed out pumpkins, cut them into slices about one inch thick, and dried the pieces in the sun. Then they tied the slices together and hung the cuttings from a rack to dry for several more days. Pumpkin slices preserved in this manner could be kept for several months.

Corn, squash, and beans, however, were not the only crops important to Indian farmers in the Northeast and Midwest. In the upper Great Lakes, the Ojibwa (Chippewa) and the Assiniboin sowed wild rice in marshes. The Ojibwa typically sowed as much as one-third of their rice crop in water averaging about 2 feet deep. The Assiniboin weeded the rice by pulling the water grasses that grew among the stalks. Since the rice stalks might rise more than 8 feet

above the water and the muddy bottom was impossible to walk on, the men poled their canoes into the rice fields, and the women gathered a bunch of rice stalks with a curved stick called a rice hoop and tied a bark string around it. By tying the wild rice into sheaves or bundles before it was ripe, the heads did not shatter into the water as the stems moved back and forth in the breeze. About a month later, at the end of September, each family returned to harvest the rice from its allotted area. Ojibwa women marked the boundaries of their rice fields with stakes about ten days before the rice was ready to harvest. Then the men poled their canoes up to the bunches of

CORN SERVED as the primary crop for Native American farmers. Women and children usually had the responsibility of driving away pests. Many farming villages erected watchtowers to make that job easier. *Author's collection.*

rice while the woman bent the heads over the center of the canoe and beat the grain out with a stick. A skilled rice gatherer would leave the unripe grain in the heads for harvest later. By so doing, these farmers gleaned a rice field several times until they harvested as much grain as possible.

Wild rice farming remained primarily the domain of women. Women harvested the grain and spread it out to dry on skins in the sun or dried the grain by parching it in a kettle or putting it in a bag hung over the smoke or heat of a fire. Sometimes they spread the rice on a mat attached to a 3-foot scaffold. Then they built a slow fire

IN THE REGION of the Great Lakes, several Native American groups harvested wild rice, but the Ojibwa and the Assiniboin sowed seed and pulled weeds from the water to raise the rice. The women had responsibility for the wild-rice harvest. *Author's collection.*

underneath the scaffold and periodically shook or turned the rice. By the end of the day, the rice had dried for threshing. To thresh the rice, these farmers dug a hole about 1.5 feet deep and 2 feet in diameter and lined it with deerskin or wooden staves. Next they poured the rice into the pit, and the men threshed the grain by treading on it or pounding it with a pestle. Once the rice had been threshed, the women winnowed it with a basket and the wind or a birch-bark fan. When the chaff had been removed, they placed the rice in birch-bark boxes or skin bags and stored it in underground pits. If the rice had been properly dried, it would keep for many years.

Some of the northern tribes also tapped sugar maple trees and boiled the sap into sugar. Among the Ojibwa, each family or group of two or three families claimed a stand of sugar trees, just as ch controlled a certain area of wild rice. The sugar-making season usually began in February or March and lasted about one month. Before the arrival of the Europeans, the Indians boiled the sap by dropping red-hot stones into wooden vessels or heating clay pots. Later, after contact with white traders, they used iron or brass kettles. The tribal groups engaged in maple sugaring commonly boiled granulated maple sugar with wild rice or corn or stirred it with bear fat to make a garnish for roasted venison. They also mixed it with water to make a sweet drink.

EAST OF THE Mississippi River, many Native American groups made maple sugar during the late winter. This practice extended from the Great Lakes into the Deep South along the Appalachians. The maple sugar provided a garnish for food and a refreshing treat. *Author's collection.*

Although these Indian farmers did not develop an agricultural system based on a combination of plants and animals, their achievements were considerable. The northern Indians showed the European settlers which plants to cultivate, particularly corn, beans, pumpkins, and tobacco, and to make maple sugar. The cultural significance of this transfer of knowledge is particularly important because Indian women transmitted the New World's agricultural tradition to white men. White farmers readily adopted the crops and techniques of the Indians and made agriculture the basis of their economy.

THE SOUTH

In the present southern United States, Indian farmers cultivated squash in the lower Tennessee River valley by 2000 B.C., and they raised squash and bottle gourds extensively for food utensils by 1500 B.C. Sometime between A.D. 500 and 1000, however, corn became the most important crop. By the time of contact with European civilization, the skill and knowledge required to raise corn had spread across the South, and the Indians had developed the concept of market centers for trade of agricultural products.

Southern Indians cultivated the flood plains of the rivers and streams where they could easily till the fertile sandy-loam soil with flint hoes and spades and wooden digging sticks. Wooden and stone tools merely loosened or stirred the soil rather than turning it over. This technology was not suitable for working the heavy clay soils of the uplands. Moreover, southern Indians farmed the flood plains because they knew these lands were the most productive.

The sandy-loam soils of the alluvial valleys in the lower elevations of the Piedmont, the Interior Low Plateau and the Cumberland Plateau, for example, contained flood-deposited silt that enriched the soil and enabled continuous cultivation without a fallow or resting period. These river-bottom soils also held sufficient moisture, even during periods of summer drought, to enable agriculture without irrigation or the wide dispersal of plants. These soils were intensively cultivated and highly productive, enabling the support of large, permanent populations. When the Spanish arrived in the early sixteenth century, they found many of the southern tribes located on fertile lands engaged in well-developed agricultural practices.

Indian agricultural practices in the American South resembled those in the Northeast and Midwest. Native American farmers cleared land by girdling trees and removing brush with the aid of fire, axes, and mattocks. The communal fields, no doubt, varied in size depending on the number of people who needed food, the work force available, and the fertility and openness of the land. Productivity varied as well, depending on soil fertility, precipitation, plant disease, and insect attacks.

Without a doubt, though, the southern Indians were skilled agriculturists who raised a substantial portion of their food. They produced a surplus to prevent starvation when the crops failed and to trade among themselves and white settlers. Certainly the Europeans recognized that the Native Americans were knowledgeable farmers because they appropriated Indian fields. Those fields needed minimal clearing, and the colonists knew that Indian farmers did not waste their time cultivating poor lands. Although river-valley soils were exceptionally fertile, southern Indians probably could not cultivate fields in the uplands for more than a decade before yields began to decline because of soil exhaustion. When that happened, they had little choice but to let their fields lie fallow or clear new lands and start over.

Indian farmers in the South used two agricultural methods—*intercropping* and *multiple cropping*—to achieve maximum yields from their small fields along the river valleys. By intercropping, they planted corn, beans, and squash in the same fields, using all the available space and enabling the crops to complement one another. Beans, for example, replaced the nitrogen the corn withdrew from the soil. Cornstalks provided a place for the beans to climb, and the squash vines covered the ground and slowed the growth of weeds. By multiple cropping, these farmers planted two crops in the same field. Indian farmers used this method primarily for early and late corn varieties. Like the Indian farmers in the Midwest and Northeast, the southern agriculturists also probably distinguished several strains of corn, based on color, shape, and size. Certainly they recognized two broad categories of corn—early and late. They picked the early corn while it was still green and roasted it on the cob. They used the late-ripening corn for bread.

In the South as in the Midwest and Northeast, corn, squash, and beans superseded all other crops in importance. The men prepared the soil, but the women had the responsibility of planting, weeding, and harvesting the crops. Moreover, women were entirely responsible for the cultivation of small fields or garden plots.

Among the Cherokee, however, the men assumed a greater role in agriculture; they commonly cleared the land and planted and harvested the crops. Although the Cherokee men served primarily as hunters and warriors, they apparently delayed their summer expeditions until after the corn harvest. The Cherokee division of labor is not entirely clear in relation to weeding, but the women and children probably performed this task, as they did in the other southern tribes at that time.

By the mid–eighteenth century, Cherokee women and children also raised hogs in pens. This livestock provided insurance against hunger if their men could not kill sufficient game. At that time, however, the Cherokee women raised few cattle because they did not fence their crops. Moreover, they associated that practice with European culture, in which the males raised cattle. The Cherokee women, however, owned the crops, hogs, and other agricultural produce. During the early colonial period, for example, English traders and soldiers purchased corn, poultry, and pork from Cherokee women and not the men.

During the late seventeenth century, some Indian farmers began raising cattle, acquired from the English colonists. They did not take particularly good care of their livestock because they did not put up hay or maintain pastures. An abundance of game may have hindered their livestock-raising endeavors, but one should not be too critical of Indian farmers in this respect because most white farmers also neglected their livestock. In any event, the integration of livestock raising into Indian agriculture took time, as did the adoption of the plow.

Without a doubt, southern Indians made an important contribution to the agricultural development of the South. Like the northern Indians, the Native Americans in the South showed the Europeans which seeds to plant on the best lands. They also taught the white farmers the importance of protecting the corn crop from rodents and the weather by storing it in cribs. One of the most lasting contributions of Indian farmers in the South to Anglo-American culture was the preparation of hominy. Certainly corn became their most important crop, and hominy served as a major food of the Cherokee, Creek, Chickasaw, Seminole, and many other tribes in that region.

Although no one is certain how much agriculture contributed to the nutrition of southern Indians, farming, together with hunting and gathering, made a major contribution to the food supply. While southern farmers relied on corn, beans, and squash, they also raised a variety of other cultigens, and after the arrival of the Europeans adopted peaches, figs, and watermelons. In time, they also adopted the European practice of raising poultry, swine, cattle, and horses, but domesticated animals and poultry did not begin to supplement wild meat and fowl in their daily diet until the nineteenth century. In the meantime, they cultivated their crops in a centuries-old manner while they resisted increasing demands for their land by white farmers.

THE GREAT PLAINS

Perhaps 1,500 years ago, several Indian societies practiced agriculture on the eastern fringe of the Great Plains. Probably they had learned agricultural techniques from the Hopewell and Woodland people in the Mississippi Valley. By no means, though, were those Plains Indians extensive agriculturists. Although they came together in the spring to plant a few crops along the flood plains of the rivers and streams, they relied primarily on hunting and gathering. When winter arrived, they disbanded their villages and scattered until planting time in the spring.

On the eastern fringe of the Great Plains in present Texas, the Caddo tilled the soil at the time of European contact. Although the Caddo were not part of the Plains culture and represented a transitional culture between the Indians of the Southeast and those of the Great Plains, they may have created the "hearth" from which Indian agriculture spread to the Great Plains. The Caddo, who collectively represented twenty-five farming tribes prior to European contact, organized their largest confederation between the headwaters of the Neches and Trinity rivers, where they were known collectively as the Hasinai.

The Caddo cleared the fields and tilled the soil with the aid of wooden and bone hoes; after the arrival of the French and the Spanish, iron hoes. The Caddo seeded two corn crops annually, which they harvested in May and July. They also planted beans, which they winnowed. The Caddo stored both crops in large reed baskets covered with a thick layer of ashes to prevent damage by insects and rats. Customarily, the Caddo kept a two-year supply of seed corn in reserve.

Similarly, about A.D. 1000, the Indians of the central Plains engaged in fairly well-developed agricultural practices. The archaeological evidence, however, is not clear about whether those activities were a continuation of earlier endeavors based on outside influences or were indigenous to the region. In any event, by the fourteenth and fifteenth centuries, the agricultural settlements of the central Great Plains had been established for a long time. By the eighteenth century, the riverine farming villages were transforming from small scattered communities to large fortified towns that may have been ancestral to the nineteenth-century settlements of the Arikara, Mandan, and Pawnee.

Indian agriculture on the central Great Plains resembled farming practices east of the Mississippi River. Planting time came when the wild plum thickets blossomed. Corn, beans, squash, sunflowers, and tobacco were the major crops. In the Missouri River region of Nebraska, farming tribes such as the Pawnee raised a dozen or more varieties of corn. Dent, flint, flour, and popcorn were common, and these farmers retained the purity of each variety by carefully selecting the seeds and planting those varieties far apart to avoid cross-pollination. The Indian agriculturists of the central Great Plains also raised fifteen or more varieties of beans and at least eight varieties of squash.

The hoe was the most important tool among these agriculturists. This tool differed from the implements the eastern and western farmers used because Plains farmers fashioned it from the shoulder blade of a bison. Because bone hoes were insufficient to break the tough, fibrous prairie sod, these farmers confined their agricultural activities to the flood plains. Their fields averaged 1–4 acres, and the women frequently traveled as far as 10 miles to tend the fields along the creeks and rivers.

The Mandan and the Hidatsa were the most agricultural people in the region. These farmers lived on the eastern fringe of the northern Great Plains in present North Dakota. They occupied permanent villages along the tributaries of the upper Missouri where the women farmed the rich alluvial soil. In this portion of the Great Plains, the Mandan began their agricultural activities as soon as the ground thawed in the spring. When the ice in the Missouri River broke up and the geese appeared, they cleared the ground with antler or willow rakes. The women, and sometimes the old men, removed the brush from the field, spread it evenly across the ground, and burned it. This procedure softened the soil and made it easier to dig.

The Hidatsa, who lived on the Knife River in North Dakota, planted corn in May. They used a digging stick, hoe, or fingers to make a hole about 4 inches deep into which they dropped seven or eight seeds. In contrast to other Indian farmers, the Hidatsa planted their hills on exactly the same spots where they had seeded the year before. Hidatsa women cultivated whenever weeds appeared and hilled the crop during the second hoeing. The women left uncut weeds on the ground unless the plants had seeded; in that case, they carried the weeds 15 or 20 yards beyond the edge of the field. During the growing season, the Hidatsa did not fertilize their fields, and they even removed horse manure from their lands because they believed it caused the weeds to grow.

When the corn ripened, Hidatsa women plucked the ears, which they piled in the center of the field. The next day, husking began, and the young men of the village aided those who wanted help in return for meat. Husking lasted about ten days, after which they braided the largest ears with the best kernels into strings. Then they dried the ears on platforms and shelled the seed corn by hand instead of sticks. Hidatsa women threshed most of the nonseed ears in skin booths beneath the drying platforms. Those booths prevented the kernels and cobs from flying about as the women threshed the ears with the sticks. Then they winnowed the chaff with baskets. The Hidatsa saved some of the stalks of green corn for fodder; after harvesting the mature corn in the autumn, they pastured their horses on the husks and stalks. By spring, almost all of the stalks had been eaten to the ground, making spring land clearing easier.

The Hidatsa planted squash in late May or early June after they had sprouted the seeds by wetting them and mixing them with the leaves of sage or buck brush. Next they spread the seeds and leaves over a grass mat placed on a buffalo hide. Then they gathered up the hide and hung it from a pole near the fire. After about three days, the seeds sprouted, and the women planted a pair of squash sprouts per hill, spacing each pair about 12 inches apart. Like the Mandan, the Hidatsa made these plantings on a hillside so that the rain would not pack the soil and prevent the sprouts from growing.

The squash harvest began just before the green corn was ready for picking. Like Indian farmers elsewhere, the Hidatsa and Mandan sliced the squash, strung it on spits, and dried it on platforms. After drying it for three days, the women strung the slices on dry grass cords and hung them in their lodges. On sunny days, they took the squash outside for more drying. When the slices were thoroughly dry, they placed them in parfleche (rawhide) bags and stored them in caches. The Hidatsa and Mandan saved the largest squash for seed, picking those squashes at the time of the first frost. The women removed the seeds, which they dried by a fire, sacked, and stored in caches. They boiled the seeds and flesh of nonseed squash for eating.

Northern Great Plains farmers planted the bean crop after the squash had been set out. These farmers planted beans like squashes, on the side of a hill. When the crop ripened in the autumn, the women pulled the vines and dried them for about three days, at which time they threshed the beans from the pods with sticks. They winnowed the beans and dried the crop for another day before pouring it into sacks. The Hidatsa recognized five varieties of beans—black, red, spotted, shield-figure, and white—and planted, threshed, and stored each variety separately to maintain purity. These women farmers were careful to save the largest and best-colored beans for seed.

Each Mandan and Hidatsa family usually planted one variety of sunflower, such as black or white, in the spring. They harvested the crop when the petals fell from the seeds and the backs of the heads turned yellow. The women cut the heads before the seeds were dry, dried the heads facedown on their housetops, and as the seeds loosened, threshed the heads with a stick. Then the Hidatsa stored the seeds in skin sacks. These farmers parched and pounded the seeds in a mortar to make a meal they boiled with cornmeal and beans.

The Hidatsa stored their crops in jug-shaped caches, which served the same purpose as cellars for white farmers. These pits, about 5 or 6 feet deep, were dug by the women. They lined the pits with grass and covered the floors with willow branches and grass. Dry willow ribbing and wooden pegs held the grass against the walls. The Hidatsa stacked the braided ears against the pit walls, then poured the shelled corn into the center and piled squash slices on top of it. Last, they dumped more shelled corn onto the squash to help protect it from moisture. The Hidatsa closed or sealed the openings with a hide cover, grass, another hide cover, and a layer of earth. They concealed the openings by raking ashes or refuse over the spots to hide the caches from enemies. These pits lasted for many years, and the Hidatsa used them repeatedly.

Corn raising became the most important agricultural endeavor, and Indian farmers on the northern Plains raised several varieties—flint, flour, and dent. The distribution of these varieties depended upon the environmental conditions which governed the agricultural practices of the specific tribes. Flint corn matures earliest and it is the heartiest type, and the most northern tribes cultivated a variety that matured in about ten weeks. Most of these farmers, though, preferred to raise one of the five or six varieties of eight-rowed flour corn. This corn was easier to grind and tasted better when eaten green. Flour or dent corns were raised primarily in the southern portions of the Plains. No family planted more than two or three varieties, and each variety was usually cultivated in a separate plot by a different family member, who spaced the plots from 60 to 100 yards apart to prevent cross-pollination.

Corn was important to the Great Plains farmers not only for subsistence but for trade. Indeed, the hunting tribes of the high Plains craved vegetables as well as corn. As a result, the agricultural villages in the northern plains became important trading centers. By trading, the agricultural tribes acquired buffalo meat, antelope hides, and flour made from the prairie turnip. Frequently, the Mandan, Hidatsa, and Arikara traveled into the high Plains to meet and trade with hunting tribes such as the Cheyenne and Arapaho. Only the northern farming tribes, however, traded substantial amounts of agricultural products with the hunting tribes.

In the central Plains, the Pawnee, Omaha, Ponca, and Otoes conducted little or no agricultural trade with the tribes of the high Plains or with their northern farming neighbors. Among the northern agriculturists, this trade was beneficial but hazardous. The Sioux, for example, good customers of the farming tribes, were temperamentally unpredictable. The agricultural tribes knew that any trade with them had to be conducted with utmost caution. With the arrival of the white fur traders and the American army, the Indian farmers of the Great Plains gained a new, more advantageous, and usually safer market for several hundred bushels of corn annually.

THE SOUTHWEST

Agriculture in the prehistoric Southwest began as early as 2000 B.C., if not before, when traders brought cultigens into this region from Mexico. By the beginning of the Christian era, the Indian farmers of the Southwest had made the seed selections and developed plant varieties best suited for the climatic conditions in that region, from the cool, moist mountains to the hot, dry desert. For example, the corn the people from northern Mexico first carried through the central Sierra Madre into the Southwest about 500 B.C. grew best in the wetter high elevations. This corn was pre-chapalote, a small-cob corn similar to the earliest domesticated corn in the Tehuacan Valley of Mexico.

Soon thereafter, the Mogollon farmers in the Bat Cave region of present New Mexico developed a new type or "race" of corn by crossing teosinte and perhaps *Tripsacum* with chapalote. The hybrid corn that resulted was more variable and drought resistant. It could be grown in the wetter high elevations as well as the dry lower regions. This new corn variety not only produced larger ears with more rows of kernels but sprouted from deep planting. Dry sandy soil necessitated this development. Planting depths of a foot or more prevented germination before late June or early July, so that the summer rains would water the new plants. If germination occurred in May as a result of occasional spring rains, the high temperatures and dry conditions that followed would ruin the crop. This corn provided the subsistence base for southwestern civilization, especially the Hohokam-Basketmaker (Anasazi) people.

About this time, the Hohokam, who lived along the Gila and Salt rivers in Arizona, also practiced agriculture. By 300 B.C., the Hohokam raised corn, a strain of which they brought into the area from along the Sierra Madre on the west coast of Mexico. Because they planted corn in the spring when temperatures frequently approached freezing and the late autumn harvest also occurred during a time of relatively cold weather, the Hohokam improved their corn to resist cool temperatures and tolerate the hot, dry summers.

This development, in part, enabled the Hohokam to seed two corn crops annually because they could plant their first crop in early spring and harvest their second in late fall. They accomplished this by letting various varieties with desired characteristics cross-pollinate to create a better genetic mixture. Then they perpetuated this new variety by selecting the best seeds for planting. Throughout the Southwest, after A.D. 950, Anasazi Pueblo farmers, for example, raised a corn variety created from a mixture of chapalote and maize de ocho. This new corn variety, exhibiting the characteristics of its parents, was especially well suited for higher elevations and more northern latitudes, and it maintained a good milling quality. As a result, these farmers expanded their agricultural practices across the entire present state of Utah.

The southwestern farmers also cultivated several varieties of squash and beans. In contrast to eastern farmers, the southwestern agriculturists did not cultivate beans among corn plants. Instead, they developed bush varieties that were self-supporting rather than vining. The development of bush beans was important because in the Southwest closely planted cultigens could not compete successfully for the limited soil moisture without irrigation. Consequently, Indian farmers seeded the bean and corn crops in separate locations where the plants did not compete with a second crop.

Besides corn, squash, and beans, the southwestern farmers also cultivated cotton. After Indian agriculturists in the Tehuacan Valley domesticated this cultigen sometime between 3400 to 2300 B.C., it spread rapidly as people learned to use it. Cotton probably reached the Hohokam settlement of Snaketown with the original settlers about 300 B.C. Since southern Arizona is well suited for cotton cultivation and the Hohokam were skilled at weaving it into cloth, they were probably instrumental in disseminating this cultigen among farmers throughout the Southwest.

After A.D. 700, the Anasazi, Mogollon, and Hohokam, among others, cultivated cotton. The southwestern Indians valued cottonseed for eating and vegetable oil. Furthermore, the dual value of cotton, food and fiber, in addition to its light weight, made it an important cultigen for domestic use and trade. No one can state with certainty, however, how much acreage the Indians devoted to cotton or how much seed the fiber produced from the annual crop, nor has anyone determined the precise range of cultivation. The cotton (*Gossypium hopi*) these farmers cultivated had a growing season that was the shortest of any variety. The bolls ripened within eighty-four to one hundred days, and it could tolerate relatively high altitudes and arid conditions.

Cotton, together with corn, beans, and squash, contributed to the development of a sedentary life. Cotton gave southwestern farmers a reliable source of fiber for weaving and thereby removed them from the whims of nature for gathering various fibrous plants. In addition to these crops, as early as A.D. 630, southwestern farmers may have sown the seeds of the tobacco plant *Nicotiana attenuata* in the ashes of burned-over areas. The tobacco variety *Nicotiana rustica,* which the eastern farmers raised, was not introduced into the Southwest until contact with Spanish civilization.

The searing temperatures of the desert meant that Indian farmers had to contend with a short growing season. Corn had to mature within sixty days, or the crop might be lost to heat and drought. Indeed, without irrigation, the combination of high temperatures and scant precipitation gave the Southwest the shortest growing season on the North American continent. Moreover, the areas of the Southwest that receive between 10 and 20 inches of precipitation annually do so in July and August during the middle of the growing season. Consequently, southwestern farmers planted their crops so that the plants could make the best use of that moisture when it arrived.

Along the flood plains, however, planting usually began in June or early July after the rivers receded from May and early June overflows. Although this procedure meant that the planting season came during the time of the highest temperatures, the moist soil enabled the crops to make a quick start, promoted rapid growth, and usually ensured a harvest, even if no precipitation fell until the crop matured. Without irrigation, though, these farmers could raise only one crop annually. When they learned the principles of irrigation, they would be able to use the long frost-free or growing season (e.g., 357 days at Yuma, Arizona) to produce two or three crops annually. Until they began irrigation, however, the areas that were nearly frost-free were little more suitable for agriculture than regions where frost restricted the growing season.

The Hohokam stored corn in baskets and pottery vessels. Pottery storage jars were particularly convenient because the openings could be sealed to prevent rodent, insect, and moisture damage. The Anasazi, in contrast, used storage pits and fashioned granaries among the cracks, niches, and overhangs in canyons. If the soil of these protected areas was too sandy or crumbled easily, they scooped out holes and lined them with flat rocks or clay in a manner similar to the lining of storage pits with bark by eastern Indian farmers. They used flat stones, clay, or mud to cover the openings. These granaries were usually about 2 to 3 feet in diameter and 3 feet deep; some had walled fronts. Indian farmers probably used granaries for temporary storage until the corn could be carried to their homes.

Although the farmers in the prehistoric Southwest were skilled in adapting plant varieties that would survive best and yield most in the hot, dry environment and storing their crops, they also enhanced the success of their agricultural activities by using irrigation. These farmers used several methods. One, the very basic use of floodwater, simply involved planting crops in areas inundated with floodwater during periods of rainfall.

Farmers like the Pueblo planted their fields at arroyo mouths; as the floodwater emptied from these small gullies or canyons, it spread across the alluvial fan and watered the crops. Some farmers planted their crops on the broad floodplains of river valleys. When a river overflowed its banks, it gave the soil a moisture supply that enhanced the planting process or irrigated the growing plants. In either case, the overflow had to saturate the soil but not wash away the crops. Consequently, this method required southwestern farmers to have an excellent understanding of the physiographic features of their environment and the principles of runoff and stream flow.

Among the southwestern farmers, however, the Hohokam exercised the greatest irrigation skills. In the Salt and Gila river valleys, an area known as the "land of the canal builders," they constructed major irrigation canals that stretched for 150 miles or more. These canals rivaled those Old World farmers made along the Tigris and Euphrates rivers. The longest of these canals in the Salt River valley extended approximately 14 miles, measured 30 feet from crown to crown of each bank, and reached depths of 10 feet or more. The Hohokam, however, did not build reservoirs to store water. Therefore, they had to channel overflow or stream flow from the rivers into the canals. Although they were able to control water flow over a vast area, they evidently never learned that a reservoir would enable more systematic irrigation over even greater areas at specific periods. Nevertheless, their technical feat in constructing such large canals is remarkable, and no one surpassed their engineering techniques until the nineteenth century.

The Hohokam did not have modern surveying equipment, of course, but they apparently had an understanding of the principles of irrigation when they arrived in the Salt and Gila valleys, because they began the construction of irrigation canals almost immediately. No one can be certain about the details of that construction. No doubt it involved a good deal of trial and error. The construction of these canals, though difficult, probably did not take as much effort or as many workers as one might expect. They probably solved the problem of determining the grade for the main canals by admitting water from the rivers into the excavating area. The water flow would have indicated the direction for further excavation and softened the soil to make digging easier.

Hohokam men and women probably dug the irrigation ditches 1 to 2 feet wide at the beginning, then increased the depth and width of the ditches. In other words, the Hohokam probably did not set out immediately to construct a canal 12 miles long that was 10 feet deep and 30 feet wide. Rather than undertaking such an incredible task, it is more likely that a small group of men and women built the canals on a more modest scale. Fifty men and women, for example, each removing 1 lineal meter of earth per day could have constructed a canal 2 meters wide, 1 meter deep, and 2 to 3 miles long, such as the Pioneer Canal near Snaketown, in about a hundred days. A labor force like this would have enabled to Hohokam to complete a small but substantial irrigation network within the first year of their arrival and in time to irrigate their first crop. This is not to say that canal building was easy for the Hohokam. It must have been excruciatingly hard work because their implements were stone hoes without handles, ironwood digging sticks, cottonwood spades, and wooden baskets.

Although the Hohokam probably constructed their canals over a long time, they apparently built this irrigation system without the aid of a highly centralized society, even though the repair of leaks and the dredging of silt probably required some sort of organized communal effort. An intervillage decision-making process perhaps determined water allocations. Archaeological evidence suggests that small family groups may have been responsible for the repair of specific portions of the canals. Certainly some system of control and maintenance would have been necessary for a canal network that spread in veinlike fashion over several hundred miles, but the precise methods of control and organization remain a mystery.

Archaeological evidence, however, clearly indicates that the Hohokam built brush and sand dikes or weirs to channel water from the rivers into the main canals. If the canals began to leak, they used adobe to patch the seepage area, then burned brush on top of it to harden the clay. The Hohokam lined part of one canal near Pueblo Grande, Arizona, with clay to minimize water loss from percolation while it flowed to the fields. Although lining the canals with clay was probably not a common practice because of the labor required to complete the task, the use of clay for lining indicates the technical ability of the Hohokam to solve problems associated with irrigation agriculture.

No one can be certain how many acres the Hohokam irrigated; only the basic engineering principles of this system are known. No one knows how they applied the water to their fields. Did they use furrows to channel the water from the lateral ditches into crop rows? Did they flood the entire field with a sheet of water? Or did they use some other method? Nor does anyone know whether they used water to wash away the accumulated salts from the fields. Probably they used some system for washing the salt away; otherwise soil fertility would have been depleted long before their civilization became extinct in the fifteenth century. No one knows how frequently the Hohokam irrigated their lands or how much water they applied to their crops at one time. Moreover, dry and silted-in canals would have affected the amount of acreage under irrigation. Thus, without knowing the method of application and the amount of water used, no one can accurately estimate the acreage under irrigation.

If the Hohokam began irrigation in the Salt and Gila river valleys at the time of their arrival, they would have practiced irrigation agriculture for more than fifteen centuries before their civilization disappeared. Archaeologists conservatively estimate that the Hohokam irrigated their fields for at least five to six hundred years, approximately from A.D. 800 to 1400. Certainly irrigation farming in the Southwest was well established long before it peaked sometime between A.D. 1200 and 1400.

In addition to floodwater and irrigation farming, the people of the Southwest developed important conservation techniques that aided agricultural productivity and stability. In contrast to the Hohokam, most farmers, such as the Anasazi and Mogollon, lived where water was scarce. To farm successfully, they had to develop water conservation techniques that enabled maximum use of the available moisture. Accordingly, they developed several conservation methods that capitalized on floodwater runoff. Indeed, water was so important that they devoted considerable time and effort to developing conservation practices that helped ensure a bountiful harvest, even during times of drought. They built walled terraces in a step fashion by placing boulders along hillsides. These walls trapped runoff and held the water until much of it sank into the soil. The plots these terrace walls formed were small, some no more than 8 by 10 feet. The Anasazi constructed similar terraces, called check-dams, in the upper parts of watercourses where the grade was not particularly steep or the runoff swift. They carefully anchored the ends of these dams to hold the structures against water rushing down the stream or arroyo bed. The Indians built these rock dikes sufficiently high so that the water would deposit silt behind them to a depth of 3 to 4 feet, thereby creating a well-irrigated garden plot with rich soil. Excess water passed down the hillside to be trapped behind other terraces until as much water as possible had been used. These terraced and check-dam plots were suitable for planting supplementary crops of corn, beans, and squash or a speciality crop.

Over time, then, the Native American farmers in the Southwest improved their crops by using great care and skill in the selection of seeds. The Hohokam, particularly, left a distinguished mark on the history of Indian agriculture. These "Canal Builders" engineered an irrigation system on a twentieth-century scale. They not only established the only true irrigation culture in prehistoric North America but also made agriculture practical where farming would have been tenuous without it. Irrigation requires greater knowledge and ability than dryland farming. These achievements enabled southwestern farmers to gain increasing control over their environment and to produce a dependable supply of food and fiber. By A.D. 500, agriculture provided an important subsistence base for the Anasazi, Hohokam, and Mogollon people.

After contact with the Spanish in the mid–sixteenth century and with the influence of the missionaries, Indian agriculture in the Southwest became more diverse as these farmers adopted European crops and livestock. Still, although Indian agriculture underwent fundamental change, southwestern farmers maintained the best agricultural practices their ancestors had developed over the centuries before European contact. They would continue to pass that knowledge from generation to generation.

THE NORTHWEST

The agricultural practices of the Indians in the Pacific Northwest were modest compared to farming practices among Native Americans elsewhere on the continent. This is not to say that agriculture in the Pacific Northwest was unimportant but that it was not extensive. Among the Shasta, for example, tobacco cultivation involved burning brush off the site to encourage a better crop. The Klamath River Shasta thinned their tobacco plants to encourage better growth, and during times of drought they irrigated the crop with the aid of baskets. As early as 1813, the Willamette Valley Indians raised a blue-eared corn that grew on short stalks.

Prior to European contact, the Salish, who occupied the Whidbey and Camano islands where the Strait of Juan de Fuca meets the Puget Sound, raised camas. They ate this small bulb much like a potato. The Salish planted and harvested camas bulbs in prairie areas cleared with fire. By 1840, however, nearly all the women of the Salish cultivated potatoes. They may have received this crop from the Pacific Fur Company, which planted potatoes in the Astoria area as early as 1811. Because many of the company's employees married Salish women, these women may have learned to cultivate potatoes from their husbands, then passed that knowledge to their villages. Moreover, the fur company may have encouraged the Indians to raise potatoes to provide a food reserve for the traders in time of emergency. Potatoes may have been introduced into the region as early as 1827 by the Hudson Bay Company at Fort Langley on the Fraser River or by the British during the 1830s at Fort Nisqually on the upper Puget Sound. The Salish women quickly saw the superior productivity of this new crop and adopted it to replace the camas. Their experience raising the camas could easily be applied to the cultivation of potatoes. In any event, potato cultivation was an activity for women who used the traditional digging stick to loosen the soil for planting and harvesting.

The potato patches in the Pacific Northwest were small, usually located in natural clearings called prairies, on river islands, and along upper river valleys. In 1854, the Duwamish cultivated about 30 acres of potatoes near the outlet of Lake Washington. From that acreage they produced 3,000 bushels, an average of 100 bushels per acre. They sold a portion of that crop to the white settlers in the area. The potato crop was important for the Indians in the Pacific Northwest because it provided a cash crop and was a tasty, starchy food they needed. As white settlers came into the region in increasing numbers, they apparently encouraged the cultivation of potatoes because they commonly purchased this vegetable from the Indians. In time, however, white settlers acquired the "potato prairies" for themselves, and their grazing livestock ruined the crops of these Indian farmers. Overall, however, Indian cultural groups in the Pacific Northwest remained reliant on hunting and fishing rather than agriculture until the reservation period, which began during the late nineteenth century.

LAND TENURE

In the northern United States the Indians adopted two forms of land tenure. First, villages claimed sovereignty or exclusive ownership over an area, which other bands recognized. When the soil, firewood, or game became depleted near the village, the Indians moved to another location in their territory. They might not return to that village site for generations, if ever. Second, in contrast to communal ownership of a large area of land, another concept of land tenure involved individual control of the gardens and fields within the general territorial boundary. Among the Iroquois and Huron, for example, the family lineage controlled and cultivated the land. Uncultivated land separated the fields of each lineage. The eldest woman of each lineage exerted overall control of the land, although the other women had usage rights to particular plots. Each lineage retained the right to use those fields as long as the village remained on the site and the women cultivated the fields. When the village moved, the cultivation rights to a specific parcel expired, and the village chiefs or headmen allotted new fields to each lineage based on the size of the lineage.

If an Iroquois family felt safe from attack, its members might not live in the village but reside outside its confines on the land they farmed. In this case, each family also controlled the land it cultivated, and the village probably claimed adjacent land the family intended to farm at a later date. Although this system of land tenure—that is, private control of land—could present problems of conflicting claims, land was plentiful, and disputes over control were probably rare.

Thus, ultimate land tenure depended upon village sovereignty over a particular area, and immediate individual control of a field depended upon actual occupation and use. If a plot was cleared of trees and brush and planted with crops, it was automatically removed from the communal domain as long as the family continued to use it. One might call this principle *squatter's rights.* If a field was abandoned, however, it reverted to communal or village control, after which it could be freely allotted to another lineage or claimed and farmed by someone else. All village members could use communal lands not under cultivation for hunting, fishing, berry picking, or wood gathering. Generally, land could not be sold or inherited. This land-tenure concept can be termed *use ownership,* because land belonged to an individual farmer for as long as he or she cultivated it. The effect of this system was to ensure an equitable distribution of property so that every family would have enough space to meet its needs while it prevented the development of land hoarding for wealth and status.

Land tenure among southern Indians resembled the land-use pattern in the Northeast and Midwest. There were no fixed rules concerning the size of the field that an individual or family might cultivate. Each village member could clear and farm as much land as he or she desired. As long as the land was cultivated, village custom protected it from encroachment by others. No one would think of seizing the land of another, but if a field was abandoned, anyone could claim it and begin to farm. No one could claim absolute right or title to a portion of land because it ultimately belonged to the village, which owned it communally. In the South as well as the North, women had immediate control of the fields they cultivated.

Among the Plains tribes in general, land was inherited from mother to daughter. In the case of the Hidatsa, if relatives did not cultivate the land after the death of the owner, anyone was free to claim it. Protocol required, however, that the woman request permission to do so from the deceased's relatives. The dead woman's son, mother, daughter, husband, or sister could grant that permission, although the entire family probably consulted about the matter. If another woman took the land, the transfer was not a sale but a gift.

In the desert Southwest, tenure differed slightly from that in the northern and southern portions of the continental United States. There, individual males could acquire land in four ways. First, the headman of a village might give land to an individual as compensation for work on an irrigation canal that brought new lands under cultivation. Second, a man could apply to the headman of the village for a plot of land. Third, he might develop a piece of unclaimed land for himself without a specific grant or allotment. Last, he might acquire land through inheritance. Inheritance and individual effort were the usual methods of obtaining land.

In contrast to eastern farmers, the women in the Southwest usually did not control the land. Their labor in the fields, however, gave them a right to a portion of the crop and the freedom to dispose of it without permission from their husbands. Besides control, inheritance was an important aspect of the Pima and Papago tenure. Since title to the land was vested in the men, it was inherited through the male line of the family. Theoretically, the entire family inherited it, and the land remained undivided as long as the widow lived. During that time, the eldest son was in charge of the farming operation. When the widow died, the male heirs divided the land, although an unmarried daughter had the right to claim a portion of it. Daughters who married poor farmers might be given a portion of the land to aid them. If a couple remained childless, the nearest relatives of both divided the land among themselves. If a woman returned to her family after the death of her husband or a divorce, however, she forfeited all rights to his land. When Pima and Papago women married, they lost their rights to share the produce from the land because this culture assumed that husbands would provide for their wives.

Among the Hopi and Tewa, the headmen divided the arable land in the territory the village claimed and apportioned it among the clans—that is, groups of closely related families. These allotments might vary from several hundred square yards to more than a square mile. The lands of each clan were specifically marked with boundary stones placed at each corner or at the point where the fields adjoined the land of another clan. These stones or rock slabs might have the clan symbol carved on their faces. No markers were used to define family fields within clan lands. Disputes were rare because each clan claimed several fields not contiguous with those of the other clans. This landholding system also helped reduce the risk of crop failure; if one field failed, another might provide a harvest.

In contrast to the Yuman-speaking farmers, Hopi women controlled the land, inheriting it through a matrilineal system. In addition, whenever a woman's daughter married, the mother gave her a portion of land. Still, absolute control always resided with the clan; individual women had merely the right to use it. Although a woman's daughter inherited the land, the men, who took the primary responsibility for cultivating the fields, owned the livestock and passed that property to their sisters, brothers, or other relatives at their death. Land disputes were rare. A man would not demand the recognition of his rights to cultivate a field if another infringed upon it. The reluctance to do so, in contrast to the Yuman-speaking farmers, was based upon a superstition that if they pressed their right to the land, they would shorten the lives of their children or themselves. If a dispute did occur, the senior woman of the clan who controlled the lineage and its lands settled it. Large families were given the right to cultivate surplus lands or unneeded fields of smaller families. No rent or interest was required for that use.

Sometimes Hopi men owned land in their own right. They could do so if they cleared wasteland and planted fruit trees or crops on it. Since wasteland was considered part of the village territory, it was not under the control of a particular clan. Upon the death of the owner, this land reverted back to the village or passed to a son or the dead man's relatives. If this land was not cultivated continuously, another person might ask permission to farm it. The owner, however, could renew his right to cultivate that plot at any time. If the owner never reclaimed that land, however, everyone considered it to belong to the farmer who now cultivated it. If a clan became extinct, a related clan took control of the land and merged it with its own, or the fields might be passed to the woman who cared for the last survivor of the clan as payment for her services. Thus, although Hopi tenure was based on rigid principles of matrilineal inheritance, in practice it was flexible, meeting the needs of a family or clan.

Among the Zuni Pueblo, the concept of land tenure based on female ownership and inheritance also prevailed, but as among the Hopi, this principle was not an absolute rule. A man worked the fields of his wife's parents, but he might also cultivate land he had obtained from his relatives. If a man did own land, his daughter's and sister's children inherited it. Usually a man who owned land gave a field to his sister's son when the young man was old enough to cultivate it. He also gave a field to his sister's daughter at the time of her marriage.

Zuni women, however, always had the first claim to the land because they were expected to remain with the clan all of their lives. Matrilineal inheritance kept the family lands together under one household and prevented fragmentation. Occasionally, the Zuni family might lend land to another family in need of more crop production. These loans did not require payment or reimbursement. In contrast to most Indian farmers in the East or West, the Zuni recognized the sale of land among their own people and transferred it from one family to another permanently for payment in sheep, horses, blankets, or turquoise beads.

Among the Navaho the control of land was also based on use. Whenever a man or woman cleared a tract of land, he or she was ensured the right to use it thereafter, and that property passed to the clan. When a man died, his sister's son inherited the land, but if he did not have a nephew, the land went to a brother or a sister. If a woman held the right of use, it passed on to her mother, sisters, or brothers at her death. The first user of the land, or anyone who inherited it, had the right to farm it as he or she pleased. If the farmer did not plant the land or moved away for a period of years, the first user had the continuing right to cultivate it upon return, whether or not someone else was farming it at the time. Generally, a farmer would ask a relative to cultivate the land if he was going to be absent from his fields for a considerable time.

Several generalizations can be made concerning the nature of Indian land tenure. Title to a general territory was a group right, not an individual right. Usually the Indians of North America did not think of private property as an absolute individual right or consider land a commodity that could be bought, sold, or permanently alienated (transferred) in some fashion. Simply put, the community owned the land, and the individual created a control or use claim by cultivating a specific plot. If arable land was plentiful, the individual's claim lapsed whenever the land became exhausted or abandoned. This characteristic of tenure was common in the present eastern United States. In the Southwest, however, where the climate limited arable land, an individual's claim to the fields remained valid even when the land lay fallow.

Moreover, tenure or control was not vested in an individual but with the lineage or household. This control unit could be either patrilineal or matrilineal, depending upon the particular culture. Although a family or individual could claim additional land by clearing wasteland for cultivation, the lineage or clan had a paramount right to it. In this sense, tribal property was much like corporate property. Every village member was a landowner, but he or she did not have absolute control of it. An individual could not sell it, and among certain people, an individual could not lose his right to the land if he left it untended. If an individual cultivated the land, however, he or she could do so indefinitely without interference.

After contact with white civilization, various individuals or groups in a village occasionally gave white settlers permission to occupy their territory and to use certain lands. Those transactions usually involved payment in some form. Although whites almost always considered such transactions sales in which they obtained exclusive ownership, the Indians invariably regarded such proceedings as nothing more than temporary permits to use the land, pending compliance with the terms of the agreement. Since the Indians did not recognize individual rights to land other than the right of use or occupancy, thus making absolute ownership by individuals an impossibility, the individual could not sell or alienate the land in any fashion. Moreover, because the village did not have sole ownership of the land any more than past generations had absolute ownership, the tribal group could not alienate it.

Although the Indians did not recognize permanent land transfers to whites, they seldom acknowledged land transfers from one Indian to another through sale. Usually, land was transferred through inheritance or gift. Moreover, Indian farmers neither developed a system of land tenancy nor recognized easements or mortgages. Consequently, they did not need to develop rules concerning the relationships between mortgagors and mortgagees or landlords and tenants. Still, the Indian farmers recognized such contemporary aspects of land tenure as trespass, squatter's rights, waste, ejectment, or reversion and escheat. Indeed, the concept of trespass sometimes caused war among the tribes. *Squatter's rights,* a white American term, was practiced in principle by those who cleared land and claimed it for their use and occupancy.

Among the Navaho, the concepts of *ejectment* and *reversion* were recognized since a person could reclaim his land after a lengthy absence, even though someone else was then cultivating it. They also recognized the principle of *escheat;* in the absence of heirs, the land reverted to the village for redistribution and use by others. The Indian farmers, however, with the possible exception of the Havasupai, apparently did not recognize the concept of *eminent domain;* the village did not interfere with individual or family tenure.

The Indian concept of tenure enabled villages to make the best use of the land to meet their specific needs. Moreover, each people or culture developed a rational system for transferring land after the death of the user. The major problem with the American Indians' concept of tenure, at least for whites, was one of degree or scale. It was not that the Indians did not have a rational system for land use; their agricultural practices were sensible to them. Indian farming was sufficiently extensive and intensive to meet subsistence needs by supplementing hunting and gathering. After they produced a surplus, they traded with other tribal groups for meat and furs and with whites for a host of goods such as cookware, clothing, knives, and food products.

The problem with Indian lands was one of philosophy and scale. For the Native Americans, the land was a gift from the Great Spirit. Only he could take it away. Indian land could not be sold because it did not belong to the present generation. The present generation acted only as a trustee of the land for the generations unborn. In this respect, then, the Indians could neither understand nor accept the whites' concept of land sales, absolute ownership, and speculation.

Another problem of Indian land ownership was that they did not use enough of it for farming. Anglo-American farmers believed their agricultural system, which emphasized raising diverse crops and livestock, superior to Indian farming practices. Moreover, the new immigrants maintained that since the Indians had not improved (that is, farmed) all of the land they claimed, the Native Americans had no legitimate title to it. In reality, the Indians did use all the land they claimed. By hunting and fishing, they were able to provide the protein necessary to complement the vegetables they raised. The white newcomers could forgo hunting and fishing because they provided protein in the form of cattle and hogs. In addition, European and American technology (plows, cultivators, seeders, and planters), together with the aid of draft power from oxen and horses, meant that white farmers could cultivate more acreage than Indian farmers.

In addition, the white farmers were able to sell their produce in a market system that artisans, merchants, and others created at home and abroad. Market demand meant that commercial agriculture was possible, and white farmers wanted more land to earn profits that would improve their standard of living. White farmers, then, believed that the subsistence farming and limited agricultural trade of the Indians resulted from an inadequate or improper use of the land. Since Indian lands seemed undercultivated, whites were convinced that those lands should give way to a new tenure system that would make the land more productive. Soon after contact with white civilization, Indian farmers along the eastern seaboard learned that the ownership among the newcomers depended not on use but "pen and ink." Moreover, for the new farmers, the possession of land was exclusive, and all transfers were understood as irrevocable sales. The methods by which this new civilization acquired Indian lands ultimately ended traditional or indigenous agriculture as a way of life for Native Americans between the Atlantic and Pacific.

In retrospect, the history of Indian agriculture is the story of supreme achievement. Nearly three millennia before the arrival of white settlers, Native American farmers learned to cultivate plants of local and Mesoamerican origins. They discovered how to select the seeds that would yield maximum harvests in local soil and climatic conditions. By so doing, they made great strides toward farming in harmony with nature. While Indian farmers improved their agricultural methods, they also developed rational systems of land tenure. By placing ultimate control of the land with the tribe and allocating specific plots to individuals, the Indians were able to meet their communal and personal needs culturally and physically in the most efficient manner. In contrast to white civilization, group rights superseded individual claims, and property rights of women were far more extensive than under English common law.

The transmission of New World agricultural techniques from the Indians to the European settlers, however, eventually had a tragic effect on Indian civilization. Although white farmers readily adopted the crops and techniques of the Indians and made agriculture the basis of their economy, and while many tribal groups also adopted European agricultural practices, white farmers sought to become commercial agriculturists as quickly as possible. Profits required the extensive use of Indian lands. As a result, white civilization, with a market-oriented agricultural economy, together with an ever-increasing population, dictated that more lands come under cultivation. As white farmers moved westward, they seized not only tribal lands that the Indians held communally but the privately controlled cultivated fields. Thereafter, the farming culture of the American Indians began to disintegrate. As white farmers ultimately directed their efforts to economic gain, Native American farmers could not command the capital, science, technology, and land base necessary to compete effectively. By the late twentieth century, few American Indians could be considered farmers, a position far removed from the time when their ancestors were the most skilled agriculturists on the North American continent.

SUGGESTED READINGS

Castetter, Edward F., and Willis H. Bell. *Pima and Papago Indian Agriculture.* Albuquerque: University of New Mexico Press, 1942.

______. *Yuman Indian Agriculture.* Albuquerque: University of New Mexico Press, 1951.

Cronon, William. *Changes in the Land: Indians, Colonists, and the Ecology of New England.* New York: Hill and Wang, 1983.

Ford, Richard I., ed. *Prehistoric Food Production in North America.* Museum of Anthropology, Anthropological Paper 75, University of Michigan, 1985.

Forde, C. Daryll. "Hopi Agriculture and Land Ownership." *Royal Anthropological Institute of Great Britain and Ireland* 61 (1931): 357–405.

Hatley, Thomas. "Cherokee Women Farmers Hold Their Ground." In *Appalachian Frontiers: Settlement, Society & Development in the PreIndustrial Era,* ed. Robert D. Mitchell, 37–51. Lexington: University of Kentucky Press, 1991.

Hurt, R. Douglas. *Indian Agriculture in America: Prehistory to the Present.* Lawrence: University of Kansas Press, 1987.

Kinney, J. P. *A Continent Lost–A Civilization Won: Indian Land Tenure in America.* Baltimore: Johns Hopkins University Press, 1937.

Nabhan, Gary Paul. *Enduring Seeds: Native American Agriculture and Wild Plant Conservation.* San Francisco: North Point Press, 1989.

Parker, Arthur D. *Iroquois Uses of Maize and Other Plant foods.* New York State Museum Bulletin 114, 1910.

Smith, Bruce D. "Origins of Agriculture in Eastern North America." *Science* 246 (December 1989): 1566–71.

Struever, Stuart. *Prehistoric Agriculture.* Garden City, N.Y.: Natural History Press, 1971.

Trigger, Bruce G. *The Huron Farmers of the North.* New York: Holt, Rinehart and Winston, 1969.

Wessel, Thomas R. "Agriculture, Indians, and American History." *Agricultural History* 50 (January 1976): 9–20.

White, Richard. *The Roots of Dependency: Subsistence, Environment, and Social Change Among the Choctaws, Pawnees, and Navajos.* Lincoln: University of Nebraska Press, 1983.

Will George F., and George E. Hyde. *Corn Among the Indians of the Upper Missouri.* St. Louis: William H. Miner, 1917; repr., Lincoln: University of Nebraska Press, 1964.

Wilson, Gilbert L. *Buffalo Bird Woman's Garden* (Originally *Agriculture of the Hidatsa Indians: An Indian Interpretation.* Minneapolis: University of Minnesota Press, 1917). St. Paul: Minnesota Historical Society Press, 1987.

Yarnell, Richard A. *Aboriginal Relationships Between Culture and Plant Life in the Upper Great Lakes Region.* Museum of Anthropology, Anthropological Paper 23, University of Michigan, 1964.

2 • THE COLONIAL YEARS

AGRICULTURE BECAME THE PREEMINENT FEATURE OF THE ECONOMY IN the British American colonies. Nearly all the first colonists engaged in farming, and approximately 75 to 90 percent still practiced agriculture by the time of the American Revolution. Rural life and agriculture influenced everyone's existence, and farming provided a relatively high standard of living. Although farmers produced a variety of crops such as corn, wheat, barley, oats, and rye in addition to garden vegetables, and raised cattle, hogs, sheep, and horses, they were never totally self-sufficient, nor did they want to be subsistence farmers. Profits, wealth, and the improvement of living standards became their chief goals beyond meeting the basic subsistence needs of their families. Early in the colonial period, these farmers developed a "commercial *mentalité*," producing commodities for local, regional, and foreign markets whenever possible. Sometimes they exchanged their produce for goods in kind, that is, through barter, but they also sold their commodities for cash. In either case, this exchange became the key to agricultural profitability and expansion.

With land cheap and readily available from the Crown, proprietor, or New England town, or by squatting, colonial farmers could easily acquire 100 acres, of which they could clear and plant about a quarter. The uncultivated land remained in pasture or woods, to be cleared or plowed whenever the fertility of the cropland became exhausted. Although few farmers fertilized their land, they rotated cultivation of their fields rather than crops. This practice met their needs and enabled depleted lands to regain fertility. It was cheaper

THE AGRICULTURAL TOOLS that farmers used during the colonial and early national periods were hand-held or animal-powered implements. The plow, harrow, spade, hoe, sickle, fork, and ax were essential farm tools. *Smithsonian Institution.*

to wear out land and plant new fields than to hire expensive labor to farm old fields properly. In the mid–eighteenth century, Peter Kalm, a Swedish botanist traveling in the middle colonies, noted that "the grain fields, the meadows, the forests, the cattle, etc. are treated with equal carelessness. . . . [The farmers'] eyes are fixed upon the present gain, and they are blind to the future."

In contrast, German immigrants in Pennsylvania gained a good reputation for careful husbandry by using crop rotation, manure for fertilizer, and a fallow or resting period in a systematic manner. Moreover, the Germans usually stayed where they settled. Among them it was axiomatic that "a son should always begin his improvements where his father left off." In contrast, other immigrants often settled on land only long enough to earn a speculative profit after they increased its price by making modest improvements such as building a cabin and clearing a few acres. Whether they stayed or moved on, a majority of colonial farmers owned their land. Although poor farmers customarily rented their land, most farmers viewed tenancy as a temporary condition to endure only until they acquired enough capital to purchase their own acreage.

Just as colonial agriculturists sought markets and profits whenever possible, most farmers and village residents in colonial America were never totally isolated. Rather, farmers and other rural inhabitants were inextricably linked economically, socially, and culturally. The small towns and villages created local markets and provided avenues for commercial gain and social contact. As the population increased, towns like Boston, New York, Philadelphia, Baltimore, and Charleston became regional markets and ports for international trade in tobacco, rice, wheat, flour, and beef as well as cultural centers.

Overall, then, the majority of the early American farmers were careless, often mobile and profit-minded. They would remain so throughout most of the nation's history. Although carelessness would give way to conservation and the westering movement would end with the acquisition of arable public domain, American farmers were always governed by the potential for economic gain. Many colonial farmers agreed that America was the "best poor man's country in the world."

LAND POLICY

Agriculture dominated the use of the land in the British American colonies. Farmland provided the basis for subsistence, wealth, and status. It gave the owner a stake in society, respectability, independence, and often the right to vote. The ownership of land and its distribution to settlers depended on whether the colony was a royal, proprietary, or corporate entity. In the royal colonies the Crown established land policy; the proprietor or the people determined land tenure elsewhere. (*Land tenure* means the way property is held.)

Fundamental to the feudal land system of England was the principle that all land belonged to the king and could be granted to subjects and among subjects in a relationship of landlord and tenant. Usually the tenant owed the landlord certain services for the privilege of farming the land. In time landlords accepted a *quitrent,* money that released the tenant from services to his feudal lord. Simply put, this rent meant the farmer was "quit" of his feudal responsibilities. Although the Crown and various proprietors attempted to transfer this system to the American colonies, it never worked as planned. The land grants from the Crown and proprietors to settlers required a quitrent to generate income, but these rents were difficult to collect because of an inadequate bureaucracy and the distance of the colonies from government agents or proprietors. Colonial farmers studiously avoided paying this obligation.

In colonial America an individual could acquire land in primarily six ways: (1) through the *headright* system; (2) purchase of Crown lands; (3) New England survey of townships for a church congregation or corporate group; (4) a grant from the Crown to supporters, friends, or others to whom it owed a debt; (5) squatting; or (6) rent. With the exception of squatters and renters, who had no legal or only a contractual right to the lands they farmed, colonial farmers considered their ownership of land to be in *fee simple,* despite quit-rent responsibilities or inheritance laws: a farmer had full right to his land, subject only to the rights reserved by society, such as taxation, policing, and eminent domain. Land owned in fee simple could easily be sold by placing the appropriate words on a piece of paper. As the paper deed changed hands for a price, subject only to register with the appropriate government official, the land title transferred. In the colonies, the pattern of easy purchase and sale of land, unrestricted by inviolate inheritance laws, became a hallmark of American agriculture. The nearly effortless transfer of land helped open the frontier for family farmers and speculators, and enabled the accumulation of wealth and the foundation of subsistence and commercial agriculture.

In Virginia, New York, Maryland, Pennsylvania, Delaware, New Jersey, Georgia, and the Carolinas, the *headright* system became the usual route to land ownership. This system granted land to those who paid passage for themselves or others to a colony. In Virginia, the Crown usually provided 50 acres to anyone who paid the passage of an immigrant. Individuals who paid their own way also received 50 acres. Families were entitled to 50 acres for every member if they paid their own passage. These lands could be selected from acreage that had not yet been claimed or settled.

In 1705, Virginia settlers also gained the right to purchase Crown lands. Individuals with considerable capital could buy extensive holdings of relatively cheap land and create vast estates, such as those established by Robert "King" Carter, who purchased 300,000 acres, or William Byrd, who bought 179,000 acres. Land companies also bought public land for resale to settlers. Virginia did not survey or plat its public lands, and a settler had to search, locate, and ensure that no other claims were attached to land that he had purchased. Unfortunately, Virginia did not require settlers to file their claims, that is, officially register them in Williamsburg or Richmond. As a result, the haphazard system of claiming land by "metes and bounds"—the surveying method of running boundaries as one saw fit by using trees, rocks, and streams rather than rectangular surveys with the corners marked precisely on mathematical principles—kept the courts busy with farmers contesting overlapping claims.

By 1642, overlapping claims had become such a problem that the Virginia House of Burgesses passed a law that gave some protection to squatters. Without a system of survey before sale, a farmer could easily and accidentally settle on land that belonged to someone else. Although the law permitted the removal of illegal settlers, an owner had to pay the squatter for his improvements, such as a cabin, fences, and the costs of clearing the land, before eviction. The Virginia system for claiming land produced a patchwork of irregular fields that can still be seen in the late twentieth century.

The New England settlers provided a different system for surveying and gaining land. Families rather than individuals settled most of the land in New England on a community basis after a group petitioned the legislature for a grant. In Massachusetts Bay Colony the provincial government conveyed land to corporate groups, usually a church congregation that petitioned for a grant to permit settlement. This tract, known as a *town grant,* usually varied from 6 to 8 square miles. Before settlement of the land, however, the grant had to be surveyed in a rectilinear manner with acreage reserved to support the school and church from the rents derived from its use. This New England town system provided for the systematic location of town lots, streets, arable lands, meadows, woodlots, and common fields. The surveyed fields were then allotted by the General Court or town council, according to need, ability to use the land, and sums a settler had invested in the enterprise. Unallotted land, land not assigned to a family, could be used as a "commons" for pasturing livestock or cutting firewood. Until 1725, when the town grant system ended, it fostered the settlement of ordered, cohesive, and compact communities. It also created an agricultural region of small-scale owners and family farmers. The regulations of the New England towns prevented the accumulation of lands on the plantation scale found in the southern colonies. The town grant system also prevented much of the speculation in land that became common in the southern colonies because land could not be sold without the permission of town officials.

Land could also be acquired from proprietors like William Penn in Pennsylvania or the proprietors who controlled the Carolinas. These proprietors sold parts of their grants from the Crown and also earned large revenues from settlers who leased (rented) those lands.

In addition, the proprietors granted headrights to encourage development and create an economic base from which they could profit via taxation and further development for more tenants.

Some settlers, particularly the Scots-Irish, who settled from Pennsylvania to North Carolina, adopted neither the headright, purchase, or town system for acquiring land. Instead, they believed that so much uninhabited land was contrary to the laws of God and free for the taking, and they did not hesitate to settle on land that they did not own. Although Virginia recognized limited squatter's rights, squatters wanted the right to buy lands on which they had settled. Ultimately, Virginia's Land Act of 1779 permitted the preemption of 1,000 acres for settlers who had occupied public lands before January 1, 1778, and 400 acres after that date. *Preemption* meant that squatters had the first right to purchase their land. This act provided the basis for the American farmer's perennial demand for a national preemption policy after the creation of the new nation.

Despite the preference of colonial farmers for fee-simple title to their lands, many colonial farmers without sufficient capital to purchase land rented it from a proprietor, speculator, or land agency. Tenancy served as a stepping-stone to ownership. It did not mean poverty and oppression but calculation and design by those with little capital. Some farmers eventually purchased cheap land from savings acquired by hard work and thrift. Still, many tenants were locked in poverty because they farmed poor land, kinship ties kept them from emigrating to the frontier, or migrating west did not appeal to them. The land, then, provided opportunity, but culture, ability, and capital determined whether a farmer could or would take advantage of it.

THE SOUTHERN COLONIES

When the Virginia Company of London planted the first successful British colony on the North American continent on May 6, 1607, agriculture was essential to its plans. Indeed, the company intended the settlers to produce their own food, trade with the Indians, and send home various commodities for processing and manufacture. Although these colonists soon learned to raise corn (maize), starvation prevailed, and agricultural production did not improve until Governor Thomas Dale granted each settler 3 acres for personal rather than communal cultivation. The incentive of immediate economic gain through private ownership and production, as opposed to the promise of gaining 100 acres for each colonist at the end of the seven-year joint stock period, worked, and the food situation improved, particularly from raising corn.

Corn became the essential frontier crop. It could be grown on partially cleared land, and it could be harvested late in the autumn after other agricultural tasks had been completed without damage or loss of the grain. Moreover, corn could be processed easily by hand with a husking peg, mortar, and pestle, while European grains such as wheat, barley, and rye required special threshing and milling equipment. Quickly the colonial settlers learned the importance of corn and made it their first crop on the new land. In Carolina, John Lawson reported that corn was "the most useful Grain in the World" and that "had it not been for the Fruitfulness of this Species, it would have proved very difficult to have settled some of the Plantations in America." By 1616, farmers were raising an abundance of corn, and new settlements were spreading along the banks of the James River toward the interior.

While the colonists struggled to provide food and shelter and endure difficult living conditions, they never abandoned their hope for commercial gain to improve their standard of living. They did not, however, begin cultivation of tobacco (*Nicotiana rustica*) until 1612. But this native variety had a "byting tast," and they soon adopted a new, smoother-smoking variety (*Nicotiana tabacum*) from the Caribbean to replace the harsh tobacco they adopted from the Indians. In 1613, John Rolfe raised the first good crop of this tobacco and shipped a sample to England. Soon many farmers planted this variety because it had a higher value per acre than any other crop.

By 1616, the residents of Jamestown had caught tobacco "fever" and even planted it in the streets. In 1628, these Virginians exported nearly 553,000 pounds of tobacco to England. Although some Englishmen believed tobacco to be the "herbe of hell," most Virginia farmers considered it an "esteemed weed" that would earn a considerable profit. As a result, Rolfe and other planters of Jamestown began the trade of a commercial crop that would dominate the agriculture of the Chesapeake region for more than a century. They also began the tradition of one-crop agriculture that dominated the South until the mid–twentieth century and provided the basis for a social relationship between whites and blacks that would influence the course of American history.

Tobacco farming expanded rapidly with the emphasis on quantity rather than quality. In 1619, the House of Burgesses attempted to improve Virginia's tobacco production by requiring burning all tobacco that graded "mean" at the storehouse in Jamestown. The legislators intended this law to ensure that only the best quality and highest-priced tobacco reached the market. Production remained high, however, and in 1621 Governor Francis Wyatt decreed that each family could raise only 1,000 plants per person, but despite this allotment, the overproduction of poor quality tobacco continued. In 1632, the Virginia legislature again attempted to limit production by restricting a farmer to 1,500 plants and fixed the price at 6 pence per pound. The problems of overproduction and low prices were not resolved, but still this act helped expand tobacco culture into the interior because planters sought the best land to improve the quality of their tobacco and abandoned worn-out tobacco lands or planted them in other crops. By 1680, most Virginia farmers raised tobacco. At that time, one man could produce from 1,500 to 2,000 pounds of tobacco on 1.5 to 2 acres. In 1688, a bumper crop brought more than 18 million pounds to the market, and the price soon dropped to a penny per pound.

During the first half of the eighteenth century, extensive cultivation and high production kept prices low, and the large-scale planters sought political relief. They advocated guaranteed fair prices, reduced production, and required quality controls to keep their earnings as high as possible and to ensure that only the best tobacco reached the market. These solutions, however, were detrimental to the small-scale planters who held less fertile lands and produced poor grades of tobacco. Between 1723 and 1730, the large-scale planters, who controlled the assemblies of Virginia and Maryland, also attempted to limit tobacco production, but these regulations were difficult to enforce.

The Virginia Inspection Act of 1730, however, became the most comprehensive inspection legislation in colonial America. It mandated grading, packing, and marketing of only the best grades of tobacco at legally established warehouses operated by agents appointed by the governor. This action proved effective, and prices stabilized during the mid-1730s. Maryland provided similar legislation in 1742. By the mid–eighteenth century, then, Virginia and Maryland regulated tobacco production and provided for the inspection of the crop at market time. These regulations proved most beneficial to the wealthy planters, who raised high-quality tobacco on good lands, but the poor farmers had to destroy most of their low-grade tobacco.

During the early seventeenth century, tobacco planters usually sold their crops on consignment to English or Dutch traders who docked at the planters' wharves. These buyers served as suppliers or agents for the planters. They provided credit, sold the crop abroad for a fee, and applied the proceeds to the planters' accounts. Although these dealers provided a necessary service, the planters usually believed that their agents cheated them because they did not know the price per pound or the total value of their tobacco for many months after shipping. British duties, freight rates, and insurance charges, together with the agent's commission, usually returned less money than the planters expected.

After 1740, Scottish purchases helped increase the price when the French gave the Scots access to the Continent. Scottish merchants increasingly replaced the English consignment trade by purchasing directly from the planters for cash. By 1775, Scottish merchants bought nearly 50 percent of the colonial tobacco crop marketed in Great Britain. Throughout most of the seventeenth and eighteenth centuries, then, tobacco dominated the social, political, and economic life of the Chesapeake. Indeed, tobacco became the "life and soul" of the region.

Despite the commercial importance of tobacco to farmers in the Chesapeake region, this crop quickly depleted the soil. Tobacco farmers rejected fertilizing their fields with manure because they believed it gave a peculiar taste to the leaf when smoked. Instead, they raised tobacco for three or four years until the soil became exhausted. Then they planted corn, wheat, or barley until these crops also failed to prosper. When their fields became depleted, the tobacco farmers let them lie fallow, often for twenty years, for the land to regain sufficient minerals for commercial production. Farmers who could not clear new lands to provide 20 to 50 acres for various crops per worker moved beyond the coastal plain to the Piedmont and even farther.

Below the Chesapeake region, rice became important to the agricultural economy of South Carolina and Georgia. Although some farmers began raising rice in South Carolina as early as 1671, the crop did not show much promise until the importation of a better variety, perhaps from Madagascar, in 1685. By 1690, rice had become an important crop, which the proprietors accepted for rent. At the turn of the eighteenth century, South Carolina planters exported 394,000 pounds. A quarter century later, in 1726, the planters along the coast had become large-scale producers who exported nearly 10 million pounds annually. The rice trade remained important until production peaked in 1770 at nearly 84 million pounds. The planters sold much of their rice to Caribbean buyers and slave traders during

the seventeenth and eighteenth centuries. This trade was so profitable and important to the South Carolina economy that it continued despite an act by Parliament that placed rice on the "enumerated" list of articles that could be shipped only to England. Although the British eased the restrictions of the Navigation Acts in 1731, regulation of the rice trade remained. South Carolinians submitted or disobeyed this inconvenience, depending on which course of action offered the best economic return.

The rice planters primarily used slaves from Ghana and Sierra

THE COLONIAL TOBACCO planters along the James River often acquired large estates and built beautiful mansions. They used slaves to raise tobacco, which they exported to England from their own wharves along the river. William Byrd II began work on this Georgian home in 1709. The center portion of the mansion was completed in 1730. The wings are twentieth-century additions. *Virginia State Library and Archives.*

Leone, a portion of the latter area called the Rice Coast of Africa. These slaves knew how to cultivate rice upon arrival, and they may have been responsible for introducing that crop to South Carolina. Until about 1720, these slaves and planters sowed rice on dry upland fields and hoped for rain. After that date, they began to practice irrigation, or "wet culture," in the inland swamps near the coast. By the late 1720s, rice had become the leading staple in South Carolina, providing the economic basis for the most extensive plantation system in colonial America.

Slave labor proved particularly suitable for rice cultivation in South Carolina and Georgia because the Africans had some immunity to malaria and yellow fever. Even so, their work was brutally difficult under horrifying working conditions. Arthur Young, an observer reporting to the British public on slavery in South Carolina, wrote that the bondsmen used for rice production often stood ankle deep in water and worked under conditions he likened to a "furnace of stinking, putrefied effluvia." In this and other respects, the development and expansion of American agriculture frequently came with a high price that men and women paid with their lives. By 1708, Governor John Archdale of South Carolina contended that the colony produced "the best rice in the known world."

In Georgia, the exact beginning of rice cultivation remains unknown, but planters exported rice from the colony as early as 1735, and lands along the Savannah and Ogeechee rivers became the major rice-producing area by the late 1740s. A decade later, McKewn Johnstone, a planter near Winyah Bay, began using the tidal flow and a series of dikes and locks to capture fresh water and hold it on his rice fields as a stream rose and fell with the tides. By 1758, the tidal-flow technique had come into common use. This method enabled better weed control; the water killed unwanted plants, and less time was required for hoeing. It also enabled the extension of rice culture about 20 miles inland and permitted the enlargement of the rice fields.

Rice plantations were large, and the enslavement of fifty to one hundred Africans for each became common. Before the perfection of tidal flooding, slaves or indentured servants cultivated between 3 and 5 acres of rice and produced about 800 pounds per acre. By 1748, tidal flooding enabled each slave to produce 2,200 pounds of rice annually, or about 50 bushels, although each bondsman cultivated only 1 or 2 acres. By 1775, rice ranked third behind tobacco and flour as the most valuable American export, and a host of traders, bankers, millers, shippers, and planters depended on it for their livelihood.

Rice cultivation required extensive intensive labor, and although it returned high profits, it was an expensive enterprise. As a result, some farmers preferred to raise indigo, which by 1748 offered an important alternative for rice planters. Although the first settlers had planted indigo upon arrival in 1670, it did not produce a high-quality dye that could compete with the product of the West Indies or the earnings from rice. Eliza Lucas introduced indigo to South Carolina in 1739. When she married Charles Pinckney in 1744, her indigo seeds became a partial dowry, and her husband distributed the seeds to many friends and acquaintances. By this means, she wrote, "there soon became plenty in the Country."

Indigo production grew and developed for two important economic reasons. First, King George's War (1739–1748) interrupted shipping and increased freight and insurance rates, and British ships to haul rice cargoes became scarce. The rice planters began to turn to indigo for another export crop. Indigo was cheaper to ship than rice, whose price had declined precipitously. South Carolina now became, in the words of Henry Laurens, "Indigo Country." Second, the British textile industry created a high demand for the plant's rich purple dye. Parliament encouraged production with its mercantile legislation, and in 1748 provided a bounty of 6 pence per pound for all indigo produced in the British American colonies. This bounty enabled planters to double their capital every three or four years. Indigo planters marketed their cubes of purple dye at Charleston. The profitable cultivation of indigo also depended on slave labor. Each bondsman was usually expected to cultivate about 4 acres, with 30 pounds of dye per acre the average return. By the mid-1770s planters exported 150,000 pounds.

In 1774, the Continental Congress prohibited exports to Great Britain, and the American Revolution soon ended the cultivation of this staple crop because the new American nation was beyond protective British mercantile policy after the war. Without a bounty, planters could not compete with producers in the British Empire. As a result, they shifted to other crops, such as rice and sea island cotton along coastal South Carolina and Georgia. Although the bonanza for indigo lasted only thirty-five years, it facilitated the shift to extensive slave-based cotton farming because indigo planters already had considerable capital and slaves at their command.

Although tobacco remained the preeminent crop in the upper South during the colonial years, corn and wheat provided the essential grains. Usually colonial farmers fed corn to their livestock and ground it into meal for family consumption, although they also marketed some of this crop abroad. In Virginia, wheat became an important commercial crop. Although the Virginia settlements struggled with agriculture at first, by 1634, Governor John Harvey proclaimed that the colony had become the "granary of all His Majesty's northern colonies." By the end of that decade, Virginia's merchants also marketed wheat in the West Indies. In 1739, they sold more than 58,000 bushels of wheat and 83,000 bushels of corn, primarily in the Caribbean, from the farms along the James, York, and Rappahannock rivers. Wheat enabled Virginia farmers to diversify and provided an alternative to single-crop tobacco agriculture. During the 1750s, crop failure in Europe and increased population growth and food consumption in the colonies continued to make wheat a profitable crop in Virginia.

Farmers on Maryland's Eastern Shore also began to diversify by raising wheat during the mid–eighteenth century, increasing their profits and decreasing their economic reliance on tobacco. After 1750, wheat replaced corn as the second most important cash crop. New England and Caribbean merchants bought this wheat to feed artisan and slave populations. Annapolis and Philadelphia now became major ports for the export of wheat from Maryland. Farmers in the upper South raised wheat with the same slave labor force that they used for tobacco because the planting, cultivating, and harvesting occurred at a different time for each crop. Wheat helped slaveowners keep their "hands" busy throughout the year. The planters also readily adopted wheat when they learned that lands planted to this grain needed to lie fallow for only one year following two crops, in contrast to the lengthy period necessary to restore exhausted tobacco lands. Wheat, then, not only brought additional money into the household but also enabled more efficient use of the land. After 1750, the demand for American wheat increased rapidly, and farmers in the upper South responded to market demands. By 1775, a "wheat belt" stretched from the Eastern Shore northwest into the Shenandoah Valley.

The early settlers in the South also raised cattle, usually within a decade of settlement. By the late seventeenth century, cattle raising had become important to the colonial economy, particularly in South Carolina where by 1674 the gentry commonly held large herds of 500 to 1,500 head. By 1710, it was "very common" for small-scale farmers to graze 200 head of cattle. Cattle and hog raising became important in Georgia and North Carolina about the same time, but South Carolina developed the most important livestock industry. There, cattlemen employed the open-range system because too much labor was required to build fences.

Colonial cattlemen did not provide feed or shelter for their livestock. Although farmers fenced their crops to keep cattle out and to ensure the "peace and harmony of every neighborhood," they allowed their livestock to graze freely on public land, the "King's land." In the fall and spring, they held "cow hunts" to brand and mark the newborn calves or separate the cattle that would be driven to market. In contrast to the Spanish cowboys to the West, Anglo stockmen were known as *cowkeepers* in the British tradition. Many of the herdsmen, however, were African slaves. Some, particularly those from Gambia, had experience raising cattle in their African homeland. Black herdsmen were most common in South Carolina where primarily large-scale landowners or gentry raised cattle rather than small-scale or yeoman farmers.

South Carolina became the heart of the British-American cattle industry, the principal markets developed in the Bahamas, Jamaica, and Barbados by 1682. There, traders exchanged barrels of salt beef for sugar, slaves, and cash. Tallow, candles, leather, and cowhides also reached foreign markets. Colonial beef reached these international markets packed in barrels rather than on the hoof because the shipment of packed beef was easier than handling live cattle on the high seas and more beef could be sent to market. Moreover, the selling price of packed beef was higher than that for live animals.

By 1712, South Carolina had developed a major trade in packed beef with the West Indies. In the mid–eighteenth century, South Carolina contained approximately 100,000 head of cattle, with 12,000 head slaughtered annually for packing. During the colonial period, this colony remained unrivaled for cattle production. The mild climate, abundant grazing lands, proximity of the West Indies market, ample supply of Gambian slaves, and merchants in Charleston, who developed a hide market, made South Carolina an ideal colony for cattle raising. By 1720, South Carolina exported more livestock products than rice. After 1750, cattle disease and deteriorating range conditions irreparably damaged the open-range cattle industry in the colony, and Pensacola, Mobile, Natchez, and New Orleans soon replaced Charleston as important cattle towns as the industry spread west along the Gulf Coast.

By the mid–eighteenth century, farmers in Virginia along the western reaches of the Shenandoah Valley had also developed an important cattle-raising industry. The mountain valleys and hillsides proved ideal for grazing livestock. These farmers, who engaged in mixed agriculture, also began the practice of fattening their livestock for market with corn, and they were among the earliest importers of purebred cattle. They too played an important role in the diffusion of the cattle industry into the transappalachian region of Kentucky and Ohio. The herds of these farmers averaged less than 50 head, and few Shenandoah settlers could be classified as large-scale cattlemen by 1775. Some Scots-Irish, Welsh, German, and Dutch farmers in this region, however, specialized in cattle raising, but most stockmen remained small-scale producers who engaged in other agricultural practices. Still, this region became a major area for cattle grazing during the colonial era, and Philadelphia served as an important cash market for Shenandoah stockmen and drovers.

Overall, then, the plantations, commercial farms, both large and small, dominated all aspects of southern agriculture. Based on plentiful land, cheap labor, staple crops, and demand at home and abroad, the plantations drove the economy, shaped the social structure, and determined political power in the colonial South. Although planters sought economic gain from the production of staple crops, they also tried to achieve self-sufficiency. Plantation agriculture, however, slowed the regional development of a merchant and artisan class, discouraged investment opportunities other than land and slaves, fostered rural isolation, and slowed the growth of urban areas. This agricultural legacy would affect southern life into the twentieth century.

THE NORTHERN COLONIES

In the spring of 1621, agriculture began in New England when the Pilgrims at Plymouth learned from the Indians to plant corn. These Separatists from the Church of England were more concerned with nourishing their souls in the New World than meeting physical needs. As a result, they gave little thought to agriculture. Although the Pilgrims planned to farm in a communal fashion, lack of harmony among them destroyed this agricultural system. In 1623, however, Governor William Bradford allotted 1 acre to each man to till for his family, the first cattle arrived, and agricultural production expanded. It further improved after each head of a household received 20 acres in 1627.

By clearing land and planting corn and English grains such as wheat, rye, and oats, the Pilgrims gradually ended their food problems. In the autumn of 1631, Governor Bradford reported a "plentiful crop." The Pilgrims also bred and sold cattle to the newcomers in the Massachusetts Bay Colony. Livestock production flourished; in 1660 one New Englander reported: "It is a wonder to consider how many thousand neat beasts and hogs are yearly killed and have been for many years past for provision in the country and sent abroad to supply Newfoundland, Barbados and Jamaica."

The Puritans, who settled not far to the north in the Massachusetts Bay Colony, gave more thought to survival than the Pilgrims. In 1629, a preliminary expedition, designed to prepare the way for the great migration that would follow the next year, brought cattle, horses, hogs, and goats. When eleven ships set sail from England on March 4, 1630, with 840 passengers, their holds carried 240 cows and 60 horses. In 1635, two Dutch schooners brought the first sheep into the colony at Boston Harbor. Soon Massachusetts Bay became the major livestock-producing colony in British North America, surpassed by South Carolina during the late seventeenth century.

Because of the irregular coastline, the New England colonists could keep their livestock on many peninsulas and necks with relatively little fencing. Their cattle could graze with a minimum of care and in relative safety. In time these northern farmers confined their milk cows in their houses or barnlots and let their beef cattle roam unattended through the woodlands until branding and market time. As the population expanded during the eighteenth century, however, farmers became legally bound to fence their livestock to protect the fields. Each town also had a keeper responsible to collect the milk cows each morning from their owners and drive them to a communal pasture. In the evening the keeper brought them home. During the winter, these farmers kept their beef cattle in nearby lots.

By 1650, Boston and New London had become major cattle markets for livestock producers in Massachusetts, Rhode Island, and the Connecticut Valley. Beef not needed for local consumption was packed for shipment to the West Indies. A decade later, pork packing had become an important economic activity, farmers driving their hogs long distances to market for regional and foreign consumption. Rhode Island also became the center for raising saddle horses, called Narragansett Pacers, the Indies again providing the chief market. Boston traders sold horses to Caribbean buyers as early as 1648. During the late seventeenth century, William Harris, a wealthy landowner, reported that more sheep grazed in Rhode Island than in the rest of New England. For him, Rhode Island was the "Garden of New England." The cool climate and often stony soil of the New England colonies made livestock raising the most profitable form of agriculture. This commercial enterprise developed rapidly during the seventeenth and eighteenth centuries along with seaports and interior towns. Indeed, if New England had staple crops, they were the annual crops of beef, pork, and wool. Specialty agriculture such as pasturing beef cattle, raising hogs, and grazing sheep helped alleviate the need for some farmers to migrate westward after their soil became infertile from repeated plantings.

New England farmers wanted to produce surplus crops and livestock for the market to make money and improve their standard of living. By the mid-1640s, commercial farmers, especially those in Connecticut and Rhode Island, raised crops as well as livestock for sale. They emphasized wheat and corn, which they exported by sea in great quantities in exchange for cotton, tobacco, indigo, sugar, and wine. In 1647, merchants from the West Indies, particularly Barbados, sent ships to Boston to trade for "provisions for the belly," and the merchants responded with grain and cattle. English ships also arrived to purchase grain with goods or bills of exchange. In addition, Boston became a provisioning port for fishing, merchant, and naval vessels, thereby increasing the market demand for agricultural commodities such as wheat, cattle, sheep, hogs, cheese, and butter, supplied by the farmers throughout New England. This trade also provided employment for butchers, meat-packers, coopers, millers, and bakers in the port cities and towns that drew upon the agricultural hinterland. Merchants, shippers, and farmers were inextricably linked to one another and to the broader commercial world beyond. The prosperity of each depended not only on the fertility of the land but the Atlantic trading network.

As early as 1636, farmers in the Connecticut River valley produced tobacco for export, a crop that they adopted from the Indians. Although colony leaders in Massachusetts and Connecticut opposed the use and cultivation of tobacco, smoking soon became a common habit, and the demand increased. In 1660, John Winthrop, Jr., governor of Connecticut, reported that "tobacco has brought some good crops." Twenty years later, the governor told a Parliament committee on foreign trade and domestic plantations, "We have no need of Virginia trade, most people plantgn [sic] so much tobacco as they spend." The West Indies became the principal market, although northern tobacco reached the English market by 1750.

By 1775, New England farmers raised wheat only on the rich soils of the newly developed frontier. It had nearly disappeared from the cropping practices of farmers in the older, settled areas, where pastures and orchards predominated. Although wheat was the most important grain and its flour the most highly prized for bread, few farmers raised it in abundance beyond the middle Connecticut Valley. Soil depletion and black stem rust, known then as "wheat blast," decreased wheat production as early as 1660, and farmers began to raise more corn and rye. A century later, New England farmers met the problems of depleted and naturally poor soils by emphasizing livestock production rather than cash grain crops.

Although wheat production declined in New England during the mid–seventeenth century, it became a staple of the middle colonies, particularly New York, Pennsylvania, and New Jersey. Wheat became so important that it soon earned the sobriquet "Bread Colonies" for the region. Although the Dutch, who settled New Netherland (New York) from 1624 to 1664 and William Penn in Pennsylvania attempted to replicate the feudal system of land tenure, wheat rather than rents provided the agricultural wealth for these colonies. Indeed, wheat became the "grand article" in Pennsylvania, where in 1700 one observer noted that it had become the "farmer's dependence," as it was in the other middle colonies. Although yields might reach 40 bushels per acre on newly plowed lands, these colonial farmers did not use fertilizer or practice crop rotation. Consequently, their lands depleted rapidly, and about 10 bushels per acre became a good harvest on old lands.

Wheat farmers in the middle colonies marketed their grain through local merchants, who in turn dealt with flour millers in Philadelphia, New York, and Baltimore. During the 1750s, flour milling began in Baltimore, and in 1758 Henry Stevenson sent the city's first shipload of flour to the West Indies. The international flour trade quickly became important and profitable to colonial farmers and millers. The flour milled in Philadelphia or Baltimore reached the West Indies markets in about a month, about half the time required for shipment from England. Flour from the middle colonies was well suited for the West Indies trade because English flour, which had been milled in a cool climate, often spoiled before arrival or soon thereafter in the tropical heat. Moreover, Jamaican planters earned greater profits from planting sugar cane, indigo, coffee, and cacao than from raising wheat for flour. Consequently, West Indians became dependent on American farmers for their foodstuffs, particularly flour. In addition, disease that ruined wheat farming in New England helped create an American market for grain from the Bread Colonies.

The farmers in the middle colonies also raised dual-purpose or "neat" cattle, that is, cattle used for both meat and milk. By 1750, however, some farmers were beginning to distinguish between "milch" and "beef-store" cattle among the breeds that descended from cattle of Swedish and British ancestry. Commonly, cows that produced a gallon per day were good milkers, but a quart per day was most common. With low production like this, farm women converted any surplus milk into butter and cheese for home consumption rather than sale. The quality of the butter and cheese, however, remained low because of poor feeding and breeding practices. By the mid–eighteenth century, milk production increased, perhaps because of improved breeding and feeding, and farm women in the middle colonies churned more butter, which they sold at local markets. This income helped meet family expenses and became crucial to the farm enterprise. By 1769, merchants in Philadelphia had developed a lucrative butter trade with the West Indies.

During the eighteenth century, farmers in the middle colonies shifted from oxen to horses for plowing and pulling wagons. These farmers gave careful attention to horse breeding and by 1726 preferred stallions at least 13 hands tall with "comely proportions." In Pennsylvania, German farmers developed big-boned, strong-muscled "Conestoga" horses from English stock. These horses became recognized as the best breed of draft animals during the colonial period, farmers preferring them to oxen because they worked with greater speed. The increased use of draft horses also required farmers to raise oats and hay and clear additional woodlands for pasture. Farmers in the middle colonies raised swine in a haphazard manner, but by 1740 Pennsylvania farmers fed corn to their hogs to fatten them for market, a procedure that midwestern farmers in the corn belt would perfect two centuries later.

Despite the prevailing interest of farmers in the middle colonies in a market economy, they too used the land carelessly. With the exception of the Germans, these Pennsylvanians gave little attention to crop rotation, fertilization, or proper tillage procedures. The land was something to be used as extensively as possible. When production and profits declined on worn-out fields, new lands could be cleared. When farmers ran out of productive land, they shifted to a rotation system by which they let a field lie fallow for three or four years before planting it again. Without seed improvement, proper crop rotation, or manuring of the land, yields remained low and never regained the level of the first year or two after the land had been broken. Few farmers tried crop rotations using nitrogen-rich clover or grasses until the mid–eighteenth century, and this farming practice did not become common until the 1780s.

The market was more important than soil conservation for farmers in the middle colonies. Low yields were not a problem if quantity could be increased by raising more acres of wheat. With labor expensive, farmers preferred to get as much return from their own work and hired labor as possible. As a result, extensive rather than intensive agriculture prevailed. Moreover, new and improved agricultural techniques often required additional labor. With land cheap and labor dear, including "victuals in the bargain," many farmers could not afford the expense of hired workers unless wheat prices were high. As long as wheat and flour exports remained profitable, these farmers saw no reason to change their cropping patterns. They would take risks if the profits merited, but dependable returns kept them attached to wheat production throughout the colonial period.

LABOR

Although most farmers worked their lands with their own labor and that of their wives, sons, and daughters, many needed more labor than was readily available. Throughout the colonial period, land remained relatively cheap, particularly if a settler squatted on Crown or proprietary lands, and the scarcity of labor became an annual problem for both craftsmen and farmers. Labor shortages drove up the prices of wages. Unable to hire field hands, some farmers met their labor needs by using indentured servants. By 1750, not counting the Puritan migration of the 1630s, more than half the immigrants south of New England were indentured upon their arrival, particularly to planters and farmers. The tobacco economy of Virginia and Maryland was based on indentured servitude during the seventeenth century.

Indentured servants contracted their labor for as many as four to seven years to individuals who paid their passage and other costs. Indentured servants could be supported for about £14 per year, but they could return £50 to their masters from their labor. In the plantation South, indentured servants worked and lived alongside the slaves, although they had higher status and greater privileges. On the smaller farms throughout the colonies, their duties usually were lighter and their treatment better.

Indentured servants were entitled to adequate food, clothing, shelter, medical care, and physical protection. They could own property, but they could not vote or engage in trade, nor could they marry without the consent of their master. At the end of their term of service, indentured servants were entitled to "freedom dues," which in some colonies, such as South Carolina, meant 50 acres. Masters also offered additional cash if their servants stayed the full term of the contract period to encourage them to remain rather than run away. Most planters preferred to employ men during the seventeenth century, particularly for tobacco production, but women also came in large numbers. Women often worked as domestic servants, but family farmers and small-scale planters used women in the fields. The men and women who sold themselves into indentured servitude usually wanted to improve their economic fortunes, but most remained poor after their servitude. Few became profitable landowning farmers. They often sold their land rights and remained poor whites who lived in rural areas and earned their living by working for someone else. Upon their freedom, women had a great choice of marriage partners because men outnumbered them in the colonies. Women who became pregnant before the end of their contract period, however, usually had their service extended from one to three years to make up for time lost before and after childbirth. This discriminatory and abusive practice enabled masters to gain many months of domestic or farm labor to recover the loss of only a few weeks.

Improved economic conditions and a declining birthrate in Great Britain decreased the number of indentured servants in the colonies between 1680 and the turn of the eighteenth century. Not long after 1700, the servant trade virtually ceased in the Chesapeake region. Planters, however, still needed extensive labor, but tobacco returned less profit for the hire of field workers. As a result, planters turned to slave labor because it was cheaper than indentured servants over time and more readily available. Southern planters also preferred slaves because they were servants for life.

In August 1619, bonded servitude began in the southern colonies when a Dutch ship deposited about twenty Africans in Virginia, but agricultural slavery developed slowly. Although slaves increasingly provided plantation labor in the southern colonies after 1680, masters held only about 28,000 Africans throughout the colonies by the turn of the eighteenth century. Even so, slavery had been firmly established by law in every colony, and few people questioned the morality or legality of bonded servitude. By 1770, however, nearly 22 percent of the population, or 459,000 blacks, resided in the colonies. In the South, most slaves worked on the tobacco, rice, and indigo plantations from Maryland to Georgia. Nearly two-thirds of these bondsmen labored in the tobacco fields of Maryland and Virginia. Slavery also became so important to the agricultural economy of South Carolina that bondsmen composed 60 percent of the population by 1760.

By 1700, slaves produced most of the tobacco in the South. During the early colonial period, most slaves worked on family farms alongside their owners rather than on large-scale plantations where overseers directed the work. In the northern colonies, the absence of a staple crop and a short growing season enabled the family to do the work. Slavery was not profitable in that region, and indentured servants tended to be acquired by craftsmen in the towns. When a small-scale northern farmer owned slaves, he usually employed only one or two in the absence of free white labor. By 1758, the Narragansett planters of Rhode Island used more than 5,000 slaves, the largest concentration of bondsmen in rural New England.

Southern planters used both the task and gang systems to work their slaves. In the *task system,* most commonly employed in the rice areas, a slave was assigned a specific amount of work during a day, such as threshing 10 bushels of rice, splitting 100 fence rails, or processing a certain amount of indigo. When the bondsmen completed that work, they were free to return to their quarters and tend gardens or enjoy some leisure time. In the tobacco region, planters primarily used the *gang system,* whereby a group of slaves conducted an assigned task such as preparing the soil for planting, seeding the crop, or hoeing the weeds. The group worked together throughout the day. Yeomen who owned only one or two slaves invariably worked alongside their bondsmen at whatever tasks needed to be done. During the eighteenth century, the ownership of slaves not only enabled planters to reap profits from the production of staple crops such as tobacco, rice, and indigo but also became a symbol of social status. As a result, the slave trade expanded rapidly.

TECHNOLOGY

Extensive hand labor and the use of few tools characterized colonial agriculture. Farmers in the seventeenth century primarily used an ax to clear the land and a hoe or mattock to prepare the seedbed. If farmers owned plows, they were usually British imports or crudely fashioned homemade plows. Most of the early colonial farmers went without plows for a considerable period. The Pilgrims, for example, who landed at Plymouth in 1620, engaged primarily in fishing and trade with the Indians for their livelihood, and they did not have plows until 1632, twelve years after their arrival. In 1637, only a few miles to the north, the Puritan farmers around Boston in the Massachusetts Bay Colony had only thirty-seven plows. And as late as 1642, Rhode Island farmers still used hoes and spades to turn the soil.

Later, British mercantile policy intentionally restricted the development of American industry to make the colonists reliant on England for manufactured goods. Consequently, colonial farmers could either import expensive English plows or fashion their own. These plowmen, however, always demanded an implement that required little draft, the amount of power needing to pull it, one that ran at a uniform depth, turned over the furrow, and pulverized the soil. But these demands were seldom met before the standardization of design and the perfection of interchangeable parts during the early nineteenth century.

With the general absence of plows in colonial America, farmers who owned one often tilled their neighbors' fields. Or the town paid a bounty to any farmer who purchased a plow and used it to prepare fields for planting. Those farmers who could neither afford a plow nor hire their plowing fashioned plows of their own design from the wood and metal available. By 1700, blacksmiths plated the wooden moldboard with iron strips to reduce friction and scour—that is, prevent the soil from sticking to the moldboard. With these walking or swing plows, two men using two or three horses or four to six oxen could plow 1 to 2 acres per day.

After a field had been plowed, farmers often used log rollers or A-frame harrows with wooden teeth to break the clods and smooth the seedbed. Then they planted their corn with a hoe or seeded small grains like wheat, barley, or rye by walking across the field and sowing the seed before them. They weeded row crops, such as corn and tobacco, with a hoe or a small light plow that stirred the soil rather than turned a deep furrow. At harvesttime, they cut their grain with a sickle, the most common reaping tool until the late eighteenth century. An axiom of the age was that no one learned to reap properly until he or she had cut the little finger of the hand that held the bunched stalks. At best, a reaper could harvest only 1 acre per day with the sickle.

By 1750, farmers in the Middle Atlantic states, particularly wheat farmers in Pennsylvania, Maryland, and Virginia, used a cradle scythe for the grain harvest. Although the scythe was a European invention that the colonists began using early in the seventeenth century for mowing hay, these farmers changed the design to make it more suitable for cutting grain. The cradle scythe consisted of a grass scythe that had a frame with four or five wooden fingers fixed to it. The frame attached above the blade and the wooden fingers ran parallel to it. As the reaper made the cut, the grain fell onto the fingers. The reaper then tilted the cradle and allowed the grain to fall into a pile, where it could be raked into a bundle and bound into sheaves. A skilled reaper could cut approximately 3 acres per day. The cradle scythe enabled the reaper to maintain an upright stance, but the 10–12 pound tool required skill to manipulate, and cradlers commonly received more pay than other harvest hands.

Although the cradle scythe achieved quick popularity with Middle Atlantic farmers, it was seldom used in the South before the early nineteenth century. Southern farmers preferred the sickle because it did not shatter the grain from the heads as much as the cradle scythe and they could manipulate it easier in heavy stands of grain. Many southern farmers preferred the sickle because it left more straw in the fields. When plowed under, this stubble helped fertilize the soil, and the more straw a farmer left in the field, the faster his threshing and stacking time. New England farmers preferred to use the sickle until the late 1830s. Most of the farmers who raised grain for sale, however, adopted the cradle scythe, and it became the standard reaping tool until about 1860.

Whether colonial grain farmers used the sickle or the cradle scythe to harvest their crops, these tools limited the acreage they could expect to cut unless they were prepared to spend large sums of money to hire harvest hands. In the case of wheat, a farmer might have a maximum of ten days to complete the harvest before it began to shatter out of the heads, even less time if a weather change threatened to slow or ruin the harvest. Until the farmer could speed the harvest mechanically, he had little hope of cheaply expanding his grain production. With land cheap and labor expensive, the cost of hiring a large number of harvest hands often remained prohibitive; many preferred to own or rent their farms rather than work for someone else. Only horse-drawn machinery would free the grain farmer from dependence on hand tools and hired labor. But until the invention and perfection of an efficient horse-drawn harvester, the individual farmer's grain production was severely limited.

Once the grain had been cut, it had to be raked and bound into bundles or sheaves for "shocking." This task involved stoop labor similar to cutting with a sickle. Generally, three binders for every two cradlers were expected to prepare about 1,000 sheaves per day for shocking over a 10-acre field. Farmers then had to thresh and winnow their crops. These jobs were left for the winter months when cold weather forced them into their barns and the fields had become too wet or frozen for plowing. Cold, dry weather was the best time for threshing because damp grain did not beat out of the heads properly. Nevertheless, threshing and winnowing required long hours of hard work.

Colonial farmers, particularly those in New England, used the flail to thresh grain. Wielding a flail required considerable skill, and beginners usually knocked themselves about the head and shoulders before they learned to use the tool properly. On the average, approximately 7 bushels of wheat, 18 bushels of oats, 15 bushels of barley, and 8 bushels of rye could be threshed and winnowed during the course of a ten-hour day. At this rate, most of the winter was needed to thresh the crop. In the middle colonies of New York, Pennsylvania, and New Jersey, where grain production was greater than in New England, the flail was too slow to enable completion of threshing in a reasonable time. There, to speed the threshing process and keep the number of hired laborers to a minimum, farmers used oxen or horses to tread the grain from the heads. A man and a boy using three horses could tread about 30 bushels of grain per day. Overall, however, threshing grain by treading with horses or oxen was not as efficient as flailing because the animals did not tread evenly and much of the grain was wasted. Nevertheless, if a farmer had a large acreage of small grain, treading was more economical than flailing. The larger volume of grain threshed by treading lowered labor costs because fewer workers were needed.

No matter which threshing method a farmer used, he had to winnow the chaff from the grain before storing it or carting it to market. He did this by using a shallow basket to toss the grain into the air so that the breeze could blow away the chaff. Some farmers preferred to use a "riddle" to remove the chaff from the grain. This riddle, or sieve, held the larger pieces of chaff as the grain fell through the screen when the farmer shook it back and forth. Once the grain was winnowed, the farmer stored it in bins or put it into bags to facilitate handling and transport to market. Winnowing by hand remained common on farms that produced small grains well into the nineteenth century.

Although some farmers raised wheat, virtually all raised corn for their families and livestock. Generally, they preferred to use both the stalks and the ears. To do so, they allowed the corn to ripen until the ears were mature and fairly firm, then cut the stalks at ground level with a sharp hoe or long-bladed knife. This was back-breaking drudgery, and workers could cut only about 1 acre per day. Then they gathered the stalks into bundles and placed them in shocks until the farmer removed the kernels from the cobs in a process known as shelling. This task was one of the most time-consuming aspects of the corn harvest. Although corn produced the greatest yield to the amount of seed planted, it also required more hand labor for harvesting than any other cereal grain, in part because it was the largest crop. In contrast to wheat and other small grains, corn production depended on the acreage that a farmer could cultivate rather than the amount that he could harvest. If weeds choked the crop, he wasted both time and money.

BACKCOUNTRY REBELLIONS

Colonial farmers occasionally endured periodic low prices and crop failures, suffered from inadequate credit and large debts, complained about high taxes, and confronted resistance from Native Americans as they expanded their farms into the interior. In 1676, all these problems compounded on the Virginia frontier, which, partly because of political repression by Governor William Berkeley, exploded into Bacon's Rebellion.

Although periodic conflicts with the Indians had plagued Virginia's farmers almost from the founding of the colony, a serious outbreak of hostilities occurred in the summer of 1675. By January 1676, farmers above the fall line lived in fear. They demanded help from the militia, but Governor Berkeley believed this armed body should take only defensive action and convince the tribesmen to talk peace. By March, at least thirty-six Virginians had been killed. When the farmers along the frontier threatened to form an expeditionary force and attack the Indians, Berkeley called them "fools and loggerheads" and promised severe punishment for anyone engaged in this activity.

In April 1676, Nathaniel Bacon, a scion of the gentry class in England who moved to Virginia in 1674, organized 300 volunteers, who refused to accept Berkeley's dictate and took direct military action against the Indians. Although one opponent claimed that Bacon attracted the "scum of the country," his supporters included a number of wealthy planters. Berkeley immediately declared Bacon and his band of farmers "rebels and mutineers." Because of Bacon's success against the Indians, his backcountry constituents elected him to the House of Burgesses. Meanwhile, Bacon recruited a new army of 1,300 men to continue his attacks against the Indians on the frontier. Berkeley responded again by declaring Bacon and his men in rebellion against the king. Bacon then drove the governor from Jamestown and burned the capital on September 19, 1676, but he died on October 26 from the "bloody flux," dysentery. Berkeley reasserted control of the colony and hanged twenty-three rebels. Significantly, Bacon's Rebellion caused the government of Virginia to develop an aggressive policy against the Indians. Within a generation, they no longer menaced agricultural expansion in Virginia.

Meanwhile, Chesapeake farmers suffered from low tobacco prices, high taxes, burdensome quitrents, political oligarchy, hostile Indians, and proprietary government. In 1680, a large tobacco crop compounded the surplus problem and threatened financial ruin for tobacco planters in Maryland and Virginia. During the spring of 1682, vigilantes under the direction of the large-scale planters took direct action to limit the tobacco crop and thereby increase prices by sending vigilantes to cut down tobacco plants of small-scale farmers. By the end of the summer, approximately 10,000 hogsheads of tobacco and the plantings of five thousand workers had been destroyed before the militia put down this lawlessness. Soon thereafter, prices increased, and order returned to the countryside.

Agrarian insurrection also occurred in New Jersey between 1735 and 1754. Farmers who claimed they had purchased their land from the Indians or received their grants from the governor of New York fought the proprietors of East Jersey, who claimed title to their land via a grant from the Duke of York in 1664. The East Jersey proprietors, a fluctuating group of land speculators, offered land for sale and lease subject to the payment of annual quitrents. They also refused to recognize land titles purchased from the Indians. They demanded that the settlers buy or lease all land from them or face confiscation of their property and eviction.

The settlers refused to purchase their land twice, particularly because they had improved the value of their property by clearing the land, planting crops, and building houses. With approximately 500,000 acres at risk, the stakes were high for both farmers and proprietors. When these proprietors had some small-scale farmers arrested and jailed for trespassing and judges under proprietor influence issued judgments against them, violence occurred. Crowds armed with clubs freed the jailed farmers and threatened the property and lives of the officials, who supported the proprietors. Although serious violence had ended by the early 1750s, the problem of determining who owned what land had not been resolved before the American Revolution drove the proprietors away.

Land riots also occurred in New York during the 1750s, culminating in the "Great Rebellion" of 1766. There too, settlers from New England contested the right of the large-scale owners to claim unused lands, and colonists squatted on the great estates. When the landlords attempted to evict the squatters, these farmers fought back by cutting down thousands of trees that belonged to the landlords and preventing legitimate tenants from occupying and cultivating the lands. These squatters also used physical intimidation and destroyed crops and fences in retaliation against the landlords. The landlords responded by hiring armed men to remove the insurgents, tear down their houses, destroy their belongings, and ruin their crops. Competing land claims by New York and Massachusetts in the contested area complicated the problem.

Until 1766, the agrarian rebellion in New York involved matters of land ownership rather than lease terms. In that year, however, the tenants demanded better leases and reinstatement of the evicted farmers from estate lands and refused to pay rent. At the same time, many squatters also refused to become tenants. Above all, the tenants and squatters sought fee-simple or freehold property of their own because property meant profit. The proprietors responded with more evictions, arrests, and intimidation. This armed conflict ended when British troops arrived from New York City and routed the insurgent farmers. In the end, this rebellion challenged and undermined provincial authority and helped shape the political and social attitudes of colonial farmers on the eve of the Revolution. All of the agrarian rebellions during the colonial period reflected the economic and social tensions that affected agricultural life in America.

THE SPANISH WEST

The European agricultural heritage of the American West began when the Spanish organized farming activities among the Indians in present New Mexico and Texas during the late seventeenth century. In 1680, the Pueblo revolt forced the Spanish back to El Paso from Santa Fe, and the Comanche drove them from Texas by 1692. The Spanish returned to Texas early in the eighteenth century, established presidios and missions, taught agricultural methods to the peaceful Indians, and worked to convert them to Catholicism. These efforts, however, did not significantly expand agriculture because the Spanish settlements remained isolated without the economic support necessary for growth and development. Moreover, the Plains Indians soon adopted the horse, became nearly invincible, and prevented further agricultural settlement until the U.S. Army forced them onto the reservations during the late nineteenth century.

As early as 1706, the Spanish began colonization of the upper Rio Grande Valley, an area known as Rio Arriba, by raising livestock and practicing a limited irrigation agriculture. In 1718, the Spanish also established the presidio and settlement of San Antonio, where cattle raising became the most important agricultural activity. By the mid–eighteenth century, the diamond tip of Texas (an area formed by San Antonio on the north and the Gulf Coast on the south, the upper sides of the diamond shaped by lines drawn from San Antonio to Laredo on the Rio Grande to the west and to the Gulf Coast on the east) had been settled by large-scale ranchers, and thousands of cattle and sheep grazed the land.

As the herds grew, the ranchers needed markets for their cattle because they could not earn sufficient profit from slaughtering their livestock for tallow and hides. For that market they turned east, and by 1750 Spanish ranchers had developed an export cattle trade with New Orleans. At that time, the Spanish also established ranches along the Rio Grande and soon founded others along the Nueces River in Texas. Crown favorites received these lands, and their ranches remained relatively untouched by the Apache and Comanche because the tribesmen could meet their food needs with the buffalo farther to the north. The isolated, widely scattered ranches in Texas did not offer adequate returns from raiding. As a result, the cattle industry became firmly established in the diamond tip of Texas.

In present California, the Spanish based their colonial expansion on military power, the Catholic church, and agriculture. There, the Indians remained peaceful, and the Spanish mission and land-grant policies proved more successful. On November 29, 1777, Spain established San Jose, the first pueblo or town in California. And in 1784, Spanish soldiers gained the right to own land. Retired soldiers could hold land near a presidio, pueblo, or mission to raise livestock and protect the inhabitants. The Spanish pueblo land grants favored private citizens and groups that promised to establish communities. After farming four years, the grantee received full title from the Spanish government. The regional Spanish governors authorized rancho (individual), pueblo, and presidio land grants, although the viceroys technically made all grants in the name of the Crown. Spain continued to grant lands for the support of presidios, pueblos, and missions until the Mexican War for Independence in 1821.

Many rancho and pueblo grants ranged from 1 to 11 square leagues, a league containing 4,438 acres. The Mexican government continued this generous land-grant policy to promote colonization and ensure its claim to empire. Under the Spanish, some mission grants reached 133,000 acres; individual families often claimed more than 100,000 acres and the Pico family secured 532,000 acres. In all, 750 land grants encumbered between 13 and 14 million acres.

Spanish officials did not survey these grants. Consequently, a host of boundary disputes developed between those who held Spanish or Mexican grants and American citizens who sought access to the public domain after the United States gained control of California at the end of the Mexican-American War in 1848. A year earlier, William Tecumseh Sherman reported from Monterey that "not a poor devil, native, Indian or foreigner, has a paper to show his title to land and houses. Moreover, no person knows the limit of his own property, so that the ranches overlap and several claim the same hill and valley. This is to be expected and will offer plenty of employment for lawyers, though it will produce distress in the land." Sherman could not have been more prophetic.

Although Spanish supply columns brought seeds of all kinds to California, cattle raising became the most important agricultural activity. Spanish ranchers in California, like those in Texas, primarily slaughtered their livestock for the hides and tallow. They did not have a market for fresh meat, and they could not adequately preserve their beef for shipment to foreign markets. By rendering the fat for soap and the tallow for candles and tanning hides for leather, these early cattlemen capitalized on opportunities for international trade. Export markets, especially in New England, became important about 1810.

In 1779, the Spanish also introduced viticulture to California at the mission of San Juan Capistrano from cuttings on the Baja Peninsula, and by 1784 the friars produced wine. The Spanish also introduced the *ard,* or plow with an iron point, sickles, and reaping knives for use with their field crops. Yet despite these agricultural improvements, profit was not the overriding motive for agricultural development among the Spanish in California and the Southwest. The Spanish based their colonial policy on the premise that land should be used to support subsistence agriculture. Farming would strengthen Spanish authority, promote colonization, encourage religious conversion, and aid the defense of the frontier over an increasingly expansive empire. With the exception of the large-scale landowners who grazed cattle, most settlers cultivated only small plots and lived in poverty. They did not develop a market economy for commercial gain.

After contact with the Spanish explorers, missionaries, and soldiers, Native American agriculture changed dramatically. During the seventeenth century, Spanish missionaries introduced a variety of new crops such as wheat, oats, barley, onions, peas, watermelons, muskmelons, peaches, apricots, and apples. They also brought cattle, sheep, goats, donkeys, horses, and mules—all of which the Indian farmers adopted over time. Although European crops became important, no one is certain about the exact varieties that the Spanish introduced. At the missions, the priests distributed seeds, iron hoes, and axes; taught the Native Americans to raise cattle; and maintained oxen and plows for community use. The effect of Spanish influence on the agricultural Indians in the Southwest, such as the Pima, was significant. Wheat, for example, provided a staple that soon replaced corn in the Pima diet. In turn, wheat required a new technology, the plow, which enabled them to produce large crops to meet the demands of newly developing markets at the missions and white settlements. New crops gave the Native Americans a more diverse and nutritious diet.

The Spanish, then, changed the traditional characteristics of Indian agriculture. Wheat and barley required different planting, cultivating, and harvesting techniques, which the Native American farmers learned from the Spanish or other Indian agriculturists the missionaries had taught to farm in the European manner. The Pueblos and Navaho adopted the Spanish practice of grazing sheep and used the wool for blankets. The adoption of cattle, horses, and sheep required crops for feed and forage. Pack animals such as donkeys and horses stimulated agricultural trade and increased the value of the corn crop. When the Indian farmers in the Southwest learned that they could sell or trade their crops to the Spanish, they gave increased, though still limited, attention to farming for profit.

RURAL LIFE

Colonial farmers lived lives governed closely by the rhythm of the seasons. The rituals of plowing, planting, cultivating, harvesting, and preparing crops or livestock for market occurred at predictable times of the year. In the southern colonies, planters prepared their tobacco lands for seeding in February. In the North, they harvested their wheat in July. Early autumn brought the corn harvest and plowing the land for winter wheat. Late autumn and early winter was hog-killing time. Most farmers conducted many of the same tasks, but the climatic characteristics of their region altered the schedule by a few weeks or months. In New England, for example, tobacco farmers harvested their crop in the late summer, but southern farmers did not cut their crop until autumn.

Regional differences also developed in the daily pattern of farm life. In the southern colonies, women conducted the milking, but in New England this was a male responsibility. Moreover, New Englanders "husked" an ear of corn, but southerners "shucked" it. Northerners picked the ears of corn from the stalks in the fields, while Southerners shocked the stalks for removal of the ears later. No matter where a farmer lived, however, sons followed the examples of their fathers in tilling the soil, and mothers taught their daughters the domestic arts. Farm women had less variety in their daily lives than their husbands. During the spring and summer, however, the kitchen garden provided some diversion as well as more work. Rural women broke the drudgery of their lives by attending church services and ritual gatherings at marriages and funerals. Still, most farm men and women lived relatively isolated lives.

Although one contemporary called the British American colonies a "paradise for women" because of the high ratio of males to females and the ease of obtaining a husband, rural life in colonial America was exceedingly hard for most women. Yet women were essential to the success of any agricultural endeavor, whether a yeoman's farm in the North, a plantation in the South, or a squatter's claim on the Appalachian frontier. Male farmers needed the labor and skill of women to process raw agricultural commodities such as flax, milk, meat, and vegetables into subsistence and commercial products.

The agricultural work of colonial women was difficult and physical, and they had few options by law and custom. A woman's property belonged to her husband at marriage. On the death of her husband, she was legally entitled to one-third of his estate, although the husband usually willed her more land than the law required because he trusted her judgment and ability to provide for his children. Most women who remained on the land relied upon their sons to operate the farm or quickly remarried from economic necessity rather than love.

A woman's world was also intellectually narrow, her life bound to domestic needs—preparing meals, milking cows, washing clothes, tending gardens, making soap and candles, and bearing children. Colonial women also carded wool, dug potatoes, picked fruit, and made cider, cheese, and butter—all bringing income into the home. Women who married small-scale farmers worked alongside their husbands in the fields, particularly at planting and harvest. Occasionally they hired themselves out to perform this work for other farmers. By the mid–eighteenth century, women often conducted a wide variety of agricultural labor, and farm work was highly interdependent for men and women. In the late eighteenth century, J. Hector St. Jean de Crèvecoeur, French essayist and observer of rural America, correctly noted that a farmer who had a "good wife" was "almost sure of succeeding."

Slave women worked in the fields, but the common assumptions about a woman's place in society did not apply to them. They were expected to perform a man's work in the fields, and their only rest came with illness or pregnancy. Despite the work expectations of their masters, slave women brought a lower price on the market than prime male field hands, although their value increased if they were exceptionally fertile. Slave women did not enjoy the sanction and respect of marriage, and they had no independent legal or social status. While most southern colonies prohibited intermarriage between whites and blacks by 1700, miscegenation was common.

Although colonial farmers emphasized subsistence agriculture in the beginning, they did so no longer than absolutely necessary. In fact, few colonial farmers relied entirely on their own production for a livelihood. Subsistence farming might mean independence to anyone with a romantic notion of rural life, but to the colonial farmer it meant only hardship. Life in a drafty, cramped log cabin with glazed paper or waxed cloth for windows, a stone hearth for cooking, and standard fare of cornmeal and pork did not appeal to farm families. They endured such hardships only until they could improve thier fortune by producing some quantity of food or fiber for sale at a market.

Although inadequate roads often relegated commercial agriculture to river areas where farm products could easily be shipped to market and limited production to nonperishable commodities such as tobacco, rice, wheat, corn, and salt meat, farmers near the towns and cities had better access to markets where perishable products met a high demand. Indeed, colonial farmers had a keen eye for the bottom line. Profit remained their goal because it alone enabled improvement in one's standard of living—a better house, furnishings, improved tools, education, and security. Profit depended on the markets, local, regional, and international. They became "jacks-of-all-trades" by necessity, but their economic interdependence became increasingly complex throughout the colonial period.

By the Revolutionary era, after the Treaty of Paris in 1763 ended the French and Indian War and provided the motivation for Great Britain to recast its economic policy for the colonies, American agriculture had become increasingly profit-oriented whenever possible. Slave labor characterized the work force on the tobacco, rice, and indigo plantations in the South, and small-scale farmers in the North raised grain and livestock, churned butter, and pressed cheese for a variety of markets. Although overproduction and economic recession periodically caused prices to fall, most colonial farmers remained committed to commercial agriculture.

Commercial agriculture was not a new economic experience for English immigrants in the American colonies. A market economy for agriculture had flourished in Great Britain for a long time, and those settlers who had been farmers at home knew how to raise grain, livestock, and other agricultural products for sale. In the colonies, however, relatively cheap and easy access to land gave American farmers freedom to produce for themselves and the marketplace and leave subsistence farming behind within a relatively short time after settlement. A merchant and artisan class needed food for sustenance and trade, which meant profits for farmers. The high cost of labor, however, forced them to adopt innovative technology, such as the swing or walking plow to replace the heavy-beamed, two-wheeled plows of Europe. If a local blacksmith could craft an affordable plow that lasted a few years, so be it. Another could be purchased relatively cheaply. Time equaled money, and the technology that sped the farmer's work and made it easier and more efficient rather than serving as a long-term investment met a ready demand.

Where high prices for labor and an absence of technology prevented the expansion of commercial crops, planters turned to slave labor and created a caste system in American society. Where farmers met economic frustration, Indian or agricultural policies that they considered unwise and unfair, they often took the law into their own hands to gain by force the rights, privileges, and protections they took for granted. In addition, farmers who could not afford land in settled areas had little choice but to leave for the frontier, but this process broke family ties. The prosperous farmers who remained behind, however, formed a gentry class that wielded economic, social, and political power.

The best and worst features of colonial commercial agriculture would spill across the Appalachians, sweep along the Gulf Coast, and extend into the interior beyond—profits, private property, economic independence, political power, and a respected social status as well as soil exhaustion, political and social elitism, ruthless speculation, and slavery. Above all, farmers who sought commercial production to profit from local, regional, and international markets, no matter what access or commitment to a market economy might be, carried the "culture of capitalism" into the countryside. During the revolutionary era and early national period, the "commercial *mentalité*" of the American farmer remained supreme.

SUGGESTED READINGS

Almaráz, Félix D., Jr. *The San Antonio Missions and Their System of Land Tenure*. Austin: University of Texas Press, 1989.

Bidwell, Percy Wells, and John I. Falconer. *History of Agriculture in the Northern United States, 1620–1860*. New York: Peter Smith, 1941.

Bridenbaugh, Carl. *Fat Mutton and Liberty of Conscience: Society in Rhode Island, 1636–1690*. Providence: Brown University Press, 1974.

Briggs, Charles L., and John R. Van Ness. *Land, Water and Culture: New Perspectives on Hispanic Land Grants*. Albuquerque: University of New Mexico Press, 1987.

Carr, Lois Green. "Emigration and the Standard of Living: The Seventeenth Century Chesapeake." *Journal of Economic History* 52 (June 1992): 271–91.

Carr, Lois Green, and Lorena S. Walsh. "The Planter's Wife: The Experience of White Women in Seventeenth Century Maryland." *William and Mary Quarterly* 36 (October 1977): 542–71.

Chaplin, Joyce E. "Tidal Rice Cultivation and the Problem of Slavery in South Carolina and Georgia, 1760–1815." *William and Mary Quarterly* 49 (January 1992): 29–61.

Clemens, Paul G. E. *The Atlantic Economy and Colonial Maryland's Eastern Shore: From Tobacco to Grain*. Ithaca, N.Y.: Cornell University Press, 1980.

Coclanis, Peter A. *The Shadow of a Dream: Economic Life and Death in the South Carolina Low Country, 1670–1920*. New York: Oxford University Press, 1989.

Countryman, Edward. "'Out of the Bounds of the Law': Northern Land Rioters in the Eighteenth Century." In *The American Revolution: Explorations in the History of Radicalism*, ed. Alfred H. Young, 37–69. DeKalb: Northern Illinois University Press, 1976.

Dethloff, Henry C. *A History of the American Rice Industry, 1685–1985*. College Station: Texas A & M University Press, 1988.

Gates, Paul W. *History of Public Land Law Development*. Washington, D.C.: Government Printing Office, 1968.

Gray, Lewis Cecil. *History of Agriculture in the Southern United States to 1860*, vol. 1. Gloucester, Mass.: Peter Smith, 1958.

Henretta, James A. "Families and Farms: *Mentalité* in Pre-Industrial America." *William and Mary Quarterly* 35 (January 1978): 3–32.

Hurt, R. Douglas. *American Farm Tools: From Hand Power to Steam Power*. Manhattan, Kans.: Sunflower University Press, 1982.

Jensen, Joan M. *Loosening the Bonds: Mid-Atlantic Farm Women, 1750–1850*.

New Haven, Conn.: Yale University Press, 1986.
Jordan, Terry G. *Trails to Texas: Southern Roots of Western Cattle Ranching.* Lincoln: University of Nebraska Press, 1981.
Kulikoff, Allan. *Tobacco and Slaves: The Development of Southern Cultures in the Chesapeake, 1680–1800.* Chapel Hill: University of North Carolina Press, Institute of Early American History and Culture, Williamsburg, Va., 1986.
_____. "The Transition to Capitalism in Rural America." *William and Mary Quarterly* 46 (January 1989): 120–44.
Lemon, James T. *The Best Poor Man's Country: A Geographical Study of Early Southeastern Pennsylvania.* Baltimore: Johns Hopkins Press, 1972.
Littlefield, Daniel C. *Rice and Slaves: Ethnicity and the Slave Trade in Colonial South Carolina.* Baton Rouge: Louisiana State University Press, 1981.
McMannis, Douglas R. *Colonial New England: A Historical Geography.* New York: Oxford University Press, 1975.
MacMaster, Richard K. "The Cattle Trade in Western Virginia, 1760–1830." In *Appalachian Frontiers: Settlement, Society, & Development in the Preindustrial Era,* ed. Robert D. Mitchell, 127–49. Lexington: University of Kentucky Press, 1991.
Menard, Russell R., Lois Green Carr, and Lorena S. Walsh. *Robert Cole's World: Agriculture and Society in Early Maryland.* Chapel Hill: University of North Carolina Press, 1991.
Myres, Sandra L. *The Ranch in Spanish Texas, 1691–1800.* El Paso: Texas Western Press, 1969.
Pruitt, Bettye Hobbs. "Self-Sufficiency and the Agricultural Economy of Eighteenth-Century Massachusetts." *William and Mary Quarterly* 41 (January 1984): 333–64.
Purvis, Thomas L. "Origins and Patterns of Agrarian Unrest in New Jersey, 1735 to 1754." *William and Mary Quarterly* 39 (October 1982): 600–627.
Rothenberg, Winifred B. "The Market and Massachusetts Farmers, 1750–1855." *Journal of Economic History* 41 (June 1981): 283–314.
Russell, Howard S. *A Long Deep Furrow: Three Centuries of Farming in New England.* Hanover, N.H.: University Press of New England, 1982.
Schlebecker, John T. *Whereby We Thrive: A History of American Farming, 1607–1972.* Ames: Iowa State University Press, 1975.
Schumacher, Max George. *The Northern Farmer and His Market During the Late Colonial Period.* New York: Arno Press, 1975.
Vickers, Daniel. "Competency and Competition: Economic Culture in Early America." *William and Mary Quarterly* 47 (January 1990): 3–29.
Washburn, Wilcomb E. *The Governor and the Rebel: A History of Bacon's Rebellion in Virginia.* Chapel Hill: University of North Carolina Press, Institute of Early American History and Culture, Williamsburg, Va., 1957.
Westphall, Victor. *Mercedes Reales: Hispanic Land Grants of the Upper Rio Grande Region.* Albuquerque: University of New Mexico Press, 1983.
Wood, Peter H. *Black Majority: Negroes in Colonial South Carolina from 1670 through the Stono Rebellion.* New York: Knopf, 1975.

AGRARIANISM

AGRARIANISM is the belief that farming is the best way of life and the most important economic endeavor. Agrarianism also implies that farmers willfully sought to avoid commercial agriculture and preferred a "moral economy" in which they produced for subsistence purposes rather than the market and economic gain. The agrarian tradition has long been recognized as central to the American experience.

Agrarianism springs from the writings of Thomas Jefferson, who believed that country people were morally virtuous and superior to city dwellers. For Jefferson, the ownership of land was a natural right and made the small-scale farmer the bastion of freedom and independence, the vanguard of American democracy. Jefferson believed that all wealth and virtue derived from the land and family farmers. Others accepted and perpetuated this belief.

During the presidential campaign of 1896, William Jennings Bryan, Democrat by party and populist by heart, said: "Burn down your cities and leave our farms, and your cities will spring up again as if by magic, but destroy our farms and the grass will grow in the streets of every city in the country." During the early twentieth century, Liberty Hyde Bailey, chairman of the Country Life Commission, also accepted the tenets of agrarianism when he wrote: "The city sits like a parasite, fanning out its roots into the open country and draining it of its subsistence. The city takes everything to itself—material, money, men—and gives back only what it does not want." In contrast, the farmer, engaged in subsistence or self-sufficient agriculture, remained independent and conducted the nation's most important economic activity.

The alliance of agrarianism and democracy created an "agricultural fundamentalism," a belief in the social, economic, and political superiority of rural citizenry. The achievement of universal white manhood suffrage in the age of Jackson, the rise of corporate power during the late nineteenth century, and the development of agribusiness in the twentieth century, however, eventually shifted economic and political power from the countryside to the city. Still, farmers and their supporters, unable or unwilling to accept the transformations of agriculture and rural life, continued to express an ideological belief in the merits of agrarianism.

It is one thing to believe in agrarian values and another to live the agrarian tradition. Agrarianism in American history is more fiction than fact, a classic example of the intrusion of myth into history. It is an example of the ease with which myth has often replaced reality for the explanation of American agricultural history, and it is an example of how some historians have recreated a past that they prefer rather than the one that existed. Agrarianism is a myth that dies hard, and by the late twentieth century it had not disappeared from the American mind. As late as 1987, a poll indicated that the public believed farm men and women had the moral mettle to keep the nation on a proper, steady course. The public still equated agriculture and rural life with the family values of hard work, thrift, honesty, and a neighborliness far removed from the impersonal life fostered by the cities.

Certainly agrarian values are eminently respectable. Freedom, democracy, and economic security based on landowning small-scale subsistence farmers have had an appealing charm, particularly for those who see American society threatened by industrialization and urbanization as well as those who would use class as an ideological historical tool to explain the past.

Yet the myth of the agrarian tradition in American agricultural history is no more clearly evident than in the South. There, planters, small-scale farmers, sharecroppers, and tenants planted tobacco, cotton, sugar cane, and rice for the sole purpose of making money. They were businessmen, although operating on different levels of ability, sophistication, and success, who produced for a market economy. Their reward was not measured by the food that came from their fields and pastures to the family table or moral righteousness but the cash or credit they earned for the purchase of more land, daily necessities, and, for a time, slaves.

Southerners emphasized staple crops for commercial gain. One-crop agriculture enabled them to exploit the soil and the labor supply to their best advantage and to apply their capital where it would gain the highest returns. In addition, sharecroppers and tenants in the South after the Civil War had no freedom of production or independence born of landownership. They looked not to themselves for their existence but to the landlord and the furnishing merchant, who dictated their daily activities and controlled their lives by contract. For them, the agrarian life meant hardship, poverty, and subordination so that the landowner could enjoy wealth, independence, and leisure. Moreover, until the mid–twentieth century, southern planters and farmers were more concerned with producing fiber for the textile industry than food for the family table. In these respects, the reality of capitalism rather than the romanticism of agrarianism governed the development of southern agriculture and history.

Similarly, in the West, the railroads, speculators, and land companies took the first claim to much of the land, not the small-scale independent farmer, who often purchased his acreage at high prices. Although family farmers would buy land or claim public domain, they had to emphasize commercial agriculture as quickly as possible to pay their mortgages, meet their tax obligations, and purchase items of daily necessity that they could not produce. Moreover, although agrarian life offered prosperity and civil rights to men, it promised only hard work and continued subservience for rural women, white and black, slave and free. Agrarianism meant that women could work in the fields alongside their men, but they would seldom own the farm, and the land gave them no political voice or social status except through their husbands. At best, agrarianism extended to women the obligation of serving as a moral presence on the land.

Since Jefferson's time, the concept of agrarianism has

so captured the American mind that it continues to transcend the boundaries of agriculture and rural life, even though farmers dropped below 50 percent of the population between 1870 and 1880 and the rural population fell below 50 percent by the 1920 census. During the 1920s, for example, Bernard Baruch, one of the nation's leading businessmen, proclaimed: "Agriculture is the greatest and fundamentally the most important of our industries. The cities are but the branches of the tree of national life, the roots of which go deeply into the land. We all flourish or decline with the farmer."

Now, agrarianism also applies to nonfarm contexts of social understanding. It has been used to explain everything from low crime rates in rural areas to the suburbanite's love of a well-manicured lawn. Politicians have championed the merits of agrarian values, particularly when fighting reapportionment or a campaign requires a forceful commitment to something that almost everyone can support. During the late twentieth century, one Kansan perpetuated Jefferson's concept that rural residents were a chosen people when he said, "In the cities you have machines so you vote for whoever is put up; in the country we vote for the man, not the party." Clearly, the concept of agrarianism has maintained its adherents for two centuries of American history.

Yet by the late twentieth century, independent subsistence agriculture had been left in the distant past. Farmers produced for a market economy, usually by specializing or emphasizing a mix of one or two crops and livestock. They purchased their food in town at supermarkets like everyone else. Never truly independent yeomen, they were now businessmen who produced for sale and purchased a host of goods. Moreover, the land-grant universities provided agriculturists with appropriate farm-management systems to monitor their activities and balance sheets efficiently and accurately as well as new forms of science and technology to improve their productivity. With this aid, any notion that agriculture was a basic independent industry became absurd, particularly when farmers sent their children to the city to earn a better living than they could enjoy on the farm.

In addition, agribusiness has so enlarged the vocation

of agriculture that small-scale farmers now involve non-household members such as bankers, food-processing companies, and outside investors in the decision-making process. Since the New Deal, the federal government has been an active player if not a partner on the farm, further alienating the traditional agrarian virtues of rugged individualism, economic independence, and personal control of the farm enterprise.

Near the century's end, some of the strongest supporters of the social values of agrarianism were not farmers but presidents, congressmen, state legislators, editors, and bureaucrats. Although most farmers gave tacit recognition to the importance of agrarian values, most understood that the size and efficiency of an agricultural operation rather than rhetoric were necessary for a good life on a farm. Although family farms grossing less than $50,000 annually remained sacred in the American mind as a symbol of democratic government, social stability, and agricultural abundance, their economic viability became increasingly uncertain. This uncertainty caused the supporters of the family farm to defend it not just on economic and social grounds but as the moral keystone of American society, even though thousands fled the farm each year for a better life in the cities.

Despite the failure of agrarian values to keep people on the farm, despite sentimental attachments to the rural past and nostalgic views of farm life, agrarianism has succeeded in becoming a volatile and sometimes effective political force in the American experience. Its most common expression occurs when farmers demand state and federal support. By the late nineteenth century, agrarianism usually meant the use of political means to achieve economic ends for farmers. Since that time, agrarianism as a political force has sought increased government regulation or control of the economy to enable farmers to earn an adequate return on their investment and labor and enjoy a comfortable living.

Today, many Americans hail the virtues of rural living based not on Jefferson's concept of an independent yeomanry anchored on the farm but for safety, privacy, and a slower-paced lifestyle. Yet the boundaries between rural and urban communities are often arbitrary and unclear. If the flight to the countryside for the benefits of rural living is agrarianism, it is a new agrarianism far different from the fundamentalism of the past. Although the "refugists" to the countryside view the city as unnatural, they seek rural life for escape or relatively cheap living rather than farming. Farming no longer offers virtues superior to other lifestyles and occupations.

Thus, while the agrarian tradition originally meant that the farm provided more than bread alone, agriculture in the late twentieth century often did not furnish enough food for sustenance or ensure economic viability. Small-scale, family-farm agriculture declined as a fundamental industry.

Although the belief in the agrarian tradition remains, it is colored with myth. It paints a mental image of a past that never was while denying the reality of contemporary agricultural life. Agrarianism now, as in the past, remains more myth than reality.

---

3 • THE NEW NATION

GREAT BRITAIN CHANGED ITS POLICY OF ECONOMIC BENIGN NEGLECT toward the American colonies following the conclusion of the French and Indian War in 1763. Quickly a host of policy changes occurred that would drive the colonists to armed rebellion a dozen years later at Lexington. While political and economic relationships changed dramatically between Great Britain and the colonies, the commercial bent of the American country people continued to shape colonial agriculture. The American Revolution benefited farmers because the armies of both friend and foe required great quantities of food. Farmers with access to military markets eagerly sold surplus commodities for economic gain. Moreover, during the late eighteenth century, England could no longer supply grain to other European countries because it had a burgeoning population to feed.

With periodic crop failures and war in Europe, American agricultural commodities increased in demand and price. By 1788, Thomas Paine's observation that American agriculture always would flourish as long as eating was the "custom of Europe" proved correct. After American independence, market demands in the West Indies for flour, corn, beef, and pork further stimulated commercial agriculture. Farmers who raised both grain and livestock earned profits and improved their standard of living. Increased grain production also encouraged urban growth, based on the processing and shipment of agricultural commodities such as flour to the markets beyond.

During the revolutionary era, then, as well as the early years of the new republic, farmers produced for a market economy whenever possible. Regional climatic and geographical conditions, however, continued to determine the nature of commercial agriculture. A short growing season, poor soil, and wheat rust prevented extensive grain crops in New England. Cattle raising, dairying, and vegetable and fruit production for local markets prevailed. In the mid-Atlantic states, wheat farming flourished. Farmers in Maryland and Virginia also planted wheat to diversify and reduce their reliance on one-crop tobacco agriculture. Thomas Jefferson reported that wheat production "diffuses plenty and happiness among the whole" and that tobacco farming was "a culture productive of infinite wretchedness." Jefferson opposed the planters' emphasis on tobacco because it tied them to European buyers and agents for credit and supplies and kept them in perpetual debt. For him, diversified agriculture offered the rewards of economic independence, a higher standard of living, and a better way of life.

At this time, southern agriculturists and farmers on the western frontier cultivated more corn than any other crop. Still, corn was not a cash but a household crop because farmers primarily used it for home consumption, livestock feed, and a recreational beverage. In the Deep South, upland or short-staple cotton became a viable crop that soon spread to the interior. During the late eighteenth and early nineteenth centuries, then, Americans remained a country people closely tied to the soil. Farmers first met household or subsistence needs, but complete self-sufficiency remained beyond their ability or desire.

Although access to markets might be limited and the surplus production of farmers scant, they eagerly sold agricultural commodities in order to purchase the things they could not produce efficiently or affordably—shoes, tea, sugar, nails, and tools. The sale of agricultural products also helped farm families acquire capital for the purchase of more land or to improve their acreage. Farmers at first moved beyond subsistence by exchanging or bartering products with local merchants. Tobacco, butter, and pork, for example, could be exchanged for flour, coffee, and cloth. This rudimentary trade was a form of commercial agriculture; farmers exchanged surplus commodities, in the absence of an adequate circulating currency, for gain. American farmers saw commercial agriculture as a goal and a reality.

THE AMERICAN REVOLUTION

After the Battle of Lexington on April 19, 1775, the newly emerging states confronted the formidable tasks of defeating Great Britain and feeding the Continental Army and the American people as well as maintaining the important international trade in agricultural commodities. To meet the immediate needs of an army, the Continental Congress offered land bounties to encourage enlistments. The government had little money to pay recruits for risking their lives against the British, but it had abundant lands to offer for their services. This land bounty policy was not new. Several colonial governments in New England had used it to recruit militiamen during the seventeenth century for service against the Indians. The Continental Congress adopted that policy and used it extensively, even though the western lands that it pledged for service technically belonged to the Crown or the new state governments.

In September 1776, Congress passed the first general bounty act, which provided 100 acres to privates and as many as 1,100 acres to major generals. If the government lacked access to the necessary lands at the conclusion of the war, the states were required to meet these obligations. Virginia also used a land bounty to encourage enlistments in its Continental Army units, offering its vast claims in present Kentucky, West Virginia, and the Old Northwest. Generosity marked Virginia's recruitment land policy; depending on rank, field officers could receive as many as 5,000 acres, although privates could claim only 50 acres. In addition, Congress lured Hessian soldiers to desertion with offers of 50 acres of land and tempted British soldiers by providing 50 to 800 acres, depending on rank, as a reward for desertion. When the war ended, the veterans frequently sold their bounty certificates to speculators for a few cents on the dollar rather than move beyond the Appalachians and claim their property. Some speculators in the Ohio country made fortunes from the acquisition and sale of these bounty lands.

During the Revolution, farmers produced sufficient agricultural commodities to keep the Continental Army and the populace adequately fed. In New England, they provided cattle, hogs, sheep, fruits, vegetables, and cider for the troops, and mid-Atlantic farmers furnished wheat, flour, bread, and beef. When food shortages occurred during the winter of 1777–78 at Valley Forge, the troops suffered hunger because of transportation problems, particularly the shortage of horses and bad roads and an inattentive Congress, rather than failure of farmers to produce enough food. Supply problems occasionally prevented Washington's army from acquiring adequate hay and oats for the horses. Local food shortages also occurred where British troops occupied or contested areas such as Boston and New York City.

Slow-moving ox-drawn wagons transported agricultural supplies over poor roads, and the army invariably moved faster than its supplies. Although water provided the quickest and most efficient means to move farm commodities, the British controlled the coastline. Farmers adjusted by shipping their products, particularly grain, on rivers and streams whenever possible. Still, most farm products reached market by land, in sacks on packhorses and wagons or on hoof.

Overall, farmers profited from the war. Although inflation quickly became a major problem when the Continental Congress and the states began printing paper money to finance the war, farmers did not mind because it caused agricultural prices to increase. Some patriots complained about price gouging, but they directed their criticism primarily at merchants and middlemen rather than farmers. Merchants contracted with them for produce to sell to the army or the Continental Congress, which took over the business of provisioning the troops.

With gold coin in short supply and the paper money issued by the Second Continental Congress nearly worthless, farmers preferred to sell their commodities for bills of credit—promises to pay with sound money at a later date. In time, bills of credit circulated like currency. British and Hessian soldiers sometimes bought food and forage from Tory farmers, such as those on Staten Island, and they paid hard money. Often, however, they took what they needed and burned grainfields. Patriot farmers tried to hide cattle, sheep, poultry, and other agricultural commodities rather than lose them to enemy confiscation. Early in the war, General Nathanael Greene reported that, "The country all resounds with the cries of the people—the enemy plunders most amazingly."

But requisitioning agricultural produce by the Continental Army also encouraged "plundering and licentiousness" rather than fair treatment of farmers. Early in 1778, to protect patriot farmers as much as possible from over enthusiastic foraging parties from his army, Washington ordered the soldiers to "leave as much forage to each farm as will serve the remaining stock 'til next grass, as much grain as will support them 'til harvest, some milch cattle and a reasonable number of horses." Little wonder, then, that patriot farmers willingly sold to British troops, who paid cash.

Although the major markets in Great Britain largely collapsed, farmers and merchants continued to trade with the British and French West Indies. This market for flour, wheat, tobacco, indigo, cattle, and barreled pork had developed before the Revolution, and neither the sugar planters in the West Indies nor the American farmers wanted to lose it. American merchants traded farm products particularly for weapons, ammunition, and European textiles. After France joined the war in 1778, the French navy provided a new, lucrative market for New England farmers. The French established headquarters at Newport, Rhode Island, and contracted with merchants for large quantities of beef, flour, and other foodstuffs, and they paid with coin.

The war did not cause a substantial decline in labor or productivity on farms because women took the place of the men who joined the army. The war, however, changed the nature of some farm activities because of the shift from manpower to womanpower. Farm women preferred to raise corn and potatoes because bread grains required more physical labor. Corn continued as the most important crop because it could feed both humans and livestock. Moreover, it was easier to harvest than wheat or other small grains that required a sickle or cradle scythe. Corn also could be shelled with less labor than wheat, which required threshing with a flail. In addition, corn could remain in the field well into the late autumn or early winter without spoilage. Wheat had to be harvested quickly, within a few days of ripening.

Women preferred to raise potatoes because this crop produced large harvests for relatively little labor, and the crop stored easily. Moreover, the soldiers often returned home during plowing, planting, and harvest to help their wives with the heavy farm work, particularly if they were part of the militia rather than Continentals. This shared toil benefited American agricultural productivity but seriously hampered Washington's ability to maintain a functional army.

Commercial livestock production increased during the Revolution. The army demanded large quantities of beef, preferring live animals to guarantee freshness and to prevent the purchase of barrel-packed horse or rotten meat from merchants. During the war, farmers also began to emphasize the number rather than the size of cattle. In the North, the average weight at slaughter fell from 1,100 pounds to 900. As a result, farmers could produce more beef with the same amount of feed. In the Shenandoah Valley, however, slaughter weights for beef cattle averaged about 550 pounds, an increase from the cattle sold to the army during the French and Indian War.

Farmers raised sheep primarily for wool rather than mutton. Indeed, it became unpatriotic to eat lamb, and the Committee of Safety in New York declared those who ate lamb "enemies of their country." The army needed wool for clothing, and farmers met that demand. They also shifted flax production from an emphasis on seeds for oil to the fiber, which could be mixed with wool for a relatively cheap hard-wearing cloth called linsey-woolsey.

Wartime demands for ropes, cordage, sacks, and sailcloth also increased the price and production of hemp, particularly in Virginia. The intensive labor required for this crop created the first demand for slaves by family farmers in the Shenandoah Valley. Scots-Irish farmers dominated the valley's hemp production during the war. In addition, the Revolution increased demands for wheat in the form of flour and cattle and hogs in the form of beef and pork to feed the army. The production of these commodities necessitated the increased production of corn, hay, and oats to feed this livestock as well as the horses used for mounts and the oxen yoked for draft power. In some areas, such as Virginia, farmers could not meet the military and civilian demands for pork and bacon, and prices remained high.

Farm life and agricultural activities remained basically unchanged during the Revolution, the rural population increasing approximately 40 percent between 1775 and 1790, compared to an urban increase of only 3 percent. Clearly, the Revolution did not restrict agricultural development or ruin farm life. Moreover, farmers became more commercially oriented than ever before, even though war disrupted some markets for the sale of butter, cheese, and milk in New England and New York, and Virginians sometimes had trouble selling tobacco.

The American Revolution most profoundly affected Loyalist or Tory farmers, whose property was confiscated. In May 1775, Congress authorized the first confiscations of Tory land in Massachusetts, a policy quickly followed in the other states and continued throughout the war. Massachusetts seized the land of anyone who fought against the United States or the property of anyone who moved to British-controlled territory without the permission of the state or central government. By 1782, New York had confiscated approximately $2.5 million in real property. The state governments usually sold those lands to wealthy patriots. Confiscations, however, did not result in a major property redistribution to landless farmers or small-scale farmers because this resource remained cheap and readily available and most farmers held more of it than they could cultivate. Even so, the Tories lost their lands forever, and many fled to

Canada or the British West Indies. Still, these developments did not affect agricultural production because relatively few Tories were farmers.

The Revolution clearly influenced the institution of slavery. Although few slaves worked on the farms of New England and the mid-Atlantic states, Massachusetts and Rhode Island organized black fighting units and offered freedom to slaves that enlisted. In the upper South, Maryland and Virginia farmers planted less tobacco because of price declines, and planters hired out their bondsmen to work in the fields and tend livestock or repair fences. In the Deep South, the rice and indigo plantations continued to require a large amount of slave labor.

In 1780, although the war moved to the South with the British capture of Charleston, Crown troops stayed close to their supply lines along the coast. Many slaves used this opportunity to flee, either to the more lenient Northern states or British lines. This flight caused serious problems for many plantation owners. White southerners feared arming blacks to strengthen defenses against the British, and slaves who were hired from owners for state militia service regularly tried to escape. Whenever Continentals captured the slaves of Tories, these bondsmen were treated as contraband and often used as military laborers or sold, the proceeds divided among the captors. The British, however, tried to weaken the American agricultural economy by offering freedom to any slave who escaped to British lines and bore arms against former masters. One Philadelphian responded to British emancipation policy by writing, "Hell itself could not have vomited anything more black than this design of emancipating slaves." Patrick Henry, Virginia patriot, feared the worst and urged "early and unremitting Attention" to the control of slaves. He had reason to worry; more than three hundred blacks joined Lord Dunmore's Ethiopian Regiment in Virginia.

Some Americans offered their slaves freedom if they would join the Continental army or navy. By contrast, Virginia and South Carolina offered a slave to any citizen who enlisted in the army and served for the duration of the war, although neither state had sufficient resources to continue this effort for long. Most white southerners believed that the British lure of freedom would incite slave rebellions, and they established special military patrols to quell possible insurrections and harshly punished fugitive slaves.

Even so, tens of thousands of slaves fled to the British. If these bondsmen belonged to Tories, they were returned to their owners. If not, they were put to work building fortifications such as those at

Yorktown. Others served as guides, general camp laborers, and servants. When the British fled at the end of the war, many slaves escaped with them. Long after the Revolution, southern planters and small-scale farmers who had lost slaves demanded compensation from the British, and southern states cited this grievance in refusing to pay their prewar debts. American losses were substantial. Virginia lost approximately 30,000 bondsmen to the British; South Carolina's losses totaled about 25,000 slaves. These bondsmen were not returned to the planters, nor were their masters compensated. Still, the planters survived this loss, and the institution of slavery quickly rebounded from the war, particularly with the invention of the cotton gin, which made short-staple cotton economically profitable.

Overall, the American Revolution, like all wars, benefited farmers, especially those who were well established, provided their lands were not laid waste, their homes and buildings burned, their crops and livestock destroyed, or their lives lost. Although the ravages of war hit some farmers hard, most profited, even though they lost their markets in the British Isles. Prices increased, and all governmental regulatory efforts failed. In New England, for example, farmers refused to sell cattle, hogs, and potatoes for temporarily low prices and forsook the market until they could secure a fair return. They were not unpatriotic, because they met the food needs of civilians and soldiers alike, but they were profit minded. The Revolution offered economic opportunity, and they took it. After the Revolution, most men and women remained farmers.

LAND POLICY

During the war, the large states such as Massachusetts, Connecticut, and Virginia continued to maintain vast land claims to the interior. The states without western land claims contended that the transappalachian region belonged to the central government because it had been won in a common effort and should benefit the public. Moreover, they argued that because those lands had been controlled by the British government, they should naturally pass to the jurisdiction of the Continental Congress. The smaller states, such as Maryland and New Jersey, specifically contended that the central government should sell the "back lands" to finance governmental operations. Otherwise, the states that made large claims in the interior could sell those lands and pay off their war debts, while the smaller states that did not claim such lands would necessarily be required to levy taxes to meet their obligations.

Speculators, who often had acquired nebulous titles to western lands, also believed they would have a better chance to profit from their investments if the central government controlled land policy. And they helped to convince the Maryland legislature to reject the Articles of Confederation until those western lands had been ceded to the national government. Ultimately, Thomas Jefferson helped convince the Virginia legislature to relinquish its claims to the interior, which it did on January 2, 1781. Maryland then ratified the Articles of Confederation in February. Effective March 1, 1781, Virginia's western lands came under the authority of the central government, with the expectation that new states in the interior would join the Union. The other states relinquished most of their western lands to the federal government by the end of the century.

Upon gaining title to the western lands, Congress provided for the survey and orderly sale of the Ohio country under the Land Ordinance of 1785. This statute formalized the New England tradi-

36	30	24	18	12	6
35	29 Reserved to Religion	23	17	11 Reserved to U.S.	5
34	28	22	16 School Reserve	10	4
33	27	21	15	9	3
32	26 Reserved to U.S.	20	14	8 Reserved to U.S.	2
31	25	19	13	7	1

ON MAY 20, 1785, Congress passed the land ordinance that authorized the survey and sale of the public domain. The ordinance provided for the survey of townships six miles square. Each of the thirty-six sections would contain 640 acres. Congress reserved four sections for future disposition—section 16 to support local schools and sections 8, 11, and 26 for its own purposes, such as the payment of military bounties. In 1787 the Ohio Company of Associates, who acquired more than one million acres in Ohio, also reserved section 29 in the New England tradition for the support of religion. An administrative decision made by the Board of Treasury during the Confederation era, not a provision of the law, standardized the numbering of the sections up and down, beginning in the southeast corner. The Land Ordinance of 1785 enabled anyone who purchased public lands to locate their property with relative ease. *U.S. Department of the Interior.*

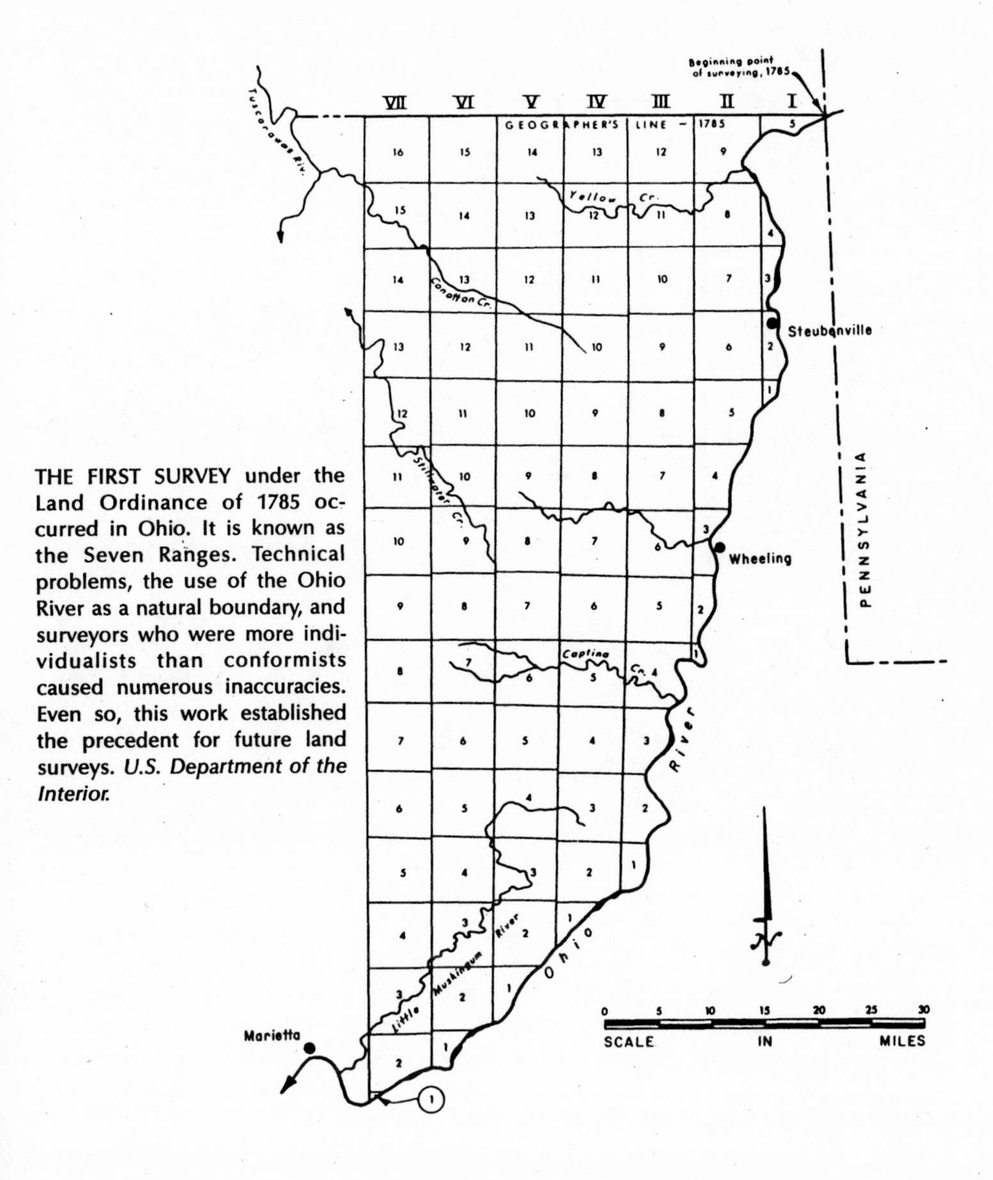

THE FIRST SURVEY under the Land Ordinance of 1785 occurred in Ohio. It is known as the Seven Ranges. Technical problems, the use of the Ohio River as a natural boundary, and surveyors who were more individualists than conformists caused numerous inaccuracies. Even so, this work established the precedent for future land surveys. *U.S. Department of the Interior.*

tion of rectangular survey before sale. The Land Ordinance authorized the survey of townships 6 miles square. With 36 lots (later called sections) in each township, section 16 was reserved to support a public school. The government then offered the remaining land for sale, with the minimum purchase of a section, 640 acres, at the minimum price of $1 per acre, at public auction. Congress, however,

6	5	4	3	2	1
7	8	9	10	11	12
18	17	16	15	14	15
19	20	21	22	23	24
30	29	28	27	26	25
31	32	33	34	35	36

Reserved Sections

ON MAY 18, 1796, Congress changed the requirements for the sale of public lands and inadvertently authorized a new system for numbering sections in the townships. Thereafter all townships had section 1 located in the northeast corner, with the numbers running horizontally back and forth across the ranges. Congress also reserved sections 15, 16, 21, and 22, although it did not designate section 16 as a school reservation. *U.S. Department of the Interior.*

reserved the United States Military Tract and the Virginia Military District to satisfy land bounties granted to soldiers who served during the Revolution. The Western Reserve also remained under the control of Connecticut and closed to public entry under federal land policy.

Most farmers, however, could not take advantage of federal land policy because they could not afford to purchase a section of land for $1 per acre. Instead, much of the newly opened public land was acquired by speculators who formed land companies that commanded sufficient capital to make large-scale purchases. Often these speculators convinced Congress to reduce the price of the public land to a few cents per acre. In 1787, Congress passed the Northwest Ordinance, which created the Northwest Territory and provided a system of government for the farmers and speculators who moved into the region.

Most governmental officials believed the western lands should be sold at a price that would generate revenue for the operation of the federal government. Others, particularly landless farmers, advocated cheap or free public land to develop the economy of the backcountry and with it the financial strength of the nation. Although this debate would not be resolved until the mid–nineteenth century, those who favored sale of public land for revenue prevailed. The Land Act of 1796 reaffirmed the provisions of the previous legislation but increased the minimum price to $2 per acre. Theoretically, this higher price would help retire the public debt and prevent too much emigration from the settled areas to the detriment of those who depended on an adequate labor force. It also provided for the consecutive numbering of township sections, from 1 to 36, beginning at the northeast corner, running to the west and back to the east, in alternating rows, to help buyers locate their land.

In time, Congress established federal land offices throughout the West to help with the survey and sale of the public domain. But public sales lagged as many settlers bought smaller tracts for less money from land companies or speculators. Although speculators charged more per acre than the federal government, they sold in smaller blocks. As a result, a settler who could not afford to pay $1,280 for 640 acres, as the government required, could more easily pay $200 for 50 acres at a price of $4 per acre from a speculator.

The increase of population and competition for land increased the prices speculators charged. Even so, no agriculturists on the frontier could farm a section of land, so speculators performed a useful service that helped expand agriculture into the interior. Moreover, many farmers became speculators. By clearing a portion of their land for crops and livestock and building fences, houses, and barns, they increased the value of their property. When the price reached an attractive level, many sold their land and used their gains to purchase other land on the frontier. Squatters, however, demanded the right of preemption, that is, first right to purchase the lands on which they had illegally settled upon survey but prior to the public auction of that property. In 1800, Congress approved the preemption of land settled before that year.

In 1800, Harrison's Frontier Land Act reduced the minimum acreage for purchase to a half section, 320 acres, but it did not reduce the price. This act, however, permitted farmers to pay in two equal installments over four years and provided for limited credit at 6 percent interest. Even so, many settlers could still not afford public land. In 1804, Congress responded by reducing the minimum acreage for purchase to a quarter section, 160 acres. Although sales increased thereafter, speculators most easily afforded the $320.

Farmers and speculators with capital quickly took advantage of this land policy, and the cotton kingdom expanded rapidly along the Gulf Coast and into the interior. Squatters went west with the buyers and claimed land of their own. In 1802, Governor Ferdinand Leigh Claiborne of the Mississippi Territory reported to Secretary of State James Madison that a "great proportion of the present population in this Territory is composed of citizens who have formed settlements

IN 1804 surveyors began using principal meridians and base lines that cross them at right angles for points of reference. By dividing areas into quadrants, surveyors could plat, that is map, ranges of townships for further division. Because meridians (lines of longitude) converge to the north, surveyors kept each section equal by adjusting their measurements every twenty-four miles from the base line and every six miles from the principal meridian. The adjusted township boundary lines, which run north and south, are called *correction lines. U.S. Department of the Interior.*

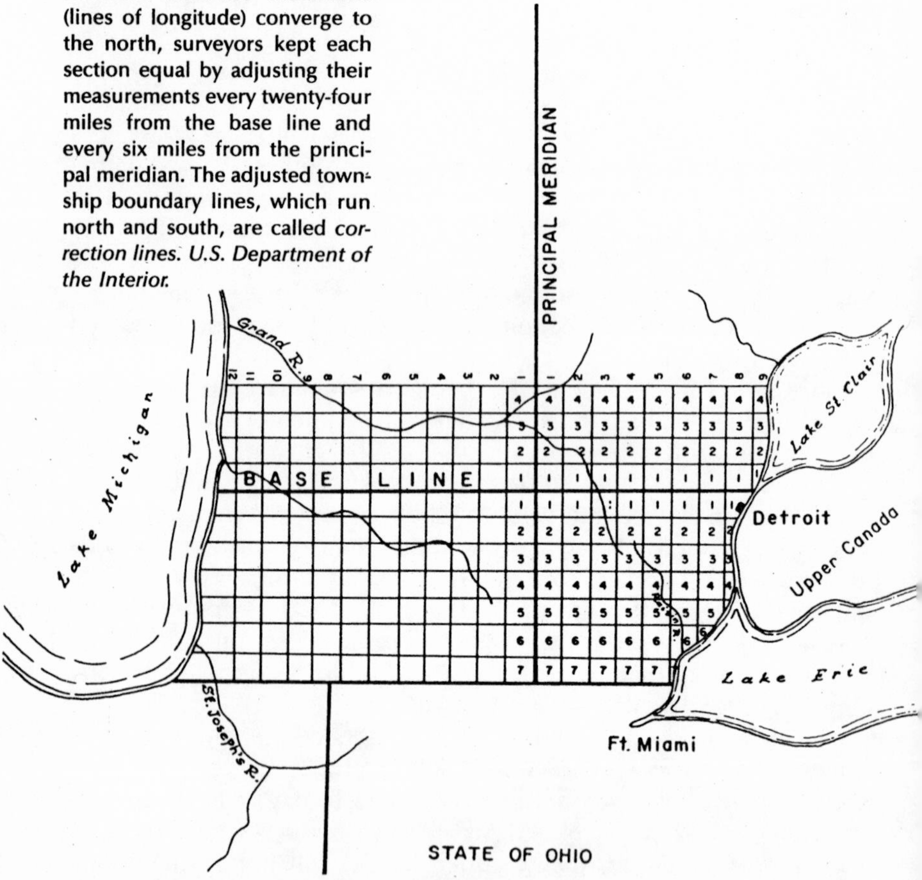

on vacant lands." The credit system established by the Harrison Land Act, however, caused incredible recklessness. Farmers and speculators acquired lands with little thought about payment. In Alabama, nearly 4 million acres were claimed on credit, and buyers in Mississippi and Missouri claimed 1 million acres in each state. When the run for land halted with the panic of 1819, much of the acreage purchased on credit under the Harrison Act was foreclosed. Even so, these westerners soon received relief from Congress with new land legislation.

THE SOUTH

After the Revolution, tobacco planters suffered from worn-out soil, surplus production, and low prices. Farmers in Maryland and Virginia continued to abandon tobacco for wheat, which caused production and exports to fall precipitously. By the end of the eighteenth century, farmers in the Shenandoah Valley had become the leading wheat producers in Virginia. There, more than four hundred gristmills ground approximately 10 million pounds of flour annually; on the eve of the War of 1812, production reached 15 million pounds. The increase of population in towns such as Alexandria and Richmond helped expand the wheat and flour markets and kept prices high. In contrast, tobacco was not a paying crop, averaging only $.06 per pound when war came again.

With the old tobacco lands largely exhausted, planters had little opportunity to maintain or expand production by breaking new lands; little good land remained to clear and plant. Until tobacco farming moved beyond the Appalachians into the backcountry of Kentucky and Tennessee during the early nineteenth century, this commercial agricultural enterprise languished. Between 1790 and 1815, tobacco returned profits only for those farmers in the Piedmont region of Virginia and North Carolina. There, tobacco farmers sent their crops to market down the rivers to buyers in Lynchburg, Richmond, and Petersburg, Norfolk serving as the chief export center for the crop.

While tobacco farming declined, the market for indigo collapsed altogether because Britain favored its own planters in the East Indies. One merchant in South Carolina reported that money from indigo came in "vastly slow." By the turn of the nineteenth century, indigo production had nearly ceased. Rice planters also suffered from overproduction and low prices. Moreover, rice farming required large acreages and considerable labor, both of which were beyond the grasp of small-scale farmers, who could not afford great tracts of irrigated land or numerous slaves.

In 1795, when the old crops of tobacco, rice, and indigo no longer offered the opportunity for great economic gain, Etienne de Boré, a Creole planter in French Louisiana, experimented with a crop of sugar cane and a granulating process for changing its syrup into sugar. Boré used thirty slaves to produce his sugar cane and earned $12,000 from the crop. Soon thereafter, French planters, who immigrated from Santo Domingo where they had experience raising sugar cane, made this new agricultural industry thrive. By 1802, seventy-five plantations produced 8.5 million pounds of sugar along the Mississippi south of the Red River. After the Louisiana Purchase in 1803, these planters had free access to the American market. In 1817, one observer reported that sugar cane plantations stretched for a hundred miles above New Orleans. These plantations were "superb beyond description," and the owners derived "immense profits," averaging annual incomes of between $20,000 and $30,000 from those lands with slave labor. Still, this new agricultural endeavor was limited to a small area along the lower Mississippi River, and American sugar never became a major agricultural commodity for international trade.

ELI WHITNEY

Smithsonian Institution.

Eli Whitney (1765–1825) is commonly known for inventing the cotton gin, thereby making the cultivation of short-staple cotton profitable. Whitney was born on December 8, 1765, and raised in Westborough, Massachusetts. He attended Yale University and in 1792 traveled to Savannah, Georgia, to teach and study law. While staying in the plantation home of Catherine Littlefield Green, he was challenged to construct a machine that would separate the seeds from the fiber of short-staple or green-seed cotton. Whitney had an inventive mind and a skillful hand. By April 1793, he had built an implement that could clean fifty times more cotton per day than any other method.

Actually Whitney did not invent the first cotton gin. In antiquity, farmers on the Indian subcontinent used cotton gins that consisted of wooden rollers attached to a frame. As a crank turned the rollers, the fiber passed between them, and the seeds squeezed out. Some cotton planters used roller gins of this type in the British American colonies as early as the 1740s. These gins also cleaned black-seed or long-staple cotton, introduced about 1786 from the Bahamas, which grew only along coastal South Carolina and Georgia. These gins, however, could not remove the seeds from short-staple cotton, which could be grown only in the uplands. To remove the seeds from the lint of this cotton, workers had to separate the fiber and the seeds by hand. At best, a worker could clean only 1 pound per day. This labor-intensive work was too expensive and time-consuming, and southern farmers could not raise short-staple cotton on a commercial basis.

Although many inventors designed cotton gins during the eighteenth century, Whitney developed a new technique to remove the seeds from the short-staple fiber. Whitney's invention consisted of a cylinder with wire teeth set in annular rows in a box. As a worker turned the handcrank, the cylinder revolved, and the wire teeth pulled the fiber through grooves in a screen and left the seeds behind. A revolving brush swept the lint from the cylinder's teeth. Whitney patented his cotton gin in 1794, but other inventors and manufacturers soon designed gins largely based on his principle. Whitney sued for patent infringement, but he did not win his case until 1807. In 1814, his patent expired, and Congress refused to renew it. Although Whitney did not earn great wealth from his invention, his cotton gin enabled the expansion of short-staple cotton production across the South and with it the institution of slavery. His cotton gin became one of the most important technological innovations in American agricultural history.

At this time, only a few farmers raised cotton. Prior to Eli Whitney's invention of the cotton gin, the black-seed or long-staple variety was the only viable cotton crop. The seeds could be removed from this variety relatively easily, but it could not be raised profitably beyond the Sea Islands of South Carolina and Georgia or more than 50 miles in the interior on the coastal plain. Short-staple or green-seed cotton could be cultivated in the uplands, but the seeds were so difficult to remove that most farmers refused to plant it except for home consumption. In 1793, cotton farmers in South Carolina and Georgia raised only about 3 million pounds. Whitney's cotton gin, which easily separated seeds from fiber, revolutionized southern agriculture, and the cotton culture soon expanded to the West. By 1811, southern farmers produced 80 million pounds.

At the same time that the cotton gin made the short-staple crop economically viable, the Industrial Revolution in Great Britain, which produced technological changes such as the power loom, flying shuttle, and spinning jenny, increased the efficiency and productivity of the textile industry, which in turn placed tremendous demands on southern farmers for cotton. The small-scale farmers and planters alike responded. By 1812, they produced 60 million pounds, three-fourths of the American crop, in South Carolina and Georgia alone. Cotton farming became a boom activity and expanded rapidly into the Piedmont of Virginia and North Carolina and along the Gulf Coast in Alabama, Mississippi, and Louisiana.

In 1800, cotton farmers could earn a profit if they sold their crop for $.12 per pound, but in that year the price soared to $.44 per pound. With good land capable of producing a 500-pound bale per acre, great profits could be earned. The cost of keeping a slave on a large plantation averaged $15 per year and from $30 to $40 on a small plantation, but a moderate cotton crop of 2,000 pounds produced by one slave returned $880, more than enough to cover this expense. Between 1799 and 1810, though, cotton usually brought between $.15 to $.19 per pound, and 300 pounds of cotton per acre became a more realistic figure for maximum production. Still, these prices and this production level remained high enough to earn a profit, after expenses for keeping slaves and buying land. Motivated by these high prices, southern farmers with capital rapidly purchased and cleared new lands, bought slaves to work the fields, and moved into the middle class. With the cotton gin, cotton and slavery became symbols of southern agriculture.

Indeed, with the invention of the cotton gin and the acquisition of New Orleans in the Louisiana Purchase, cotton farming expanded rapidly across the South. New Orleans provided an international export market for short-staple cotton, particularly for Great Britain's rapidly expanding textile industry. In 1800, one planter in Natchez estimated the cotton harvest from the region would reach 3 million pounds and $750,000. Cotton, he wrote, was "the staple of the Territory, and is cultivated with singular advantage to the planter." With ginned cotton bringing $.25 per pound in New Orleans and Louisiana planters harvesting between 500 and 800 pounds per slave, the future looked bright. One woman planter wrote that New Orleans provided a market for "all we could make." By 1803, two dozen cotton gins along the Red River pushed cotton farming still farther into the interior.

In the Shenandoah Valley of western Virginia, cattle raising remained important. Drovers assembled their herds of 100 head or more by purchasing steers from local farmers for trailing to the markets in Philadelphia, Baltimore, New York City, and Richmond. Shenandoah cattlemen fattened their livestock on corn, fed from the stack during the winter, for sale in the spring. The drovers trailed hogs to market behind their cattle herds to fatten them on the wasted feed. In 1793, a British traveler was "astonished at the multiplicity of cattle" that poured through Winchester, Virginia, from the backcountry on the way to urban markets. Cattlemen in the Shenandoah Valley continued to perfect their corn-feeding practices to fatten cattle for market and improve their breeding stock, some by importing purebred bulls from England. In the valley, most farmers raised cattle and hogs for "insurance" against crop failures. This practice in turn necessitated the widespread cultivation of corn and the maintenance of hayfields and pastureland.

Like the frontier farmer of the North, economic gain provided the motivation for southern cotton farmers to pioneer into the transappalachian West. They differed only in their extensive land holdings and their ownership of slaves. Yet with much of this new southern agriculture financed by credit and with farmers and planters reliant on international markets, profits were by no means guaranteed. Still, cheap land, slave labor, and available credit enabled commercial cotton production to spread across the black soils of the Gulf Coast and into the uplands. By 1815, commercial cotton farmers had reached the Mississippi River, and they laid the foundation for the cotton kingdom that flourished until the Civil War.

SLAVERY

The fertile lands of the South could not be exploited for the cotton culture without a large labor supply. Yet these cheap lands made labor expensive because most men and women preferred to own their farms rather than work for someone else. With a new crop that offered the potential of great profits, southern farmers and planters expanded the institution of slavery to meet their labor needs. Although slave labor had been primarily restricted to tobacco, rice, and indigo plantations during the colonial era, short-staple cotton and the gin enabled the institution to spread quickly across the South. Cotton farming now became profitable where it could not have existed before.

Cotton was well suited for the gang-labor system; large fields required extensive labor for planting, cultivating, and harvesting. Because cotton bolls ripened at different times, slaves usually picked the fields three times. With labor costs high, planters preferred slaves, who were servants for life and whose costs could be spread out over a lifetime. With each bondsman able to cultivate as many as 10 acres of cotton and 10 acres of corn and moderately fertile land producing from 300 to 500 pounds of fiber per acre, slaveowners could keep their workers busy and earn a substantial profit each year. Surplus slaves from the Upper South, New Jersey, and New York, together with bondsmen from Africa, supplied farmers and planters in the rapidly developing cotton lands.

By 1810, more than 1 million slaves worked in the cotton, rice, tobacco, and cornfields of the South. As farmers and planters sought new land in Alabama, Mississippi, and Tennessee, slaves cleared much of this land for cotton fields. While the large-scale planters used drivers who were slaves themselves and white overseers to manage their gangs of bondsmen, small-scale farmers, who owned only a few slaves or hired them from the planters, usually worked alongside their bondsmen in the fields. By 1815, Nashville, Natchez, Baton Rouge, and the lower Tombigbee and Alabama river valleys had become major cotton-producing and slaveholding areas.

Throughout the postrevolutionary era, the price of slaves increased with the demand. In 1790, prime male field hands, those 18 to 25 years of age, averaged about $200 on the Charleston market. That price jumped to $300 in 1795 after the cotton gin began to have an effect on production, and the average reached $550 in Charleston by 1812. At that time, prime male field hands brought between $600 and $700 on the New Orleans market.

THE NORTH

After the Revolution, farmers continued to migrate to New Hampshire, Maine, and Vermont, and by 1812 they had claimed almost all the good agricultural land in those areas. As worn-out fields and increasing land values pushed agriculturists away, the fertile, cheap, "wild lands" on the frontier pulled them from the older communities. In the settlements of New England, they continued to raise corn, rye, and livestock for their own needs and to supply local markets. But wheat crops still suffered from black stem rust, which also hampered farmers in the mid-Atlantic states. During the early

1790s, however, farmers in present Maine (then a part of Massachusetts) increased their wheat production for markets in British Nova Scotia, New Brunswick, and the West Indies.

The Napoleonic Wars (1795–1815) disrupted British wheat supplies, and farmers along the east coast of Maine, especially the Penobscot River valley, quickly took advantage of a profitable opportunity. By mid-July 1804, flour prices reached $26 per barrel in Puerto Rico, and other Caribbean markets remained nearly as high at a time when Maine farmers considered $10 per barrel profitable. Grain shippers in Portland also found a "good market" for wheat and corn at Cadiz, Spain, despite an American embargo in 1807 and the watchful eye of the British navy.

Although peace and bad weather ended this prosperity after 1815, for a brief time these northern farmers prospered from expanded grain production. In the mid-Atlantic area, wheat remained the "principal article for making money," averaging between 10 and 20 bushels per acre. With the exception of New England, wheat served as the major crop for farmers engaged in mixed or diversified agriculture. Philadelphia, New York, Baltimore, and to some extent Boston provided the primary markets for grain raised in the northern backcountry and served as major export cities for grain shipments to the West Indies.

However, corn remained the most important cereal in the North, production averaging between 20 and 25 bushels per acre. In New Hampshire, farmers considered 20 bushels a "paying crop." To achieve a moderate level of production, many New England farmers fallowed their land for a year or two after several consecutive plantings. Another method entailed converting depleted lands to pastures of clover or grass. Grazing cattle provided manure, which augmented the soil's nitrogen content and benefited subsequent grain crops. Still, most farmers did not rotate their crops, and soil exhaustion remained a major problem.

Between the American Revolution and the War of 1812, Connecticut became the preeminent agricultural area in New England, where farmers provided beef, pork, horses, rye, oats, and flax to the new American states. New London and New Haven served as the major ports for agricultural exports. Livestock raising remained important; farmers in New England confined their cattle to barns during the winter and tried to raise enough hay to feed their livestock until fresh grass appeared in early May. In 1774, one observer noted that farmers in Rhode Island "choose to graze their land as they think it more profitable." In New England, farmers sold their cattle to drovers, who trailed the livestock to markets in New York and Boston.

Farmers still used cattle for beef and milk and marketed surplus cheese and butter via the local general store. Farm women especially traded their butter for needed goods, and the local merchants in turn sold these commodities to other dealers in the major towns and cities such as Boston. To the south, Pennsylvania and New Jersey stockmen continued to let their cattle roam with virtually no shelter or supplemental feed throughout the year. German farmers, who built sturdy stone barns capable of sheltering livestock during the winter, were the exception.

New England farmers also increasingly raised sheep after the Revolution, although the harsh winters caused significant losses. Flock masters wanted to produce wool primarily for home consumption, although wool production became a commercial enterprise in some areas such as Martha's Vineyard, Nantucket, and Narragansett. Each sheep produced 2 to 3 pounds of coarse wool annually. Few farmers raised sheep for mutton because Americans never developed a taste for this meat and it did not preserve easily.

In 1802, Colonel David Humphrey, American ambassador to Spain, imported the first fine-wooled Merinos to the United States. The embargo of 1807 and escalating wool prices encouraged farmers to purchase Merinos and to upbreed their flocks to improve the annual wool clip. But when peace came with the Treaty of Ghent in 1815, flock masters did not have tariff protection; British merchants flooded the American market with woolens, and the price of the raw commodity declined precipitously. Many flock masters now stopped raising sheep.

In contrast, farmers continued to raise swine for home use and sale. Salt pork packed in barrels called hogsheads met a ready demand from the fishing fleets that docked in New England ports. Because hogs provided lard, the chief cooking fat of that day, fat animals brought the highest prices. When swine reached 200 or more pounds, they were ready for slaughter, packing, and sale. After the Revolution, farmers increasingly used horses rather than oxen for draft power. Oxen were slower than horses and apparently less prestigious. In 1788, one observer noted: "I know many persons who would sooner carry their articles to market on their own shoulders, than be seen driving an ox team."

New England farms in the early Republic averaged between 100 and 200 acres, a clear indication that the village system of allotment had broken down. Most of this land, however, was not cultivated near the turn of the nineteenth century. On a 100-acre farm, only 4 to 6 acres would have been planted, with an additional 8 or 10 acres in pasture. The rest remained woodland or "brush pasture," which the farmer used to graze cattle and hogs and kept in reserve for clearing and planting when his croplands became exhausted. In the middle states, where better soils prevailed, farmers cultivated more acreage. In 1790, for example, a 200-acre farm in Bucks County, Pennsylvania, averaged about 75 acres in crops and 65 acres in meadow and pasture, the remainder in woodland.

Increased agricultural trade from the major port cities not only expanded regional markets at Baltimore, Philadelphia, New York, and Boston but encouraged the development and improvement of transportation systems. In Baltimore, for example, harbor improvements accommodated more ships and helped stimulate the international trade for flour, thereby improving the market for wheat. In 1785, the Maryland legislature also authorized road construction to link the backcountry with that port city. Soon Baltimore's flour merchants entered into partnerships with rural millers, who in turn purchased wheat from local farmers, thereby creating submarkets. Other cities experienced similar market and transportation developments.

After the Revolution, farmers also moved west into the Mohawk Valley of New York and across the Alleghenies. Before the defeat of the Indians in the Old Northwest at Fallen Timbers in 1794, few farmers ventured into the Ohio country. Those who braved the wilderness and hostile Indians remained subsistence agriculturists until they gained access to markets via flatboats along the Ohio River for the sale of commodities such as butter and cheese. As early as 1789, however, the merchants in the newly founded town of Marietta, Ohio, offered shoes, shovels, and clover seed for sale. Although local farmers were "poor and proud" and "totally Voyd of Money," they traded corn and pork for these necessities. This trade moved them beyond the level of subsistence because barter required farming with a commercial *mentalité*. The production and exchange of surpluses for gain, no matter how small the production or slight the level of trade, enabled farmers to participate in a market economy.

Between 1795 and 1810, the population west of the Appalachians increased from 150,000 to more than 1 million. Invariably these settlers came for economic gain. Culturally, they drew upon their knowledge of the frontier past from their home regions. The agricultural practices of the transappalachian frontier differed little from the older frontiers east of the mountains. Traditional crops and livestock provided the means for commercial success. In some areas, these frontier farmers planted corn, tobacco, and hemp and raised hogs and beef cattle. Others became dairy farmers who produced butter and cheese for consumers on the eastern and western seaboards. All became tied to international events that affected their markets. As soon as they produced for the trade or sale of their commodities, they became linked to the wider world beyond. Jay's and Pinckney's treaties, ratified in 1795, the Embargo Act of 1807, and the War of 1812 affected the pocketbooks of farmers in the interior for both good and ill. Commercial agriculture on the frontier gave these farmers every reason to be concerned with international events.

Farmers rapidly moved into the Old Northwest after 1795. Soon commercial agriculture became the basis for all economic activities, but it did not come easily. The wilderness exacted a price in backbreaking toil from men, women, and children. In 1810, one Indiana settler wrote that land was cheap and fertile but settlers should expect to meet "hardship and expense." Still, by raising corn and hogs, these frontier farmers could put cash in their pockets or trade for commercial goods. As early as 1803, Merino grades had been driven and shipped to Ohio from Massachusetts to improve the quality of fleece among the flocks in that new state. In 1810, Ohio farmers drove as many as 70,000 hogs over the mountains to eastern markets. Dairy farmers began selling cheese and butter in Pittsburgh and various towns down the Ohio River. Packed pork soon followed, and New Orleans became the major market for the Old Northwest. In 1817, the agricultural economy improved when steamboats gained sufficient power to ply the waters upstream, bringing needed goods and carrying even more flour, corn, cheese, butter, and pork down the Ohio and Mississippi rivers to the Crescent City for transport to eastern and international markets.

Agricultural wealth and commercial production, however, did not mean the acquisition of hard cash at first. In 1800, Arthur St. Clair, governor of the Northwest Territory, wrote that tax levies exceeded the amount of money in circulation. "Money is extremely scarce in this part of the country," he noted. Debts in Cincinnati could be paid with barrels of farm-packed pork. Across the transappalachian frontier, farmers paid their debts and acquired needed goods with corn, flour, lard, butter, whiskey, tobacco, and hemp. But this commercial trade supported the development of grist- and fulling mills, slaughter and meat-packing houses, and distilleries, and with the multiplier effect, stimulated carpentry, blacksmithing, merchandising, and other trades in the towns. Economic gain lured settlers to the transappalachian frontier, and by the end of the War of 1812, it began to pay dividends, even though life remained hard and the environment unforgiving.

AGRICULTURAL IMPROVEMENTS

During the Revolutionary era, farmers continued to prefer practical to permanent agricultural tools and implements. Instead of heavy-beamed plows with wheels, the American farmer favored the light swing or walking plow that any village blacksmith could craft. These wooden moldboards with wrought-iron shares were relatively cheap, and easy to replace if broken. Still, wooden-moldboard plows were not standardized in design to permit uniform and consistent tillage of the soil. If a farmer owned a plow that cut easily through the soil and turned a furrow smoothly, it was because of accident rather than design.

Thomas Jefferson believed that a moldboard plow could be designed on mathematical principles to permit standardization and uniform reproduction. Such plows would provide maximum tillage ability and reduce the power required for pulling. Moreover, scientifically designed moldboards could be made for specific soils, conditions, and crops. Jefferson's suggestions for improving plow designs were significant, particularly his recommendation that moldboards be made from iron rather than wood. This was an important idea; standardization of design could not be achieved by using wood since each plow maker crafted it as he pleased. Only metal that was cast, wrought, or molded would permit the consistent duplication of a superior design.

Jefferson never cast a moldboard, but in 1797 Charles Newbold, a New Jersey inventor, patented a cast-iron plow. Newbold cast the moldboard, share, and landside in one piece, but it was far from practical because if any part broke, the entire plow became useless. Several years later, in 1807, David Peacock, also from New Jersey, patented a cast-iron plow with a separate moldboard and landside to which he attached a steel-edged share. Peacock's design was more practical than Newbold's because a worn-out or broken part could be replaced. About this time, the belief that cast iron poisoned the soil began to fade, and Peacock's plows gained widespread popularity in the mid-Atlantic states.

The concept of standardized, replaceable parts for agricultural implements, however, is usually credited to Jethro Wood of Scipio, New York, because he was the most successful in marketing a plow with these features. In 1814, when Wood patented his plow with interchangeable parts, he probably knew about Peacock's design because he did not claim that he had invented the principle of interchangeable parts. Wood's steel-tipped share cut through the soil efficiently and required less sharpening than cast-iron plows. His plow probably did more to eliminate the clumsy old wooden plows than any other design at that time, and farmers were quick to adopt it. Wood's plow remained popular for decades and inspired other inventors to fashion plow designs after his model. Most of the plows patented for a long time thereafter differed very little in their general principles.

By the 1790s, the farmers used two basic types of harrow—the square and the triangular or A-frame. The square harrow served best on old fields that were free from obstructions such as tree stumps, roots, or rocks. Farmers commonly used the triangular harrow on newly plowed lands that had such obstructions because it was stronger and less likely to break, and did not collect as much stubble as the square harrow. Both harrows had wooden frames with wood or iron teeth.

American farmers began using the seed drill on a limited basis by 1775. These drills were imported from England or made locally. Newly cleared fields with stumps and rocks were not suited for this implement. Furthermore, most farmers seeded winter wheat in fields where corn had recently been harvested and the stalks plowed under; the stalks and weed stubble clogged the drill tubes. These early drills also failed to plant uniformly on roughly plowed ground and were too expensive for most farmers. Moreover, agriculturists who were familiar with grain drills did not find them practical. Farmers could sow as much grain by hand as they could reasonably expect to reap; if they planted more with a seed drill, it would be wasted in the absence of sufficient labor at harvest.

Even so, in 1799, Eliakim Spooner of Vermont patented the first American grain drill. Other patents for seeders followed during the next forty years, but inventors made little technical improvement, and few farmers had any direct knowledge of drills or their use. Instead, farmers either sowed by hand or used broadcast seeders. These seeders were of three general types—handcrank, fiddlebow, and wheelbarrow. Grain drills did not become practical until the 1840s, and farmers continued to plant their corn by hand with a hoe or dibble stick until the 1850s.

Farmers cut their small-grain crops with a cradle scythe or sickle. They would not use a mechanical reaper on a wide basis until the mid–nineteenth century, although Richard French and T. J. Hawkins of New Jersey patented the first American reaper on May 17, 1803. The design of this machine is unclear, but it had three wheels, a series of scythelike knives that revolved on a vertical spindle, and long wooden fingers that extended into the grain below the cutter. A team of horses drew the machine from the side. Other experiments followed, all of which met with either failure or limited success. American inventors had begun to give their attention to the grain harvest, although success would elude them for another thirty years. Farmers continued to harvest corn by picking the ears for shelling or shocking the stalks for removal of the ears later.

After the harvest, small-grain farmers threshed their crops with a flail or treaded the grain from the heads with horses or oxen. By 1775, the fanning mill had been introduced from Great Britain to speed the job of winnowing. The fanning mill consisted of a series of wooden paddles attached to a rod geared to a crank. The paddles or fans were enclosed in a boxlike frame that housed several screens. As the grain and chaff were poured into the container at the top, the farmer turned the crank, causing the fans to draw air through apertures in the sides and blow it across the screens. The grain fell onto the screens and sifted to the bottom as the forced air blew away the chaff. The clean grain poured into a basket below. In 1791, Samuel Mulliken, a Philadelphia inventor, patented the first threshing machine in the United States. This machine was too complicated to work efficiently, and American inventors made little improvement on this implement until the 1820s.

Agricultural reformers, usually large-scale wealthy planters, also began experimenting with various cropping practices and attempted to disseminate new knowledge about the best farming methods. In 1785, John Beale Bordley, a Maryland planter, and several prominent Pennsylvanians organized the Philadelphia Society for Promoting Agriculture, and the South Carolina Society for Promoting and Improving Agriculture and Other Rural Concerns was also organized. In New Jersey, the Burlington Society for the Promotion of Agriculture and Domestic Manufacturers formed in 1790. The New York Society for Agriculture, Arts, and Manufacturers organized a year later. In 1792, the Massachusetts Society for Promoting Agriculture was established, and in 1811 the Society of Virginia for Promoting Agriculture was founded.

These organizations of planters, gentleman farmers, merchants, and professionals encouraged agricultural experimentation. The members were especially interested in plowing, fertilization, and crop and breed improvement, particularly with the importation of purebred English livestock such as Shorthorn cattle and Merino sheep. These organizations, however, did little to aid the average small-scale farmer, who not only lacked access to their publications but distrusted "book farming."

The desire to improve agriculture, however, led to the creation of a new institution—the county fair. In 1811, the Berkshire Agricultural Society in Massachusetts held what may have been the first American fair, and others quickly followed. Men and women exhibited their best crops, livestock, and home manufactures and competed for prizes. They tried to learn new and improved agricultural techniques, foster a spirit of competition, and promote social contacts.

BACKCOUNTRY REBELLIONS

During the revolutionary era and the early days of the new Republic, political, economic, and social problems as well as Indian hostilities sometimes caused farmers to rebel against civil authority. Occasionally animosity arose between the settled areas and the frontier over matters of representation. This regional schism led to a violent confrontation between Pennsylvania's rural western and more populous eastern regions. The eastern counties enjoyed greater representation in the assembly, which meant that political control resided with the Quakers instead of the largely German population in the West. Quaker control and opposition to the use of force prevented a satisfactory policy for dealing with Indian hostilities, a situation that brought sharp criticism from the men and women who lived on the frontier. Although the Quakers lost control of the assembly in 1756 and defensive measures improved, frontiersmen believed the government did not act decisively enough, particularly with the outbreak of Pontiac's Rebellion in the summer of 1763. As farmers fled their homes for shelter in the towns, the assembly continued to prohibit offensive action by the militia.

Because of the government's inaction and its inability or unwillingness to distinguish between hostile and peaceful Indians, a group of vigilantes organized near Paxton and marched toward Philadelphia in 1764 to topple the government. Some Pennsylvanians feared civil war. Although negotiations defused the situation, hostile feelings between West and East remained. The backcountry farmers remained underrepresented in the colonial legislature, their grievances unattended, although the government strengthened the British troops in the colony and continued to pay for Indian scalps. Most important, the march of the "Paxton Boys" indicated the severe rift between those who governed and those who farmed on the frontier. The frontiersmen would continue to demand increased political participation, and when the revolution came, the Paxton Boys and their supporters would champion expansion of democratic government and overthrow of a sectional aristocracy.

Soon after the Paxton rebellion in Pennsylvania, another farmers' uprising occurred in the backcountry of North Carolina, where changing economic and political relationships contributed to an agrarian revolt. A shortage of circulating currency relegated trade to barter and exacerbated property-tax payments. Dishonest sheriffs customarily seized movable property to meet those taxes, often selling it for their own profit. By 1767, the backcountry sheriffs, particularly in the western counties of Anson, Orange, and Rowan, were pocketing approximately half the tax revenue they collected. Farmers often could not seek protection or restitution from the courts located far away, and legal fees were extortionate. Economic hardship prevailed.

Many backcountry farmers refused to pay their taxes; forcibly disrupted court sessions; physically attacked sheriffs, judges, and lawyers; and attempted to make local government more democratic and responsive to public problems by electing farmers to office. At Hillsboro in October 1770, a gang of farmers, "headed by men of considerable property," closed the session of the superior court and attacked the lawyers present, whom they "cruelly abused with many and violent blows." These farmers called themselves Regulators. Government officials considered them no better than "traitorous Dogs." This Regulator movement was not suppressed until the militia, under the direction of Governor William Tryon, defeated several thousand Regulators at the Battle of Alamance near the headwaters of the Cape Fear River on May 16, 1771. On June 19, six Regulators were hanged for violating the Riot Act of 1770. This military defeat, together with the unwillingness of the colonial government to provide tax relief, caused many backwoods farmers to move farther west, laying the foundations for the state of Tennessee.

During this time, South Carolina also experienced a Regulator revolt. Backcountry farmers in South Carolina had a host of grievances. They wanted a more equitable property tax, schools, roads, and bridges, as well as additional courts so that they could avoid the long trip to Charleston for legal business. But they were most concerned about suppressing bandits, who "range[d] the Country with their Horse and Gun, without Home or Habitation" and pilfered cattle, assaulted men and women, robbed farm homes and country stores, stole or offered refuge to slaves, and endangered the social and economic order of the colony.

To contend with these problems, bands of Regulators, bent on subduing "roguish and troublesome" people who robbed and terrorized the countryside, whipped, jailed, and banished those they considered idlers and criminals. The South Carolina Regulators were far closer to vigilantes than other farmers who participated in agrarian revolts during the late eighteenth century. Indeed, the South Carolina Regulators, who were primarily commercial landowning farmers and slaveholding planters, enforced law and order in the backcountry on the eve of the Revolution. They required those deemed "reclaimable" to cultivate a certain acreage within a specific time or be whipped. Presumably, this forced labor would train vagrants to become self-sufficient farmers and civilize them so that they would no longer burden their neighbors. For the South Carolina Regulators, farming ensured social order, and they would force criminals to take up the plow or leave the countryside.

In contrast to North Carolina, South Carolina's colonial legislature supported this vigilante action by organizing two ranger companies that included many Regulators. Lieutenant Governor William Bull called these farmers an "industrious, hardy race of men." Additionally, in 1769, the assembly approved the Circuit Court Act, authorizing the creation of a judicial system and the employment of more sheriffs. By 1770, the authorities had brought many bandits and other troublemakers to justice. Some had been executed and others jailed; many fled the colony. The South Carolina Regulator movement ensured property rights, slavery, and social order in the backcountry.

In the North, the farmers in western Massachusetts also experienced economic hardship that led to civil strife, in this case after independence from Great Britain. Following the Revolution, the British terminated credit and demanded that merchants pay for their orders in specie, hard money. These merchants in turn demanded that farmers meet their obligations by the same terms rather than goods in kind such as butter, meat, or grain. Moreover, the state government imposed a heavy tax on land to help pay the revolutionary war debt. When farmers could not meet their tax obligations or debts, creditors and the government often foreclosed. The seizure of their property infuriated these farmers, who feared not only the loss of their economic security and social status but the possibility of being jailed as common debtors.

During the mid-1780s, these Massachusetts farmers sought relief by petitioning the state government to print paper money, thereby increasing and inflating the currency supply, and to pass legal-tender laws, thus permitting payment of debts and taxes with agricultural commodities. The farmers also obstructed the courts to prevent foreclosures, and they intimidated tax collectors. The creditors rejected the farmers' proposals for reform and won the support of the legislature to continue their more restrictive economic policy. In the autumn of 1786, many farmers abandoned peaceful protest and attacked the local courts to prevent foreclosures until the legislature provided relief. These "rebels," under the leadership of Daniel Shays, also prevented the state supreme court from meeting in Springfield.

The governor responded by requesting troops from the central government to quell the rebellion, and the legislature passed a riot act authorizing county sheriffs to kill the rioters if necessary to restore order. Anyone convicted of violating this statute was subject to a public whipping, jail, and the loss of property. When contingents of the state militia rather than national troops marched against the rioters, the farmers renounced the state government and pledged to overthrow it by force. Then they attempted to seize the arsenal at Springfield on January 25, 1787, in preparation for an armed march on Boston to burn the city and topple the government. The attack on the arsenal failed, however, and the troops dispersed the rebels.

Although insurrection continued until June 1787, the movement dissolved with little gained for these farmers. Shays's Rebellion, however, highlighted inherent weaknesses of the government under the Articles of Confederation and influenced reform. Accordingly, when the Constitutional Convention met at Philadelphia during the summer of 1787, the delegates were determined to endow a new federal government with sufficient power to maintain peace at home and ensure the execution of the law.

The new government under the Constitution did not have long to wait for the first test of its power. In 1791, farmers in western Pennsylvania vociferously opposed an excise tax on whiskey that Congress authorized to help pay the national debt. These farmers, who converted much of their corn crop to liquor to facilitate shipment to market, rejected this oppressively high tax of 25 percent on the net price of a gallon of whiskey. Moreover, like those who had participated in Shays's Rebellion, they customarily used whiskey as money, bartering wet goods for dry goods at local stores.

In 1794, the protests of these Pennsylvania farmers became violent when they terrorized excise officers and closed federal courts by force of arms. These insurgents even talked about attacking Pittsburgh and Philadelphia. Deploring their refusal to pay the excise tax and propensity to interfere with the lawful activities of the federal government, President George Washington ordered them to disperse but offered them amnesty in return for their pledge to obey federal law. When the farmers rejected Washington's directive, he called upon the states to provide troops to suppress the rebellion, which threatened to spread to Kentucky, Maryland, Georgia, and the Carolinas. Federal troops dispersed the rebel farmers with a show of force, and the tax remained. Washington's decisive action contributed to the development of a strong national government that became the facilitating agent for agricultural improvement more than a century later.

In 1799, the last major backcountry rebellion occurred in Pennsylvania. Known as Fries's Rebellion or the Hot Water War, it involved the violent opposition of German farmers in Bucks and Northampton counties to federal tax laws. The opposition of these farmers became so serious that President John Adams sent federal troops and the state militia to quell their rioting and arrest them. The root of Fries's Rebellion can be traced to a congressional enactment in 1798 that levied a direct graduated tax on land, slaves, and horses to help finance naval operations in the undeclared war against France. John Fries and more than two dozen others refused to pay those taxes, verbally and physically intimidated assessors, and occasionally scalded them with water as they measured houses to determine the tax. Fries and others circulated petitions, demanded the repeal of this property tax, and organized town meetings to find ways to foil the law. In addition, the local militia forcibly released prisoners who had refused to pay their property taxes from the custody of a federal marshal in Bethlehem. Federal troops arrested Fries and his supporters for this "system of intimidation." Then the government tried and convicted them for treason.

In 1800, President John Adams pardoned the rebels, perhaps realizing that these German farmers simply did not understand the purpose of the tax. The inability of the assessors to speak German or the farmers to speak English no doubt contributed to many misunderstandings. Apparently, after these farmers had the law explained to them, they readily complied with its provisions. Most important, though, the speed and force with which the federal government moved to suppress this rebellion clearly indicated that the new nation would act decisively to protect its interests, even against a band of culturally different farmers who posed no threat to the Union. After these backcountry rebellions, American farmers generally shifted their methods of protest from violence to political action. Violent agrarian opposition to federal economic policy would not recur until the Farmers' Holiday movement during the 1930s.

These backcountry rebellions were local in origin. None had links to other episodes of civil disobedience in the new nation. Yet each resulted in part from problems caused by the rapid expansion of farmers on the frontier. In each case, these farmers found themselves located in remote areas and unable to participate easily in the marketplace. Faced with economic, social, and political problems that adversely affected their agricultural way of life at a time when state and federal government increasingly infringed on their lives, these farmers wanted both government protection and noninterference. They certainly did not want higher taxes but the opportunity to meet their subsistence needs and produce for commercial sale. Consequently, they struck at anyone who sought to reduce their financial security, whether Indians, local gentry, or government officials.

RURAL LIFE

American men and women remained a country people during the Revolution and the early years of the Republic. With approximately 90 percent of the population engaged in agriculture or related work at the time of the Revolution, rural life prevailed. At the turn of the nineteenth century, perhaps 80 percent of the population still worked the land, towns and villages serving their market needs. Only Boston, New York, Philadelphia, Baltimore, and Charleston merited the designation of city. Rather than living in villages and working the fields beyond, as agriculturists did in Europe, American farmers scattered into the countryside. They often claimed or purchased large acreages in the South for the extensive production of staple crops, or they bought or squatted on smaller holdings and raised corn and hogs for subsistence and sale in the form of whiskey, salted or smoked pork, or swine on the hoof. The thump of the woodsman's ax and the pungent smell of woodsmoke were common in the backcountry, and both meant that men and women were clearing the land for agriculture.

Rural life changed with the seasons, particularly for men. Whether one lived in the North or the South or somewhere in between, planting time came in the late winter or early spring for crops like tobacco, rice, and corn. As the months passed and the seasons "circled," men had changing responsibilities for livestock, crops, and the land. Summer meant weeding, and autumn brought the harvest and processing time. A variety of farmer's almanacs provided a monthly schedule of chores, from cutting rails for fences and preparing seedbeds to hackling flax and flailing the wheat crop. When in doubt, the astrological explanations in each almanac explained why a farmer should plant tobacco or shear his sheep at a certain time. In an age when farmers had greater access to reading materials that explained the significance of the constellations than to publications that discussed the use of science to improve agricultural productivity and profits, superstition often prevailed. Indeed, superstition, along with tradition, determined most agricultural practices in America.

Although social contacts decreased between spring planting and the autumn harvest, the country stores provided convenient meeting places where farm men and women collected their mail, sold or traded produce, and exchanged gossip. With corn the preeminent crop in North and South, families and neighbors often congregated in the late autumn to husk the crop. These "affairs of mutual assistance" not only helped farmers complete a major task but also served as important social experiences that eased the burden of rural isolation.

In New England, where the Congregationalists followed the path of their Puritan ancestors by insisting that everyone learn to read in order to know God's law as revealed in the Bible, the public provided tax-supported schools to teach the basics of reading, writing, and arithmetic. Beyond New England, if boys or girls attended a school, it was often a private academy. Public school attendance was not mandatory, and few children old enough to work attended between spring planting and the autumn harvest. Only in the winter months, when cold weather made outdoor work too difficult in the North and often uncomfortable in the South, did children regularly attend school. Rote memorization and oral recitation served as the prevailing educational methodologies, and illiteracy remained high in rural America. Farm children learned the practical country arts from their parents. Vocational agricultural education would not come until the twentieth century.

Americans both rural and urban enjoyed an abundant food supply, but it seldom had great variety. On the frontier, "hog and hominy" prevailed; rural inhabitants usually dined on salted or smoked pork and a corn dish such as bread, mush, or hominy. Garden vegetables in season, as well as eggs, milk, cheese, and butter, gave some variety to their diet, but the absence of canning techniques limited food preservation to storage in root cellars, smoking, salting, and drying. Fresh meat often came only in the late autumn after the weather had turned consistently cold at hog-killing time. Then fresh pork or beef could be kept for several days without danger of spoilage. Women served salted meat for the daily fare at urban dinner tables in the North; smoked meat and bacon, together with corn bread (known as "corn pone" and "hoecake"), became common foods in the South. North of the Chesapeake, rural dwellers preferred wheat and rye rather than corn, although they mixed cornmeal with other flours for their bread. Western and southern farmers also relied on game.

By 1810, one European observer noted that Americans had undergone a revolution of diet, with a "profusion" of summer and winter vegetables. Farmers adopted other crops, such as "Irish potatoes" in the North and sweet potatoes in the South. Moreover, the expansion of American commerce at home and abroad enabled rural merchants at the crossroads stores and market towns to offer a wide variety of foodstuffs such as coffee, salted fish, fruits, and sugar. These modest changes in the America diet helped improve the quality of rural life.

If the diets of rural people improved, their clothing remained homemade. Linsey-woolsey served as the standard cloth for pants, shirts, and dresses. Wardrobes were not cluttered with clothes. Farm women might own only a "best" dress for special occasions and one or two other dresses for daily work. Men and boys often wore "tow-cloth" pants and shirts, clothing made from the shortest fibers of the flax plant, particularly in New England. These clothes were functional but far from fashionable. One observer wrote: "A man whose clothes were made at home could be easily distinguished at a hundred yards distance by his slouchy and baggy outlines." Women's dress also showed little concern for "elegance or fashion." Stylish clothes, however, would have been out of place throughout most of rural America. With the increased availability of factory-made cotton and wool cloth after the turn of the nineteenth century, rural women put away their spinning wheels, and the spinning bees that brought them together for social interaction disappeared. As a result, isolation prevailed, and social gatherings remained infrequent.

Just as the seasons determined the practices and customs of rural life, gender determined the division of labor. The world of farm women remained tied to the garden, barn and dairy, chicken house, and kitchen. Their world focused on the home and family; that of their husbands and sons turned outward to the commercial world of profit, politics, and power. Although women worked alongside men hoeing tobacco and corn, raking hay, harvesting grain, digging potatoes, cutting flax, and picking cotton, their primary work was domestic labor—grinding cornmeal; feeding chickens; making soap, butter, and candles; cooking meals; keeping house; washing clothes; and tending children. After the general adoption of the cradle scythe by the middle of the eighteenth century, fewer women went into the fields at harvesttime because this implement weighed considerably more than the sickle, which they had used with skill for reaping in the past. In the South, once farmers and planters could afford slaves, women left the tobacco and cotton fields for household responsibilities. White women on small-scale farms still might work in the fields, but European cultural tradition generally kept them from heavy work such as plowing, shearing sheep, or breaking flax and hemp.

Farm women continued to be chiefly responsible for the domestic manufactures and production that brought money into the farm home. They processed milk into butter and cheese, made brandy from fruit, raised poultry, and prepared meat at butchering time by salting, pickling, or smoking it for home consumption, commercial sale, or trade. They brought eggs, beeswax, and goose feathers to country merchants and exchanged these commodities for dry goods, flour, and ironware. The butter churn and the spinning wheel became the chief tools of country women who produced for a market economy. Men needed women for labor as well as companionship. Although the work of men and women was interrelated and often shared, gender divided many tasks and enabled the efficient management of farm and family activities. The work of women improved the quality of rural life and the economic soundness of the household.

In the plantation South, slave women generally served in domestic capacities, chopped and picked cotton, or tended other field chores. Girls began their bondage with household tasks such as kitchen work, child care, or laundry. As teenagers, young slave women began working in the fields. They remained field hands until pregnancy and childbirth temporarily took them from their responsibilities of planting, hoeing, and harvesting. Yet children increased their work load because they had to perform many of the domestic duties of white women farmers after they returned from the fields. When a slave woman became too old to work in the fields, the mistress often assigned her domestic duties around the plantation. Even so, only women cooks, laundresses, and thread spinners had any assurance that they could avoid field work, so necessary and important was this responsibility. The larger the plantation, the more likely that African-American women would work in the fields.

The American Revolution and its aftermath still touches the present wheat farmer. When German troops arrived on Long Island, they brought the Hessian fly in the bedding of their horses. This insect increased rapidly, and by 1788 it had spread to the wheat fields of Pennsylvania. By 1800, it wreaked havoc on crops in New England, New York, and west of the Alleghenies where the fly's larvae fed on the roots of newly sprouted plants. In time, farmers learned to delay planting until after the first hard frost, which killed the adult flies and prevented the larvae from devouring their crops. Today, wheat farmers still depend on soil temperature to determine planting time and help ensure a harvest.

Overall, by the end of the War of 1812, agricultural surpluses were not always large, but they were profitable. Crop failures and war in Europe increased American agricultural trade abroad, particularly for wheat. Thomas Jefferson, reflecting on the hostilities between Great Britain and France during the early nineteenth century, thought that a major new European war would bring "good prices" for wheat. If the United States could avoid that impending conflict, Jefferson believed, American farmers would profit if only British and French soldiers would eat a "great deal." British imperial policy on the high seas temporarily ruined Jefferson's hopes, but after the embargo of 1807 and the end of the War of 1812, American shippers renewed agricultural trade abroad, especially to the West Indies and South America. In 1815, Baltimore resumed its leadership as the preeminent port for flour exports.

Although subsistence agriculture prevailed on the frontier, commercial production in the settled areas determined farming activities. As the cotton culture swept along the Gulf Coast and into the Old Southwest, the international market economy encouraged its rapid expansion and the extension of slavery. Where farmers could not afford slaves or it was socially or legally unacceptable, children provided much of the needed labor, and families were large, often with a half dozen or more children. An axiom on the eve of the American Revolution held that "a daughter is worth £50; a son £100." No matter the size of a family, inadequate technology, an absence of agricultural science, and poor transportation often limited commercial production, but whenever possible, farmers produced for a market economy. They also withheld surpluses occasionally to force prices up. In 1773, for example, a Rhode Island merchant complained that "farmers having been used a long time to great prices the most of them are become wealthy, and therefore will keep back their supply unless they can obtain what they call a good price."

Throughout the nation, most farmers continued to exploit the soil. Their plowing, planting, and cultivating practices encouraged soil erosion, and continuous cropping and the absence of fertilizer depleted soil fertility. Abundant, cheap, fertile lands on the frontier encouraged farmers to move west when confronted with diminishing productivity or competition for land by a growing population. Those who could not move or who preferred not to leave adjusted by emphasizing speciality production for town and city consumers or improving their agricultural practices by crop rotation and better tillage.

During the revolutionary era and the early years of the new nation, most farmers attempted to meet the subsistence needs of their families, but they were never completely self-sufficient. They needed to exchange agricultural products or sell farm commodities to acquire the goods and services necessary for daily life. By raising corn and livestock, especially hogs; milking a cow; making butter; and raising chickens on 50 to 100 acres of land allotted to crop, pasture, and woodland, they could meet many family needs and take advantage of markets if opportunities occurred.

Only through commercial agriculture could they improve their economic condition. During the antebellum period (1815–1860) that followed, large-scale commercial agriculture would dominate the American economy. For small-scale farmers and large-scale planters, commercial agriculture remained their most important goal.

SUGGESTED READINGS

Appleby, Joyce. "Commercial Farming and the 'Agrarian Myth' in the Early Republic." *Journal of American History* 68 (March 1982): 833–49.

Baldwin, Leland D. *Whiskey Rebels: The Story of a Frontier Uprising.* Pittsburgh: University of Pittsburgh Press, 1939.

Bidwell, Percy W. "Rural Economy in New England at the Beginning of the Nineteenth Century." *Transactions of the Connecticut Academy of Arts and Sciences* 20 (April 1916): 243–399.
Bidwell, Percy Wells, and John I. Falconer. *History of Agriculture in the Northern United States, 1620–1860.* New York: Peter Smith, 1941.
Boyd, Steven R., ed. *The Whiskey Rebellion: Past and Present Perspectives.* Westport, Conn.: Greenwood Press, 1985.
Brown, Richard Maxwell. *The South Carolina Regulators.* Cambridge: Mass.: Belknap Press of Harvard University, 1963.
Clark, Christopher. *The Roots of Rural Capitalism: Western Massachusetts 1780–1860.* Ithaca, N.Y.: Cornell University Press, 1990.
Gates, Paul W. *History of Public Land Law Development.* Washington, D.C.: Government Printing Office, 1968.
Gray, Lewis Cecil. *History of Agriculture in the Southern United States to 1860,* vol. 2. Gloucester, Mass.: Peter Smith, 1958.
Green, Constance M. *Eli Whitney and the Birth of American Technology.* Boston: Little, Brown, 1965.
Gross, Robert A., ed. *In Debt to Shays: The Bicentennial of an Agrarian Revolt.* Charlottesville: University of Virginia Press, 1993.
Harris, Marshall. *Origin of the Land Tenure System in the United States.* Ames: Iowa State College Press, 1953.
Hibbard, Benjamin Horace. *A History of the Public Land Policies.* Madison: University of Wisconsin Press, 1965.
Hindle, Brooke. "The March of the Paxton Boys." *William and Mary Quarterly* 3 (October 1946): 461–86.
Hurt, R. Douglas. *American Farm Tools: From Hand Power to Steam Power.* Manhattan, Kans.: Sunflower University Press, 1982.
Jacobs, Wilbur R., ed. *The Paxton Riots and Frontier Theory.* Chicago: Rand McNally, 1967.
Jensen, Joan. *Loosening the Bonds: Mid-Atlantic Farm Women, 1750–1850.* New Haven Conn.: Yale University Press, 1986.
Jensen, Merrill. "The American Revolution and American Agriculture." *Agricultural History* 43 (January 1969): 107–24.
Jones, Donald P. *The Economic and Social Transformation of Rural Rhode Island, 1780–1850.* Boston: Northeastern University Press, 1992.
Kay, Marvin L. Michael. "The North Carolina Regulation, 1766–1776: A Class Conflict." In *The American Revolution: Explorations in the History of American Radicalism,* ed. Alfred F. Young, 71–123. Dekalb: Northern Illinois University Press, 1976.
Klein, Rachel N. *Unification of a Slave State: The Rise of the Planter Class in the South Carolina Backcountry, 1760–1808.* Chapel Hill: University of North Carolina Press, Institute of Early American History and Culture, Williamsburg, Va., 1990.
Kulikoff, Allan. *The Agrarian Origins of American Capitalism.* Charlottesville: University of Virginia Press, 1992.

Larkin, Jack. *The Reshaping of Everyday Life, 1790–1840.* New York: Harper and Row, 1988.

Levine, Peter. "The Fries Rebellion: Social Violence and the Politics of the New Nation." *Pennsylvania History* 40 (July 1973): 241–58.

MacMaster, Richard K. "The Cattle Trade in Western Virginia, 1760–1830." In *Appalachian Frontiers: Settlement, Society, & Development in the Preindustrial Era,* ed. Robert D. Mitchell, 127–49. Lexington: University of Kentucky Press, 1991.

Rohrbough, Malcolm J. *The Trans-Appalachian Frontier: People, Societies, and Institutions, 1775–1850.* New York: Oxford University Press, 1978.

Russell, Howard S. *A Long Deep Furrow: Three Centuries of Farming in New England.* Hanover, N.H.: University Press of New England, 1982.

Schlebecker, John T. "Agricultural Markets and Marketing in the North, 1774–1777." *Agricultural History* 50 (January 1976): 21–36.

______. *Whereby We Thrive: A History of American Farming, 1607–1972.* Ames: Iowa State University Press, 1975.

Schumacher, Max George. *The Northern Farmer and His Markets During the Late Colonial Period.* New York: Arno Press, 1975.

Sharrer, G. Terry. "The Merchant-Millers: Baltimore's Flour Milling Industry, 1783–1860." *Agricultural History* 56 (January 1982): 138–50.

Slaughter, Thomas P. *The Whiskey Rebellion: Frontier Epilogue to the American Revolution.* New York: Oxford University Press, 1986.

Szatmary, David P. *Shays' Rebellion: The Making of an Agrarian Insurrection.* Amherst: University of Massachusetts Press, 1980.

Whittenburg, James P. "Planters, Merchants, and Lawyers: Social Change and the Origins of the North Carolina Regulation." *William and Mary Quarterly* 34 (April 1977): 215–38.

4 • ANTEBELLUM AMERICA

DURING THE ANTEBELLUM YEARS, 1815–1860, FARMERS PRODUCED surplus crops, livestock, and domestic manufactured goods for commercial sale if at all possible. As the American population spread westward at a rapid pace, many farmers supported federal aid for internal improvements such as canals and railroads to enable the relatively quick and efficient transport of their commodities to market. Commercial agriculture, however, involved more than the production of surplus commodities for sale. It also meant that farmers became less self-reliant. They worked to produce food and fiber for money to purchase manufactured goods, and as they specialized, they bought more food.

The drive of farmers for economic gain and a higher standard of living also spurred the rapid settlement of the transappalachian frontier. It encouraged the adoption of labor-saving equipment in the North to make farm work quicker and easier and extended the institution of slavery across the South for the extensive exploitation of the land and production of staple crops. Those farmers who chose not to migrate changed their cropping practices, adopted new techniques, and improved their livestock. Agricultural production for a market economy, however, encouraged farmers to exhaust the soil.

Above all, antebellum agriculture reflected the great differences in America regarding race, class, and gender. While small-scale family farmers, called yeomen, typified farming in the North and Midwest, the planters symbolized southern agriculture. When the antebellum period began, the ax and plow served as the most common farm tools, but by 1860 horse-powered seed drills, cultivators, and reapers had replaced hand-powered tools, especially for the production of small grains. The Civil War also revolutionized the labor supply in the South and brought a technological revolution to northern agriculture. Despite rapid change, however, agriculture in every region depended on the hard physical labor of men, women, and children.

THE SOUTH

The South remained agricultural and rural during the antebellum years. By 1860, more than 75 percent of the population still practiced agriculture, and most of the remainder earned a living in areas directly related to farming, such as handling, transporting, and processing cotton, tobacco, rice, hemp, and sugar. Flour and corn milling were also major industries.

The rich lands of Alabama's black belt and the Tennessee and lower Mississippi river valleys proved a compelling lure in the Deep South. The Bluegrass and Pennyroyal areas in Kentucky and central Missouri's Little Dixie drew thousands of settlers from the Upper South. Those migrants moved westward on roughly parallel lines and brought their culture, farming techniques, and slaves with them. Virginians spread the tobacco culture across the Upper South where by the 1850s they had developed a new variety known as bright yellow and a charcoal-fire-curing process that produced a leaf that manufacturers preferred to use as wrappers for plug chewing tobacco. South Carolinans took the cotton crop into the interior uplands and along the Gulf Coast. In Louisiana, sugar production boomed by the mid-1820s, and planters in Kentucky sent their slaves into the fields to produce hemp.

Only the rice planters failed to enjoy expansion and price increases before the Civil War. Although some of the largest plantations in the South could be found along the tidal rivers in South Carolina and Georgia, the best lands for rice production had been taken. But planters along the South Carolina and Georgia coast sometimes shifted from cotton back to rice because of soil exhaustion and competition from the Old Southwest. During the late 1830s, Fanny Kemble, a Georgia planter, wrote that rice had become a "great thing to fall back upon" and that rice selling at only $.045 per pound was quite valuable, if not more so than cotton.

During the antebellum years, rice exports usually ranged from 82.7 million pounds in 1815 to 43.5 million pounds in 1860. The productive ability of rice planters peaked about 1820, although exports reached 127.7 million pounds in 1835. Competition with Cuba and Latin America increased, and cotton became essential to the economy of South Carolina after 1815. By 1820, some farmers also had begun to plant rice along the Mississippi River; the industry would move west entirely after the Civil War. During the antebellum years, the price of rice declined from about $.06 per pound in 1816 to $.03 per pound on the eve of the Civil War. Still, many large-scale planters remained committed to it with the use of slave labor. One hand could cultivate 5 acres and produce about 50 bushels per acre. With low prices, large acreages, and many slaves required, however, only the major planters could earn a profit from extensive production. Some, like Nathaniel Heyward along the Combahee River in South Carolina, called their plantations "gold mines," but rice did not compare in importance to cotton in the antebellum South.

In the New Orleans area, the rich alluvial soil of the Mississippi River and bayou lands remained well suited for sugar cane, although even there the comparatively cool climate required replanting one-third of the crop every year. Most field hands cultivated and harvested 5 acres of sugar cane. Cold weather, crop disease, and insects, such as the borer worm, always kept the harvest uncertain, but planters considered 1,000 pounds of sugar per acre an average yield. Sugar brought $0.13 per pound in 1815 before dropping to $0.035 per pound in 1840. It then rose above $0.07 per pound during the late 1850s, a price that encouraged more production. Protective tariffs kept prices relatively stable on the domestic market and stimulated investment in the sugar industry. Planters sold their sugar directly to buyers, who came to their plantations, or through "commission men." Schooners carried hogsheads of sugar and molasses to distilleries and merchants along the Atlantic coast.

The typical southern family farmer, however, cultivated fewer than 100 acres, mostly for subsistence, but raised some tobacco and cotton for sale or trade. In contrast, a midsize plantation ranged from 1,000 to 1,500 acres, and the large plantations often consisted of several thousand acres. The plantations were not always composed of contiguous acres. A planter might hold several hundred or thousand acres in one location and more acres miles away. Whether the plantations were composed of several parts or a single property, they symbolized southern agriculture and became the chief source of commercial agricultural production.

Although the plantation system provided the basis for commercial agriculture in the antebellum South, most farmers were not planters but small-scale operators without slaves. Many could not afford to purchase farms and squatted on public domain. Whether yeomen or members of the planter class, southern farmers continued their traditional methods of wasteful cultivation. One observer called them "real land destroyers" who "skinned" the soil to plant cotton and earn a profit. While the soil remained productive, the small-scale farmers raised cotton, corn, hogs, and, in the upper South, wheat to meet family needs and if possible trade with a local merchant. Yeomen practiced a mixed or diversified agriculture on lands they often owned or sometimes rented as tenants. In these respects, they differed little from small-scale family farmers in the North unless they owned slaves.

By the 1840s, many farmers and planters had exhausted their land in Alabama and Mississippi. Lacking knowledge about the benefits of crop rotation and fertilization and without adequate technology (or because fertilizer was too expensive), many of these farmers abandoned their land and moved to Arkansas or Texas to begin anew, repeating a pattern already established by farmers who moved from Virginia and the Carolinas to Georgia and Alabama from 1820 to 1840. Cotton planters commonly wore out 30 to 50 acres for each field hand. By midcentury, the value of agricultural land in the South averaged about 50 percent less per acre than land in the North largely because of poor tillage and conservation practices.

In Maryland and Virginia, tobacco farmers had also exhausted the soil, and they tried to raise wheat and corn. Low yields brought poverty where prosperity had reigned. Some farmers, however, attempted to restore their lands by planting clover, applying marl and fertilizer, and improving tillage procedures. Moreover, a spreading railroad network and an expanding manufacturing system along the seaboard lured many farmers to better-paying jobs in the cities. In the Upper South, however, diversified agriculture prevailed, yeomen raising wheat, tobacco, corn, and livestock. Throughout the South, corn remained the leading crop, providing livestock feed and meal, grits and hominy for human consumption. One person could cultivate about 10 acres of corn.

Cotton, however, became the major commercial crop in the South. It was well suited for the southern climate, soil, and slavery. Farmers needed only a shovel plow and access to a gin to raise it. With one person able to cultivate about 10 acres and good land producing at least 3 bales per acre, upland cotton became the all-pervasive crop in the South and determined the strength of the southern economy. By 1860, the cotton kingdom stretched from Virginia and North Carolina to Texas, from Tennessee to Louisiana and

Florida, the Gulf states the most important producers.

Relatively stable and usually profitable prices, together with increased foreign and domestic demand, encouraged every farmer who could do so to raise cotton. Between 1815 and 1860, production increased from 150,000 to 4.5 million bales. During that time, the value of the cotton crop increased from $28 million to $249 million. By 1850, cotton contributed 63 percent of the value of all exports from the United States. Great Britain consumed more than half the American cotton crop; the textile manufacturing industry in the Northeast purchased most of the remainder. During the 1850s, a price of $.11 per pound provided sufficient return to encourage farmers to expand production, buy or clear more land, and invest in slave labor. Cotton became the "money" crop in the South. With the South producing three-fourths of the world's cotton supply in 1860, one South Carolinian reflected regional pride and a clear understanding of the southern economy when he stated that "cotton is king."

Beyond the Deep South, the hemp planters in Kentucky and Missouri relied on the cotton culture to keep their prices high; cotton planters needed hemp rope and bagging to bind their bales. These farmers produced dew-rotted rather than the time-consuming water-rotted hemp, the latter producing stronger fibers for rope and canvas required by the navy and merchant marine. With about 10 acres of hemp cultivated per slave and production of 600 pounds per acre, farmers could earn a profit if prices averaged about $5 per hundredweight. Kentucky and Missouri became the great hemp-producing areas with slave labor during the antebellum period.

Similarly, Kentucky and Missouri became major new tobacco-producing states, the leaf shipped to market in New Orleans and St. Louis. Farmers on Maryland's Western Shore and in central Virginia also continued to raise tobacco. European markets determined the price and demand, the German principalities, Great Britain, and France the major customers. Immediately following the War of 1812, tobacco brought $.25 per pound, a price that encouraged many farmers to raise it. But overproduction soon caused the price to fall to $.04 per pound by the late 1820s, and wheat proved more profitable east of the Appalachians. By 1860, tobacco culture had disappeared from the tidewater area, where it had been a major commercial crop during the colonial period.

Throughout the South, however, cotton drove the economy. Most cotton planters, those farmers who owned at least twenty slaves and 500 acres, sold their cotton through an agent known as a factor. The factor in an export city like New Orleans received the crop that the planter sent by wagon or steamboat, located a buyer, provided marketing and agricultural advice, loaned money and extended credit, and served as a purchasing agent for items the planter requested. For these services, the factor usually charged a 2.5 percent commission on the gross sale of the crop. Small-scale farmers who did not raise enough cotton to require the services of a factor and could not afford one, marketed their crop through local merchants. With 75 percent of the farms in the South producing cotton, farmers increasingly moved toward specialization and one-crop agriculture.

The planters primarily invested their profits in slaves and land rather than commerce or manufacturing. Consequently, a diversified economy did not develop, and northerners gained control of the cotton-marketing process. As a result, money earned from cotton went north, further retarding the development of southern transportation and manufacturing. Although some critics have charged that this economic relationship between the agricultural South and the commercial North imposed a colonial status on southerners, it did not. Southerners chose to use the credit, banking, shipping, marketing, and insurance systems of the North and put their earnings back into their agricultural operations. They traded freely and equitably with the North and Europe. One-crop specialized agriculture appealed to them economically, socially, and racially. By 1860, plantation agriculture gave white southerners a higher average per capita income than anywhere in the world except Great Britain and the northern United States. Extensive agriculture, slave labor, and good management made the plantation system profitable. Moreover, by 1860, the South still contained considerable unclaimed and uncultivated land that could have been developed for the extension of cotton production and the institution of slavery.

On the eve of the Civil War, then, the South was not only the most rural region in the nation but also the most agricultural. By 1860, approximately 84 percent of the labor force in the South pursued farming compared to 40 percent for the country. While about 40 percent of the population of the United States lived in the South, farmers and planters in that region produced 60 percent of the hogs, 50 percent of the corn and poultry, 52 percent of the oxen, and 90 percent of the mules raised in the nation. The South trailed the North only in the production of wheat, oats, rye, and sheep, and it had no equal for the production of cotton, tobacco, sugar cane, rice, and hemp.

Overproduction, low prices, and economic depressions periodically checked the commercial operations of southern farmers and planters. Moreover, agricultural wealth was not evenly distributed. In the cotton South on the eve of the Civil War, 5 percent of the farmers and planters owned 36 percent of the agricultural wealth, while 50 percent owned about 6 percent and 40 percent of the population lived in slavery. The agricultural South was a land of great wealth and extreme poverty. In contrast with the North, poor farmers could not escape rural poverty in the rapidly expanding cities and manufacturing towns. Moreover, planters had no incentive to improve their agricultural methods if those changes would endanger their investment in slaves.

And after the mid-1830s, the growing animosity between antislavery northerners and proslavery southerners retarded northern investments in the South that would have encouraged agricultural diversification by providing better railroads, textile factories, slaughterhouses, and marketing services. As a result, by 1860, the plantation system had begun to stagnate as an economic system. But it had become a way of life based on race and social status rather than an agricultural enterprise solely concerned with commercial production, profits, and economic gain.

The cotton plantation system also drove small-scale farmers and tenants onto less fertile lands, sustained the institution of slavery, and created a market for midwestern agricultural commodities such as pork, corn, and beef. It encouraged the use of hand labor rather than technology, restricted settlement by immigrants, and slowed the growth of the population in the South. Although based on a captive labor force, the plantation system never became as efficient as the master class intended. Slaves had little motivation to work at full capacity, and they generally labored only enough to avoid punishment. They had little incentive to give the best care to livestock and implements. Nevertheless, once planters committed to large-scale extensive agriculture, primarily cotton, and the institution of slavery, the economic and social returns justified continuation of this agricultural system. And they would not change or give it up without a fight.

SLAVERY

During the antebellum period, the institution of slavery remained tied to agriculture. When slaveholders moved across the South before the Civil War, they took their bondsmen with them. With great demands for labor to clear land and plant cotton and the international slave trade prohibited by the Constitution after January

1, 1808, a prosperous interregional slave trade developed. Slaveholders in the Upper South, where tobacco had worn out the soil, thereby decreasing the need for bondsmen, sold large numbers of surplus slaves, usually teenagers and young adults, to traders who marketed these servants in Natchez, New Orleans, and other locations in the Deep South.

With cotton profitable, the price of slaves increased. By 1860, a prime male field hand brought as much as $1,800 on the New Orleans market. Women aged 18 to 25 usually brought prices averaging 25 percent less than males. Farmers who could not afford a slave, did not want to own one, or had only a small amount of work often hired slaves to complete special projects. When a planter hired out a slave, he earned an income and put his surplus labor to productive use. Generally, slaves hired for at least 10 percent of their value. In Missouri, for example, a male slave valued at $1,000 commonly hired at $200 to $250 annually during the late 1850s. Female slaves valued at $900 hired at $75 to $100. Hiring required far less capital investment by farmers who needed slave labor. Most hiring contracts ran from January 1 to December 25 annually.

The courts also hired out slaves on the death of their master until settlement of the estate. The earnings went to the widow and children or to the court for the settlement of the deceased's debts. When slaveowners hired out a bondsman, they protected their property with a contract that required the person hiring the slave to meet certain requirements. Contracts required the hiring party to give the slave proper medical care and provide adequate food and clothing, described the nature of the work to be performed, and prohibited the slave from leaving the county or being hired to a third party.

By 1860, most southerners still did not own slaves, and most of those who did held fewer than five, but most slaves were owned by only a few planters. In 1860, for example, 50 percent of the slaves belonged to masters who owned twenty or more bondsmen, and approximately half the slaveholders owned fewer than five slaves. Overall, slaveholders averaged 12.7 bondsmen in the Deep South and 7.7 slaves in the Upper South. The planter class, then, owned most of the slaves. The largest slaveholders operated rice and sugar plantations, where extensive labor was required year round. Slavery always remained an agricultural labor system. Only 10 percent of the slave population lived and worked in the towns and cities. Moreover, most slaves lived in the cotton states.

While most southerners did not own slaves, they aspired to acquire them. Although 27 percent of the slaveholders owned 75 percent of the slaves in the South in 1850, this figure is misleading because most southerners supported the institution of slavery. Put differently, about 25 percent of southern families owned at least one slave. More southerners had a direct financial and economic interest in slavery than twentieth-century Americans had in stocks or the employment of hired labor. As a result, the capital assets in slaves had a wider distribution than other assets anywhere in the North before or since the Civil War. Little wonder, then, that no class conflict erupted between slaveowners and nonslaveholders. Slavery offered the promise of financial success, economic security, and higher social status—all admirable goals for whites in the antebellum South. When these aspirations were mixed with the almost inherent "racist" attitudes of whites, the willingness of nonslaveholding small-scale farmers to support the planter class in the Civil War becomes more easily understood. Throughout the South, the ownership of slaves and land determined social status.

Planters employed an overseer to organize the work force to achieve maximum efficiency. The overseer managed the daily activities for the plantation. He had the responsibility of meeting the production goals of cotton, sugar, or rice as well as maintaining discipline and ensuring some subsistence production of corn and pork. Overseers occupied an essential position in the plantation's organizational structure. For their services, they received a wage, house, garden, and servant.

The slave drivers, who served much like foremen in other occupations, answered to the overseer. Drivers—dependable, loyal, strong slaves who executed the orders of the overseers—supervised the activities of the work gang, ensuring that a certain number of acres were plowed; that a specific amount of cotton, rice, or tobacco was harvested; that a definite amount of molasses was produced. They had the task of maintaining order and often meting out punishment. The drivers also spoke on behalf of their people to the overseer or planter; they enjoyed high status and better living conditions on the plantation. Farmers who could afford only a few slaves worked in the fields beside their bondsmen without suffering any loss of prestige.

Planters continued to use the gang system primarily in cotton, tobacco, and sugar cane, and the task system in rice, hemp, and sea island cotton. Most planters, however, used a combination of the two systems in which they considered age, sex, strength, and skill of their slaves in relation to the job at hand, such as plowing in a gang or hoeing a specific number of rows by task. Although gang labor harvested the cotton crop, overseers expected each field hand to pick between 150 and 200 pounds per day. No matter which system a planter employed, he expected his bondsmen to use part of their leisure time to raise gardens, chickens, and hogs and thereby contribute to the subsistence production of the plantation.

Slaveowners rated the work ability of their bondsmen by the term *hands.* A prime male field hand merited the designation of a full hand, while a woman of childbearing age received the classification of one-half hand. Usually planters referred to their bondsmen as "servants," "hands," the "force," or "my people" rather than "slaves." No matter the term, they tried to keep their slaves busy at all times. Usually the days were full. When the cotton needed cultivating or picking, a day's labor often reached sixteen hours. The sugar plantations, however, demanded the most labor; at harvesttime, bondsmen often worked eighteen-hour days seven days a week. Field work on cotton plantations began at sunrise with breakfast often brought to the fields. Overseers usually permitted two hours of rest at noon; then the slaves worked until sundown. When the crop had been "laid by" (after the last hoeing before the harvest), the work day became temporarily lighter. On small-scale farms where a slave labored alongside the master, the schedule was less regimented. Generally, planters did not require work on Sunday, and they released their slaves from duty at noon on Saturdays. Christmas remained a cherished holiday.

Planters usually divided their slaves into field and house servants. On the larger plantations, slaves also specialized by craft. Some served as blacksmiths, coopers, and carpenters. Domestic servants and craftsmen had the highest status next to the nurse and slave drivers, while the field hands had the lowest status. Old men were often assigned light duties tending livestock and farm tools; elderly women served as assistant cooks, seamstresses, and babysitters. On the other end of the age scale, when children reached 8, the planters often sent them into the fields to pull weeds. At this young age, they believed, a boy or a girl could manage without the care of a parent, gain work experience, and begin the socialization process of agricultural slavery.

Although slaves on the plantations cultivated their own gardens during their leisure time, they could not produce all of their food. Overseers distributed rations once each week. Each adult usually received 1 to 2 pecks of cornmeal and 3 to 4 pounds of bacon. Children received half rations. In the sugar cane area, the slaves sometimes received a quart of molasses and sweet potatoes, and rice planters occasionally substituted that grain for corn. Only Louisiana required masters to feed meat to their slaves. The planters also issued clothing in the summer and autumn, although women usually received only cloth to make their own dresses. Agricultural labor caused several common ailments such as tetanus, hernia, and pneumonia, all of which required home treatments with folk medicines or a call to the doctor.

The institution of slavery remained strong until the Civil War. Slave prices and valuations continued to increase as farmers and planters demanded additional labor to clear new lands and increase production. Slavery also remained profitable, the return averaging about 10 percent annually. This rate of return on investments in slaves was comparable to investments in northern manufacturing. Only the slave traders, whose profits varied from 15 to 80 percent during the antebellum years, earned more money. With a slave costing a master less than $50 annually for food, clothing, and medical care, and capable of producing 3,000 pounds of cotton, slavery paid, and southern farmers and planters had no intention of casting this labor system aside for more expensive, scarce, and sometimes inadequate white labor. By 1860, the slave population totaled 4 million, and the institution permeated every aspect of southern life, even though three-fourths of the population did not own slaves. Slavery caused the Civil War, and that great conflict destroyed the institution. Until the war came, however, slavery remained agricultural and essential to the southern economy.

THE NORTH

The small-scale family farm typified northern agriculture during the antebellum period. Although northern farmers still met most of their home or domestic needs for food and fiber, the rural population consumed approximately 80 percent of the farmers' produce, and local markets became increasingly important for the sale or trade of surplus products. A growing urban population and improved transportation in the form of canals, steamboats, and railroads expanded market opportunities for commercial production. Roads, however, remained poor; one Englishman remarked that they were "inferior to those of any other civilized country." By the 1820s, farmers on the fertile lands of the Connecticut, Hudson, Delaware, and Susquehanna river valleys commonly sold 25 percent or more of their produce for commercial gain.

Urban demands and cash payments encouraged farmers to specialize and to produce those commodities that brought the greatest economic return to the family, particularly for farmers who owned fertile, highly productive lands. Rural residents had increasingly better access to manufactured goods, such as cotton and wool cloth, factory-made candles, oil lamps, furniture, kitchen utensils, lumber, nails, iron stoves, wagons, and farm tools. Whenever possible, farmers preferred to purchase these items rather than produce flax, spin wool, or craft tools and utensils on the farm. To so so, however, they needed money, which required the surplus production of commodities that could be transported to market for sale. Profit helped transform farmers from subsistence producers to commercial agriculturists.

With the completion of the Erie Canal in 1825, the cost of shipping grain east fell rapidly, and farmers in the Old Northwest began sending their grain to market via the Great Lakes and the canal to Buffalo, Albany, and New York City. As farmers increasingly emphasized production for income, they gave less attention to the production of surplus commodities that could be sold the most conveniently and began to produce specific commodities like wheat, tobacco, pork, or beef. Farming now became a business rather than a way of life. Although some farmers balked at this specialized emphasis, most eagerly sought commercial production and adopted new technology if not improved farming practices.

West of the Appalachians, farmers primarily acquired land from the federal government or speculators, although many could not afford to purchase it and squatted until they were driven off by government officials or lawful owners or until they saved enough to buy the land they cultivated. Some settlers worked as farm laborers and saved to buy land. With wages ranging from $9 per month in the 1820s to $13 per month in the 1850s, a worker needed to save for a decade before he could accumulate at least $500 to $1,000 for the purchase of land, equipment, seeds, and livestock to begin farming. Others borrowed money from a bank to buy land and mortgaged their property, and some began farming by renting from a landowner. Tenancy became common west of the Appalachians, particularly for young men and women who lacked the capital to buy land, but it also was important in New England, New York, and the mid-Atlantic states. Tenant farmers hoped to earn enough money from commercial agriculture to acquire land of their own.

After acquiring land in the Old Northwest, farmers tried to become as self-sufficient as possible and to produce a surplus for sale. Because these farms were isolated and without adequate transportation to eastern markets, farmers at first emphasized the production of hogs, corn, and dairy products. They fed corn to hogs, which were slaughtered and the pork packed in salt and shipped down the Ohio and Mississippi rivers for sale to planters in the South or for transshipment from New Orleans to merchants on the eastern seaboard and international markets. As early as 1810, drovers took hogs over the mountains from the Ohio country to markets in Baltimore, Philadelphia, and New York. By the 1820s, the pork-packing industry boomed in Cincinnati and earned the "Queen City" of the West the new sobriquet "Porkopolis."

West of the Alleghenies, drovers also trailed cattle eastward. The early settlers in Ohio's Scioto River valley came from Virginia, where they had gained experience raising corn-fed cattle. And as early as 1805, George Renick drove a herd of cattle to the Baltimore market and sold them for nearly $32 per head, a price that justified taking corn-fed beef cattle to distant markets. After the War of 1812, western drovers took some cattle to market over the National Road, a portion of which the federal government completed from Cumberland, Maryland, to Wheeling, Virginia, in 1818. By the 1840s, Chicago had become a major cattle market in the West; Albany provided the same services for livestock producers in the eastern portion of the Old Northwest and New York. New England farmers also sent their cattle to the Brighton market, which served Boston, until the railroads brought cattle from New York via Albany and the West beginning in the 1840s.

Throughout the North, country merchants accepted butter and cheese for dry goods or paid cash for these commodities whenever possible. Farm women had the task of churning butter, which country merchants collected and periodically sent to regional or central markets. Inadequate preservation, packing, storing, and shipping, however, usually kept this butter at a low quality. Farmers close to major towns or cities, however, could send their butter to market via wagon or railroad and make fresh deliveries. By giving more attention to improved breeding, feeding, and sheltering of their dairy cattle, northern farmers also increased the butterfat of their milk and produced a higher-quality product.

Many northern farmers converted their surplus milk to cheese. Before 1850, dairy farmers made cheese at home and marketed it through local merchants. New York State became the major cheese producer at this time, while Ohio's Western Reserve earned the nickname "cheesedom." Farm-made cheese, however, like farm-packed pork, had few quality controls, and the product often suffered from improper curing, storage, and shipment. Although better care and feeding practices for dairy cattle improved the quality of cheese, standardized production with significant quality controls did not occur until the late 1840s with the development of cheese factories.

These cheeseries collected the milk or curd from dairy farmers for processing. Farmers could now sell their surplus milk and avoid the work of making and marketing cheese. With this new system, cheese production, like the meat-packing system, moved from the farm to the factory.

Most northern farmers, particularly those west of the Appalachians, still gave little consideration to soil conservation. When their fields wore out, they cleared and broke new land if they had uncultivated acreage. Some adjusted by planting smaller crops on the good land that remained. Other farmers returned their worn-out land to grass and grazed sheep, while still others, with speculative minds, sold their farms and emigrated to the frontier or moved to areas where good land remained. By the time most farmers recognized that soil exhaustion had become a serious problem, it was too late to restore those lands easily by the use of fertilizer, crop rotation, and fallowing. Farmers in New England, however, made these adjustments on a more frequent basis because they had better knowledge of improved British methods through a growing number of agricultural periodicals.

During the 1840s, farmers settled the midwestern prairie. The tough prairie lands proved difficult to plow, and farmers often hired breaking teams to turn the sod at rates of $2 to $5 per acre. Prairie farmers usually planted corn and potatoes in the newly turned sod. These crops provided food for the new settler and helped break down the fibrous root system of prairie grass so that the field could be cross-plowed for the preparation of a smoother seedbed the next year. Wheat became a primary crop for "subduing the prairies" and with corn yielded the greatest returns for the least labor and monetary investment. Farmers marketed wheat as grain and corn as pork and beef. Usually, four to five years were required to produce surplus crops for market. Before the adoption of the reaper and mower, 30 cultivated acres were the most one man could farm west of the Appalachians. Once past the pioneer phase, farmers emphasized the production of crops and livestock that ensured the "best profit." If they often moved toward commercial production slowly, they did so inexorably.

As western production increased, the surplus, especially wheat, undersold eastern producers. Eastern farmers responded by emphasizing specialty noncompetitive crops and products such as vegetables, milk, butter, cheese, and fruits that had a ready demand and brought a good price at nearby urban markets. Although northern farms remained conservative in their practices and more tradition-bound, by 1860 they had begun to adjust when necessary to better methods and improved technology, particularly if they could save time and labor and increase their profits. The Civil War era would bring even greater change in their cropping, livestock raising, and marketing practices.

LAND POLICY

Soon after the War of 1812, the Treasury Department attempted to regulate state banking practices by decreasing the currency supply. This ill-conceived policy, in addition to falling agricultural prices, caused an economic depression known as the panic of 1819. Many farmers could not meet their payments for land purchased on credit from the federal government, and others could not afford to purchase land at all. Congress responded by eliminating its credit policy, reducing the minimum bid at public-land auctions to $1.25 per acre, lowering the minimum purchase to 80 acres, and requiring full payment on the day of purchase. Congress hoped the lower price, smaller acreage for purchase, and absence of credit would eliminate problems of speculation, overextension, and indebtedness and would encourage more people to buy public land. Unfortunately, this legislation did not aid settlers because the $100 purchase price was still more than most farmers could afford and the government required payment in cash. Farmers needed credit, and the law did not provide them that purchasing flexibility.

As a result, many farmers on public lands remained squatters, and they demanded the right of preemption (the right to have the first opportunity to purchase the lands on which they lived and farmed upon survey). Often, speculators or other farmers bought public land only to find squatters on their property. Naturally, when the rightful owner attempted to remove the squatters who had cleared trees, planted crops, and constructed buildings, they met great opposition and resistance. Squatters did not consider themselves lawbreakers but rather "respectable" citizens, "sturdy" pioneers, and "hardy" yeomen.

With the pressure from westerners growing, Congress eventually aided some squatters with the Preemption Act of 1830, which gave every "settler or occupant" on *surveyed* public land the first or preferential right to buy the claim from the federal government if he had settled and cultivated a portion of the land prior to 1829. The cost would be $1.25 per acre, with 160 acres the maximum that could be acquired. Those who wanted to preempt had to file their claim and pay the $200 cost within one year, or it would be opened for public sale under the Land Act of 1820.

Squatters quickly took advantage of the Preemption Act and descended on local land offices to file their claims. One official complained: "They come like the locusts of Egypt, and darken the office, with clouds of smoke and dust, and an uproar occasioned by whiskey and avarice." However, most squatters could not afford the $200 to purchase a 160-acre claim under the law.

Although Congress intended the Preemption Act of 1830 to be a temporary measure to solve a political problem, westerners now demanded a general preemption act that would make squatting legal on surveyed land and thereby protect westward-moving pioneers from the loss of the land they occupied but did not own. Finally Congress passed the general Preemption Act of 1841, which required squatters to inhabit and cultivate their land and build a house on it to claim 160 acres. No one could make a preemption claim if he or she did not intend to farm that land or already owned 320 acres. Preemption was available to heads of households, widows, or men 21 or older who were citizens or had filed for citizenship. No one could claim more than one preemption.

Even this legislation, however, did not satisfy those westerners who demanded that the federal government make public lands available free to the American people. Those who supported this policy argued that free public land would solve the nation's immigration and urban unemployment problems by providing acreage to support individual small-scale family-farm agriculture and preventing speculators from hoarding land and slowing settlement until they could sell it for a high price. Others argued that a free-public-land policy would be unfair to those who had already purchased public domain. Some contended that such a policy would depress land values in the older settled states, promote fraud by dummy entrymen, favor foreigners, and reduce federal revenue. Southerners also opposed a free-land policy because the Constitution did not authorize the disposal of the public domain in this way and such a policy would encourage western settlement, reducing southern power in Congress.

The public and government officials continued the old debate whether federal land should be sold to generate revenue or to encourage settlement. As early as 1828, however, Thomas Hart Benton, senator from Missouri, supported a price-graduation plan for the sale of public land, the price of unsold land to be reduced annually until all acreage had been sold. This policy would generate some revenue for the federal treasury while enabling less productive lands to pass from public to private ownership. This plan would create taxable farmland and stimulate the agricultural development of the west and the nation's economy. Benton's proposal for a price-graduation program became extremely popular in the West.

Congress did not respond favorably to Benton's proposal until 1854 when it passed the Graduation Act, providing that public land not sold after ten to fifteen years would be reduced to a minimum price of $1.00 per acre, then, at five-year increments, reduced by an additional $0.25 per acre until lands not sold after thirty years would be reduced to $0.125 per acre. This generous policy created a land rush in areas where the public domain remained unsold. Although purchases were limited to 320 acres, this restriction was not enforced, and speculators acquired large tracts for little money, then sold them at higher prices. John Wilson, commissioner of the General Land Office, complained that the Graduation Act produced "fraud and perjury." Speculators, however, had great political influence in Congress, and westerners wanted cheap or free land.

The Graduation Act of 1854 did not quiet the demands of others for free public domain. Westerners believed the nation would greatly benefit if the public lands were settled and put into agricultural production. The economy of the nation would grow and benefit the federal government more than the sale of the public domain at high prices to raise revenue. Southerners, however, continued to block a free-land or "homestead" bill, and the issue remained unresolved until the Civil War.

TECHNOLOGY

Between the antebellum years 1820–1835, farmers increasingly adopted cast-iron plows, but the wooden moldboard, particularly the Carey plow, remained the favorite. Although the Carey plow differed somewhat according to the skill of each blacksmith or plowwright who made it, the general style was uniformly reproduced on a wide basis. This plow had a wooden moldboard and a wrought-iron share. Plowwrights used wood for the beam and handles and attached the various pieces with wooden pegs. Over time, these joints loosened, and the wood cracked or broke, which made plowing difficult and repairs frequent. Still, the Carey plow was popular in the North and the South. With it, a farmer could plow about 1 acre per day.

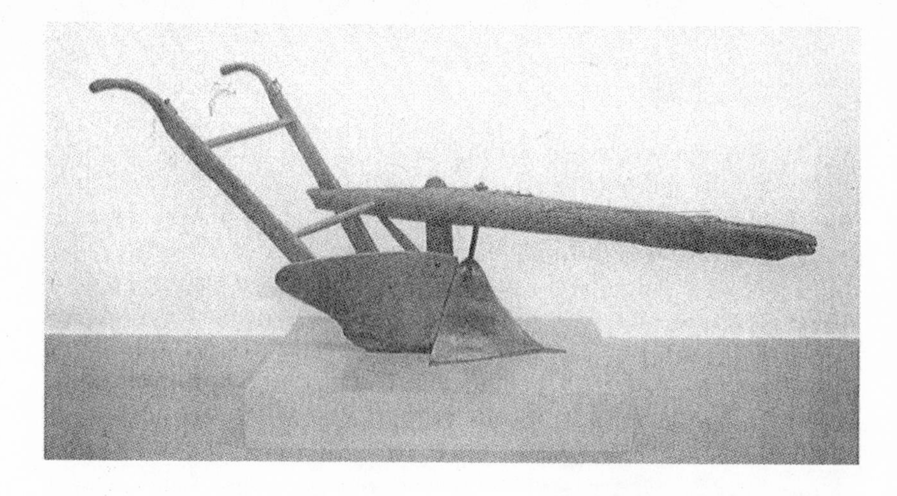

THE CAREY PLOW was the most popular tillage implement during the early nineteenth century. It had a wooden moldboard and a wrought-iron share. Wooden pegs held the parts together. Local plow makers could easily craft this implement, and farmers could plow about 1 acre per day with it. *Smithsonian Institution.*

Many southern farmers, however, preferred to use the shovel plow, maintaining their prejudice against the cast-iron plow. The shovel plow had a rough wooden beam into which another wooden piece with an iron point was attached. Two handles were pinned to the sides of the beam. The wrought-iron point, approximately 9 inches across, had a shovellike blade with the convex side turned outward. A welded loop on the back of the blade provided a place for the stock to enter. One horse or mule could easily pull the shovel plow. It cut a shallow furrow and turned the soil both ways. The nearly upright position of the handles forced the plowman to maintain a tiring erect position; nevertheless, southern farmers continued to use it for plowing and cultivating until the Civil War.

Although plowwrights fashioned moldboards into various shapes to meet the requirements of different soils, more adjustments were necessary when farmers crossed the Appalachians and entered the rich prairie land of the Midwest. There, the tough prairie sod made the wooden plow useless. It would neither penetrate the sod and cut the roots nor stay in the soil. Consequently, prairie farmers used another plow. That tool, called a breaking plow or prairie breaker, popular from the 1820s through the 1840s, was a heavy plow with the moldboard alone often weighing 125 pounds. The 14- or 15-foot beam and handles made it even heavier, but this weight kept the plow from bucking out of the furrow as it struck the fibrous root system. Two small wheels supported the beam in front, and a lever ran from the handles to the front of the beam to regulate the depth of the cut. By lifting the lever, the share would dig deeper into the soil; by depressing it, the plow could be raised from the ground.

Plowing with a prairie breaker was slow, hard work, even though it cut only 2 or 3 inches deep. Depending on the toughness of the root system, as many as three to seven yoke of oxen might be needed to pull this plow. If a farmer broke as many as 8 acres during the plowing season, he was fortunate. Hired workers called breaking teams could plow more acres because they did not have other farm chores. A two-man breaking team using three yoke of large oxen and a 24-inch plow could turn 3 acres a day.

Although plowing with a prairie breaker was slow, the cast-iron plow of the 1830s and 1840s did not give the western farmer a viable alternative because the sticky prairie soil clung to the moldboard. The cast-iron plow had two problems that made it nearly useless in prairie soils. First, it did not take a high polish. Consequently, the moldboard would not scour—that is, it tended to hold sticky soil instead of allowing it to slide or peel off as the implement traveled through the ground. Clogged moldboards caused the farmer constant delays scraping them clean with a paddle. Second, the cast-iron surface contained small cavities known as blow holes that filled with the claylike soil. The clogging caused unwanted friction that increased the required draft.

About 1833, John Lane, a Lockport, Illinois, blacksmith, made the first successful effort to design a plow that would not clog after the initial breaking of the prairie had been completed. Lane recognized that only steel, not cast iron, would scour suitably to permit the moldboard to turn a clean furrow. To produce such a plow, Lane plated a wooden moldboard and share with strips of steel cut from an old saw. Lane's innovation worked better than any plow tried in the prairie soil to that time. However, he did not patent his idea or produce steel plows on an extensive basis. Steel remained scarce and expensive until the Bessemer and open-hearth processes enabled producers to increase production and reduce the price. Even so, other implement inventors began applying Lane's discovery to their work. John Deere was one of the most successful and widely recognized of those inventors.

Actually, John Deere, a Grand Detour, Illinois, blacksmith, did not make steel plows on a wide basis until the mid-1850s. His early plows, the first made in 1837, consisted of a highly polished wrought-iron moldboard and landside with a steel share. There is no clear evidence that he fashioned his first moldboard from a steel saw blade. Deere's reputation for devising a steel plow came from his use of a steel share that was stronger than cast iron and held a sharp edge better than wrought iron. Farmers calling this implement a steel plow merely used the term to distinguish it from more traditional wooden and cast-iron plows. Deere's plow was so superior to the heavy breaking plows that farmers commonly used it as a breaking plow.

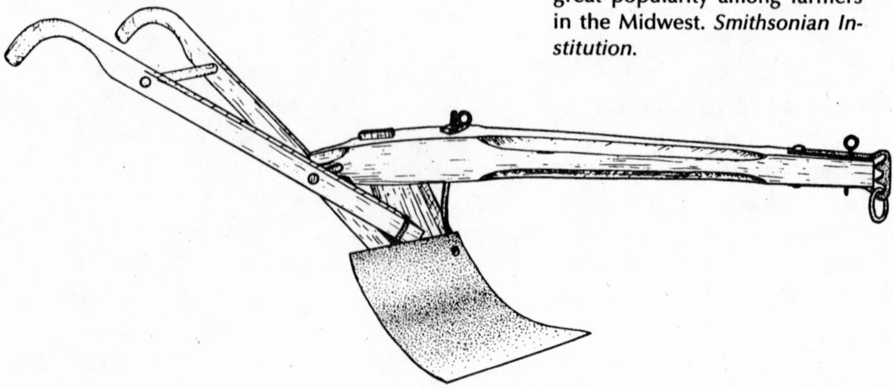

IN 1837, John Deere crafted his first plow that could easily cut through the tough root system of prairie grasses. Deere's plow had a highly polished moldboard and a steel share. Because steel was expensive, plow makers did not use it to make moldboards until the mid-1850s when new metallurgical processes reduced the price for this metal. Deere's plow could be used for breaking prairieland or general plowing. It achieved great popularity among farmers in the Midwest. *Smithsonian Institution.*

JOHN DEERE

Deere & Company Archives.

John Deere (1804–1886) stepped down from the stagecoach into the dirt street of Grand Detour, Illinois, in November 1836. He carried little more than $70 in his pocket and a bag of blacksmith tools. In Hancock, Vermont, a pregnant wife and four children worried about his safety while creditors wondered where he had gone. In the late-autumn chill, John Deere at thirty-two had only the promise of the rich prairie lands of the Old Northwest to give him hope. On the rolling prairies, farmers needed blacksmiths to hammer plowshares keen after the cutting edge wore thick and dull in the tough prairie sod. Farmers needed plows, harrows, shovels, and pitchforks; horses shod; and trace chains and clevises repaired. Soon after his arrival, he worked seven days a week at his forge and gained a reputation for producing a quality product at a fair price.

The fertile prairie, however, extracted a heavy physical and economic price from the men who tried to plow it and plant crops. The root system of the prairie grasses created a thick, nearly impenetrable sod, too tough for ordinary plows. To make the prairie lands return harvests of corn, wheat, oats, and other crops, the farmer had to break the land, and prairie breaking wore out man and beast alike. Because plows in common use could not cut through these grasslands and turn a furrow, farmers frequently hired itinerant workers with special plows. Farmers and hired plowmen had to take their plowshares to the local blacksmith about once a week for sharpening. At the rate of $2 per acre for hire, prairie breaking was not only tiresome but expensive work. John Deere thought there had to be a better way.

Sometime in 1837 Deere had an idea that would change farming in the Midwest and permanently etch his name in American agricultural history. In the sawmill of Leonard Andrus where he had gone to repair a pitman rod, he noticed a broken steel mill saw. Deere took the saw blade back to his blacksmith shop and cut off the teeth with a sledge hammer and chisel. Next, he cut a diamond-shaped moldboard from a piece of wrought iron that he then heated and bent over an anvil until it took the desired trapazoid shape. To this moldboard Deere attached a steel share that he had cut from the saw blade. Last, he gave the moldboard a high polish. Deere's moldboard cut easily through prairie sod and allowed the soil to peel away without dulling the share or clogging the moldboard. His new plow was lighter than the prairie breakers, and it required only about half the draft for cast-iron plows.

Given the state of American metallurgy, however, steel was expensive, limited in quantity, and hard to acquire. Consequently, despite Deere's technical breakthrough, he had sold only a few plows with steel moldboards by the mid-1840s. Not until the mid-1850s did prices warrant the extensive production of steel plows. Until that time, most of Deere's plows consisted of highly polished wrought-iron moldboards with steel shares. During the 1850s, his steel-moldboard plow earned the sobriquet "singing plow" because it gave a slight whine as it cut through the soil.

If Deere's plow was an overnight success from a design standpoint, as a businessman he faced an uphill battle. Major manufacturers in Pittsburgh produced nearly 34,000 plows annually in the 1840s, some of them featuring steel plowshares. Remarkably, Deere was neither absorbed by his competitors nor relegated to oblivion as a small-scale producer in a blacksmith shop. His "steel plow" quickly gained recognition far beyond the relative isolation of Grand Detour, Illinois. He increased the scale of his operations by making and selling other plows under license.

In many respects, Deere was innovative rather than inventive. John Deere was not afraid to borrow ideas from other inventors, a trait not unique to him; implement manufacturers customarily pirated ideas from their competitors. Deere had the uncanny ability to assimilate the best ideas from many manufacturers to meet the exact needs of prairie farmers. Expansion through branch houses eventually spread Deere's implements across the Midwest, the Great Plains, and the nation. His company remained in family hands until 1982, a preeminent example of a well-run family business.

Although many men and women helped settle the midwestern prairies, few can rival John Deere's immediate or lasting contribution to the agricultural development of the region. Deere never pretended to be more than a good blacksmith at a time when prairie farmers in the Midwest needed one. His ability to craft an excellent plow gained him respect, a comfortable living, and finally national prominence. At the same time, his innovation eased the burdens of thousands whose lives depended on tilling the soil and whose labor fed millions.

While many farmers awaited the perfection of the steel plow, others still needed an efficient tool that would turn a furrow and not break or dull quickly. This meant using the best cast-iron plow available. Fortunately, Joel Nourse of Worcester, Massachusetts, developed a cast-iron plow that adequately broke rough ground and turned a furrow in soils with weeds and heavy stubble. This implement, the "Eagle plow," differed in two respects from all other cast-iron plows. First, Nourse lengthened the moldboard; then he gave it additional curvature. As a result, the Eagle plow lifted the soil and turned it over more effectively than any other cast-iron plow. Outside the prairie lands, the Nourse plow became a popular implement between 1840 and 1860.

After plowing, farmers continued to sow their crops by hand or with broadcast seeders. At best, they could sow about 14 acres per ten-hour day by this method. The evenness or uniformity of the crop stand depended on the consistent turn of the seeder's crank or pull of the bow and the operator's walking speed, but frequently the seed scattered too thinly or too thickly. In 1841, however, Moses and Samuel Pennock of Chester County, Pennsylvania, built a grain drill that dropped seeds from a hopper down tubes and into the soil. The Pennock drill reportedly worked well on rough or smooth ground, and it could be regulated to drop various amounts of seed per acre. By 1850, farmers used the Pennock drill on wheat lands from Virginia to New York and from the mid-Atlantic states into Ohio, and it remained one of the most popular grain drills until the Civil War. Pennock's drill planted as many as 15 acres per day.

Although farmers began to seed small grains with a drill during the antebellum period, they continued to plant corn with a hoe. This technique limited the farmer to planting about 0.5–1 acre per day, or approximately 10 acres during the planting season. During the 1850s, however, the hand corn planter appeared on the market. This planter consisted of two wooden slats with handles and a seed canister. When the farmer thrust the point of the planter into the ground and closed the handles, seeds from the canister fell into the soil. These tools never gained widespread acceptance because farmers did not believe the corn planter saved enough time and most preferred to plant with a hoe. Although a mechanical corn planter that deposited the seed in a furrow had been patented as early as 1839, these implements did not perform adequately, especially on rough lands, and farmers rejected them. They argued that corn planted in a furrow was harder to cultivate than corn planted in hills because they could not plow both ways between the furrowed rows with their cultivators. The success of Pennock's grain drill, however, encouraged inventors to apply the same principles for seeding small grains to planting corn.

In 1853, George Brown of Tylersville, Illinois, developed a two-row horse-drawn corn planter that appealed to many farmers. This implement dropped seed into the furrow by a mechanism geared to a wheel. In 1860, he added a hand-operated dropping device that released the seed at the proper moment. Horse-drawn planters gained rapid acceptance during the 1860s in the Midwest. A farmer could plant from 12 to 20 acres per day, approximately twenty times more than he could plant with a hoe. In 1857, Martin Robbins, a Cincinnati inventor, patented the first corn planter that dropped seeds automatically in evenly spaced rows. This implement attached to a jointed rod or a chain with metal buttons that, when pulled through the seeding mechanism, tripped the dropper. The chain was staked down at each side of the field, and the planter followed the chain as a guide. This implement provided the basic concept for the check-row planter that other inventors perfected during the late nineteenth century.

Once the corn sprouted, farmers began their age-old battle against weeds, which if allowed to remain would rob the soil and the crop of moisture and nutrients. To kill the weeds, corn farmers usually hoed their crop four times. At the rate of about 1 acre per day, six

DURING THE 1860s, the self-rake reaper became popular. This implement eliminated the need for someone to walk alongside and rake the gavels of cut grain from the platform, and the driver could ride. Self-rake reapers could harvest from 10 to 15 acres per day, depending on the crop stand. *Smithsonian Institution.*

or more days' labor per acre might be spent cultivating by hand. The amount of corn that farmers could efficiently cultivate determined the acreage planted, and cultivation by hand severely limited the total corn acreage nationwide. About 1820, however, farmers began using an implement called a horse-hoe to weed their crops. This horse-drawn implement and the shovel plow became favorites among corn farmers nationwide. During the 1850s, farmers began using two-row cultivators, and by the 1860s, the sulky or riding cultivator gained widespread use in the Midwest. This two-horse implement cultivated each side of the plants as one horse and one wheel traveled down a row while the other horse and wheel went down another. By cultivating both sides of the row at once, farmers doubled the amount of corn that they could weed in a day, and they could ride at the same time. No cultivator, however, had more than a two-row capacity, and most farmers preferred single-row implements.

At harvesttime, many small grain farmers continued to use the sickle until the 1820s. But more farmers adopted the cradle scythe or grain cradle, and it remained the standard reaping tool until about 1860. Whether grain farmers used the sickle or the grain cradle to harvest their crops, these tools restricted the acreage they could expect to harvest safely during the season unless they could afford hired hands. Until the farmer could speed the harvesting process mechanically, he had little hope of expanding his grain production. Only horse-drawn machinery would free the grain farmer from a dependence on hand tools and hired labor. But until someone developed an efficient horse-drawn harvester, grain production was limited.

A new age of grain harvesting began in 1831 when Cyrus Hall McCormick tested his first reaper in Rockbridge County, Virginia. McCormick's reaper had a straight knifelike blade that he linked to the drive wheel with gearing. Projecting fingers or guards on the

BEFORE THE ADOPTION of steam power to the threshing process, farmers used horse-powered treadmills or sweeps to drive the mechanism of a threshing machine. Small-scale grain farmers primarily used the treadmill because it required fewer horses and a smaller and less expensive threshing machine than horsepowered sweeps. During the late 1870s and early 1880s, steam power replaced this threshing process. *Smithsonian Institution.*

cutter bar caught and held the stalks as the oscillating blade cut through them. The grain fell onto a platform, and a worker walking alongside raked it off, keeping the grain out of the way as the reaper made the next round. McCormick continued to improve his reaper before patenting it on June 21, 1834, but he did not place his machine on the market until 1840. In the meantime, Obed Hussey tested a reaper near Carthage, Ohio, before the Hamilton County Agricultural Society.

Hussey patented his reaper on December 31, 1833, and sold his first machines in New York and Illinois the following year. Hussey's reaper differed from McCormick's in several respects. First, it did not have a reel to gather and hold the grain while the sickle cut it. Second, the 5-foot sickle consisted of a series of triangular steel plates that he riveted to a flat iron bar. This cutter bar had a reciprocal motion between slotted spikelike fingers. As the machine moved forward, the sickle clipped or chopped through the stalks. The grain fell onto a platform where it remained until enough accumulated to make a bundle. Then a worker raked it onto the ground for binding.

With the reaper, grain farmers could harvest larger crops with less labor than ever before. Because the reaper did not shatter the grain from the heads as much as the cradle scythe, it wasted less grain. The reaper also cut the stalks cleanly close to the ground and thereby increased the amount of straw that could be saved. These features convinced many farmers that a reaper would pay for itself in a year. During the early 1850s, some northern Ohio farmers were using reapers with "success and decided advantage." For them, the

practicality of this machine was a "fixed fact." By 1855, the reaper was a common sight in the wheat fields at harvesttime, and by 1860 farmers harvested an estimated 70 percent of the wheat in the West with this implement. Reapers could harvest from 10 to 12 acres per day.

Still, reapers simply cut grain. The reaper eliminated the need to hire cradlers and sped the harvest, but hired hands were needed to rake the gavels of cut grain together, bind the sheaves, and place them in shocks. Consequently, by the mid-1840s, inventors were turning their attention to the development of an automatic raking mechanism that would remove the cut grain from the platform. By 1854, the firm of Seymour and Morgan in Brockport, New York, marketed the first commercially successful self-rake reaper. As this machine moved forward, a rake swept across the platform at intervals and deposited the gavel onto the ground where it awaited the binders. With this invention, one more worker was eliminated from the harvesting process.

During the antebellum period, farmers began using threshing machines. In the 1820s, Jacob Pope, a Boston inventor, built a threshing machine that became popular. This machine had an endless belt that led into a spiked cylinder. As a worker fed the grain into the

thresher, the cylinder with iron teeth rotated and beat the grain from the heads. Machines like Pope's simply threshed; they did not separate the straw or winnow the chaff from the grain, and they deposited the grain and straw on the ground for separating and winnowing. Farmers also complained that it was easier to wield a flail than turn the operating crank on the threshing machine, but this problem was soon resolved by attaching horse-powered gearing to these implements. In 1837, Hiram A. and John A. Pitts of Winthrop, Maine, improved the threshing machine when they patented an implement designed to thresh, separate the straw, and winnow the chaff in one operation. A two-horse treadmill powered this small portable thresher. This implement could thresh about 100 bushels per day. By the 1850s, mechanical threshers were in common use, and some threshed as many as 500 bushels per day.

About the time that McCormick and Obed Hussey patented their reapers and the Pitts brothers were experimenting with their threshing machine, Hiram Moore of Climax Prairie, Kalamazoo County, Michigan, built the first successful combine. This implement simply combined the functions of a reaper with a threshing machine and winnower. Although it is uncertain when Moore began working on the problem of designing a combine harvester, he completed a

IN 1835, in Kalamazoo County, Michigan, Hiram Moore built the first grain combine. This machine was designed to cut and thresh grain at the same time. In this early-1850s picture, Moore sits on top of his combine during a test of the machine. Moore's combine required sixteen horses for draft power. It cut a 10-foot swath and sacked the grain at the rate of 25 acres per day. This large, heavy implement did not work well in the small fields of midwestern farmers, and the cost of $500 each proved prohibitively expensive. The technical problems of the combine would not be solved until the 1880s in California. *Smithsonian Institution.*

patent model in 1834 and field-tested this implement the next year. In 1835, after making improvements on his machine, Moore's combine, pulled by twelve horses, cut and threshed 3 acres of grain. By the late 1840s, this combine, now pulled by sixteen horses, cut a 10-foot swath and harvested 25 acres per day.

Despite Moore's success, midwestern grain farmers did not adopt this implement because their fields were too small for it to operate conveniently, the damp climate prevented the grain from threshing efficiently, and the implement cost $500. Midwestern farmers would adopt the combine, but they would not do so until the twentieth century, when smaller machines became suitable for their cropping patterns.

In the meantime, inventors in California continued to work on a combine that would be practical for the dry harvest season and large wheat fields in the Far West. These inventors, however, could not solve significant gearing and power problems. They continued to rely on horses to pull the implement, gearing the cutting and threshing mechanisms to the wheels so that as the wheels turned, the machine harvested and threshed. Although some of these combines used forty-horse hitches, they did not generate consistent power to harvest and thresh at the same time. Only when inventors placed a steam or internal-combustion engine on the combine to drive its working parts would this implement become viable, but that innovation would not come until the late nineteenth century.

During the antebellum period, some cotton and sugar planters in the South began using steam engines. By 1860, steam power had almost replaced horses and oxen for powering sugar mills in Louisiana, and plantation owners used steam engines to operate rice mills and power cotton gins. A steam engine easily outperformed horse- or manpower. Three men and a cotton gin could remove the seeds from 1,000 to 4,500 pounds of cotton per day, or about one hundred times more than they could gin without steam power. Although steam engines were attached to cotton gins about 1830, they were not widely used before the Civil War. Rather, horse-powered sweeps usually drove the mechanisms. Inventors also improved the cotton gin to enable the machine to remove more dirt and trash yet reduce damage to the lint. By 1860, most gins averaged 1.5–6 bales per day.

Before the 1850s, stationary rather than portable or traction steam engines generated steam power on farms and plantations. These machines could not be moved because they were bolted to solid brick and mortar foundations. A belt linked the steam engine to the machinery such as a cotton gin or draining and milling equipment in rice and sugar cane country. Southern plantations devoted to the extensive cultivation of staple crops like cotton or sugar cane were better suited for stationary steam engines than farms that emphasized diversified crops. Steam engines did not become popular on grain farms until portable models reached the market on a widespread basis after the Civil War. The daily chores on most farms were too diverse for the convenient use of stationary steam engines.

AGRICULTURAL IMPROVEMENTS

The federal government did little to aid agriculture until 1836 when Henry Leavitt Ellsworth became commissioner of patents. Ellsworth had a great interest in agriculture, and although his office did not have a specific responsibility for promoting agriculture, he quickly began to use his office to aid farmers. Under Ellsworth, the Patent Office approved the designs for a host of new implements such as plows, threshing machines, and harrows. Ellsworth also continued to collect seeds from abroad through the consulate system and the navy. He distributed seeds to farmers through agricultural societies and congressmen. He also asked Congress for an appropriation to help pay for these activities.

Congress responded to Ellsworth's request on March 3, 1839, when it appropriated $1,000 to support the collection and distribution of seeds and the compilation and dissemination of agricultural statistics and other information relating to farming. Ellsworth used these funds to establish an agricultural division in the Patent Office, the first federal money to aid agriculture. Congress, however, did not make another appropriation until 1842, and annual appropriations did not begin until 1847.

At the same time that Ellsworth struggled to gain annual appropriations from Congress, he also championed the creation of a federal bureau of agriculture. In addition, Ellsworth issued annual reports on the agricultural activities of the Patent Office. These reports were popular with congressmen, who gave them to their constituents to assure those voters that they were interested in the farmer's welfare. Each annual report contained agricultural statistics on production, essays on methods to improve crop and livestock raising, and letters about agricultural problems nationwide.

Not everyone in Congress or in the general public believed the agricultural work of the Patent Office was worthwhile. In 1846, Senator Ambrose Hundley Sevier of Arkansas charged that the reports were "comparatively worthless" and no more than an "accumulation of newspaper paragraphs." Senator John C. Calhoun of South Carolina denounced the agricultural work of the Patent Office as "one of the most enormous abuses under this government." Senator William Person Mangum of North Carolina charged that "practical farmers turned up their noses with utter scorn and contempt" at the agricultural reports. Such criticism, however, was not strong enough to stop the agricultural work of the Patent Office. By 1849, its annual report was so large that it was published in two parts. One volume was completely devoted to agriculture, and Congressmen willingly distributed it free of charge to farmers in their districts. As a result, the Patent Office served as the focal point for the agricultural activities of the federal government.

While the Patent Office struggled to conduct and expand its agricultural work, various farm groups on the state and local levels asked Congress to create a department of agriculture to support the improvement of farming nationwide. The work of the U.S. Agricultural Society, the Maryland Agricultural Society, and the Massachusetts Board of Agriculture was particularly instrumental in achieving the eventual creation of the department by petitioning Congress again and again to create such an agency. The U.S. Agricultural Society, for example, adopted a resolution on February 2, 1853, that urged Congress to create a department of agriculture with cabinet status—that is, establish a department led by a secretary who would be an instrumental part of the executive branch of government. Congress, however, remained divided over this issue and did not act upon that proposal. During the next ten years, the U.S. Agricultural Society continued to urge Congress to establish a department of agriculture.

While the society pressed for the creation of a department of agriculture, the Patent Office increasingly had less time and fewer resources to continue its farm-related work. The burden finally became so great that on January 11, 1860, William D. Bishop, commissioner of Patents, asked Congress to relieve his office of its agricultural responsibilities. Although some governmental officials believed the agricultural work of the Patent Office should remain with the Department of the Interior, in 1861, Thomas G. Clemson, superintendent of the Agricultural Division of the Patent Office, stated that a department of agriculture should be free from the influences of a parent agency. Southern congressmen, however, who championed states' rights (the retention of all powers by the states that the Constitution did not grant to the central government), opposed that action. They continued to argue that a department of agriculture would increase the centralization of the federal government at the expense of state authority and individual liberty and that it would create an expensive bureaucracy.

While the debate continued over the establishment of a department, agricultural reformers such as Jesse Buel, Edmund Ruffin, and Solon Robinson published a host of books to help farmers improve their farming practices. The agricultural press, however, became the most important source for disseminating new agricultural knowledge, particularly for methods and market prices. The *American Farmer,* founded in 1819 at Baltimore by John Stuart Skinner, and the *American Agriculturist,* published in 1842 at New York City, became the most important periodicals in antebellum America. Other journals such as the *Ohio Cultivator, Genesee Farmer,* and the *Southern Cultivator* provided similar information on state and regional levels. Although an estimated 350,000 farmers, usually in the North, subscribed to an agricultural periodical by 1860, most farmers did not have access to this information because of isolated location, illiteracy, or lack of wealth.

Agricultural societies also organized to promote the improvement of agricultural techniques and livestock. These societies sponsored lectures and annual fairs to stimulate a spirit of competition and serve as a social occasion to lessen rural isolation. Statewide agricultural societies often embarked on experiments and demonstrations to improve agriculture. In 1819, for example, the Massachusetts Society for Promoting Agriculture imported British cattle for breeding improvements. The New York State Agricultural Society hired the first agricultural entomologist during the 1860s. Local agricultural societies also organized, such as the Cincinnati Society for the Encouragement of Agriculture and Domestic Economy in 1819 and Missouri's Boone County Agricultural and Mechanical Society in 1852. Special-interest groups based on commodity specialization also formed, such as the New England Society for the Improvement of Domestic Poultry and the Wool Growers' Association of Western New York.

Wealthy stockmen also attempted to improve their breeding practices by importing purebred or "blooded" stock. During the 1820s, for example, sheepmen imported the fine-wooled Saxony and dual-purpose Leicester breeds. A decade later, they imported Southdown and Cotswold sheep, which they bred to their native stock in a process called upbreeding. In 1822, several individuals, such as editor John S. Skinner and John Hare Powell, a wealthy farmer, imported Shorthorns to improve beef production. Because of the expense involved, cattlemen sometimes formed associations to pool their resources to purchase improved cattle. In 1833, for example, George and Felix Renick and a group of cattlemen organized the Ohio Company for Importing English Cattle. The association bought Shorthorns in England for the improvement of Ohio Valley herds. Similar though less successful cattle-importing and improvement associations organized in other states.

These efforts played an important, incremental role in upgrading beef production, particularly west of the Appalachians. Breeders also brought English hogs such as the Leicestershire, Hampshire, and Suffolk to the United States during the antebellum period. Dairymen imported the Jersey in 1817 from an island in the English Channel by that name, the Ayrshire in 1822 from Scotland, the Guernsey in 1830 from the Isle of Guernsey, and the Holstein in 1857 from that German principality. Upbreeding of dairy cows, however, did not become significant until the late nineteenth century because most dairy farmers could not afford to buy improved stock.

RURAL LIFE

During the antebellum years, few distinctions sharply divided rural and urban life. More than half the population remained farmers, and almost everyone had been born and raised on a farm or knew something about agriculture. Craftsmen and businessmen often doubled as landowners, and tradesmen frequently worked as seasonal laborers at harvesting and threshing time. The towns and villages remained important locations for marketing, blacksmithing, and various processing services such as flour milling, wool carding, meat packing, cotton ginning, tobacco manufacturing, and hemp processing and for public schools and private academies. The villages and towns also provided other special and professional services that farm families needed occasionally such as clerical, medical, and legal. Most important, rural merchants bought or traded agricultural commodities and provided cloth, ironware, and crockery for home use. The local merchants often extended credit, served as bankers by keeping the earnings of farmers on deposit in their safes, wrote letters, and collected the mail. As in the past, farmers were not entirely self-sufficient, and the village merchant gave them access to the wider world beyond.

Until villages developed, however, farmers on the frontier remained isolated and subsistence-oriented in their agricultural practices. Contact with others came infrequently, usually with a visit from a Methodist circuit rider or itinerant Baptist or Presbyterian minister. As the rural population grew, the church became important not only to nourish the spirit but to provide social contacts. The activities of the church at revivals, regular Sunday meetings, or special occasions such as baptisms and weddings contributed to the education, courtship, and communication of the members. In addition to the community of the church, farmers on the frontier met for cabin and barn raisings, harvesting, threshing, and cornhusking. Shared work was not only physically necessary but socially important. With many farms still isolated, particularly west of the Appalachians, men and women had a great "desire for society." Visitors were few, but they were always welcome.

The lives of rural women focused on the home. Their world remained domestic, their primary duties in the words of one observer relegated to "boiling and baking, turning the spinning wheel and rocking the cradle." During the antebellum period, women contributed to the farm in two primary ways. First, they worked in the home to produce food and clothing and provide the services necessary to keep the home and family organized—cleaning, cooking, laundering, and nursing. Second, they helped produce subsistence crops such as garden vegetables, but they also helped raise grain, tobacco, and cotton that could be traded or sold to provide needed items for the family. They also made butter, raised poultry, sold eggs, wove linen and wool cloth, collected feathers, and packed pork, all of which had a market value with a local merchant. Some women operated farms and held slaves, although they usually did so after the death of their husbands and with the aid of their sons.

Still, rural life for women during the antebellum period began to lose many of its common characteristics. Although certain general characteristics still applied to farm women in all sections of the country, they did so with varying degrees of emphasis. In the Northeast, for example, the rapid growth of the textile industry and other manufacturing decreased the value of home-produced goods such as linen cloth. The farm families that lived near major commercial markets in New England and along the mid-Atlantic seaboard now gave less attention to home manufactures. With access to mercantile items such as ready-made cloth, farm women moved the loom and the spinning wheel to the attic, barn, or storage shed. Manufacturing industries not only drew women away from the farm for employment but decreased the value and need for them to produce many

of these items in the home. The factories could do so cheaper and with more uniform quality, particularly cloth production. Consequently, the work of farm women became less valuable for producing farm income, and they became increasingly relegated to the same services around the home that most urban women experienced. Where the market offered the opportunity for commercial gain, however, women specialized in the production of certain commodities such as butter and eggs.

Women in the South and on the midwestern frontier continued to produce home-manufactured goods for domestic use and trade to a greater extent than their sisters in the Northeast or along the Atlantic seaboard. Often, farmers in these areas planted flax. The woman of the house harvested the crop by pulling the plants from the soil

DURING THE LATE eighteenth and early nineteenth centuries, a number of utopian groups created rural communities and practiced agriculture. Among the best known are the Shaker communities, Brook Farm, New Harmony, and the Amana Colonies. In 1817, Lutheran Separatists from Württemberg founded the colony of Zoar. Located along the Tuscarawas River in eastern Ohio, the Zoarites practiced communal agriculture. They eschewed the sacraments and looked for the imminent return of Christ. In the meantime, they raised grain, livestock, fruits, and vegetables to meet their own needs and to sell to the outside world. This photograph dates from the late nineteenth century, but their practice of harvesting grain with a cradle scythe, raking the gavels, and binding the sheaves in a communal process had remained unchanged since the founding of the colony. *Ohio Historical Society.*

and processed it by retting, breaking, swingling (or switcheling), hatcheling, spinning, and weaving it. They also carded and spun wool for the manufacture of homemade clothing. Where farmers raised sheep to help meet their clothing needs, farm women washed the wool to remove as much of the natural grease or "yolk" as possible before spinning. Then they dried the fleece and picked the fibers into a soft fluffy pile ready for carding, a task that involved running a wire comb, or card, through the wool to align the fibers in parallel rows. The wool was then ready for twisting into yarn on the spindle. The preparation of fleece for spinning was a time-consuming task that required considerable skill. When entrepreneurs established carding mills in rural communities, farm women preferred those firms to wash, pick, and card their fleece.

But they did not give it up everywhere. During the 1850s, for example, one traveler in Mississippi noted that although few women in the Midwest used the spinning wheel and hand loom and homespun clothing had nearly disappeared, "half the white population of Mississippi still dress in homespun, and at every second house the wheel and the loom are found in operation." Still, southern women and others gave increasingly less attention to these tasks through the antebellum years. Women who could afford factory-made cotton cloth or ready-made clothing purchased it because it was more cost-effective in time, effort, and quality. The spinning of flax or wool did not survive as a domestic rural craft long after the Civil War.

Although white women sometimes worked alongside their husbands and sons in the fields hoeing corn, chopping cotton, or binding sheaves of wheat if their labor was crucially needed, and although preparing substantial "well-cooked" meals at harvest and threshing time was a traditional responsibility, slave women commonly worked in the fields. They grubbed roots, spread manure, and plowed; built fences, threshed grain, and picked cotton. In the 1850s, one observer in the South noted that "the ploughs at work, both with single and double mule teams, were generally held by women, and very well held too. . . . Twenty of them were ploughing together, with double teams and heavy ploughs." No task was too difficult for their assignment, whereas Anglo-American culture usually restricted white women to the least physical tasks unless absolutely necessary. Slave women did not have this cultural protection.

Yet although masters did not hesitate to order slave women to do agricultural tasks, they did not expect as much productivity from them as men. One former slave recalled that men were required to pick 300 pounds of cotton per day; women, 200 pounds per day. Although his memory may have placed the daily harvesting requirements too high, the proportion was about right. Slave women also worked in household production to spin and weave linen and wool for blankets and clothing. White women, mistresses of their plantations, supervised the domestic work as well as some of the home manufacturing. They also managed the care of gardens, poultry, and dairy cattle. The life of a plantation mistress, however, usually involved more work than leisure. Sarah Williams, a New York woman on an Alabama plantation, reflected: "People may talk of the freedom from care of southern life, but to me it seems full of care." Even for elite plantation women, agricultural and rural life involved hard work and isolation.

The labor of women, then, remained crucial to the success of the farm for both subsistence and commercial production. Rural women worked in a society that was divided by gender, class, and race, a society in which the market drove the agricultural economy, and they became increasingly tied to it.

From 1815 to 1860, most farmers were committed to a mix of subsistence and commercial agriculture. They pursued economic gain by choice. Where subsistence agriculture prevailed, inadequate transportation and the proximity of towns rather than farm size or preference temporarily excluded farmers from the marketplace. Those farmers left behind or outside the market economy formed a "lower sort" in the agricultural community, especially in the Appalachians. At the same time, commercial farming dominated both the agricultural and national economy.

In the South, slavery made the rapid expansion of agriculture possible. It institutionalized a biracial society and for a time the plantation system that developed during the seventeenth century. Most important, slavery determined the economic and demographic patterns of the South for generations to come because it brought considerable profits to owners. Slavery and the plantation system also created profitable markets for farmers in the Old Northwest for the sale of corn and pork. Bonded labor meant that southerners could hold larger areas of land than agriculturists in other regions where labor was expensive and the supply inadequate. Without laborers, farmers in the Midwest could not cultivate as much land as those who owned slaves unless they substituted horsepower and machinery for field workers.

In areas well suited for commercial agriculture, rapid settlement occurred. In the Midwest, Indiana became a state in 1816, Illinois in 1819, and Iowa in 1846. Similarly, rapid settlement and agricultural development in the South helped bring other states into the union—Mississippi in 1817, Alabama in 1819, and Texas in 1845—while other states such as Ohio and Louisiana had joined shortly before the War of 1812 and still others would follow after the Civil War.

American agriculture spread rapidly between the Appalachians and the Mississippi River between 1815 and 1860, but pioneering still occurred in New England and upstate New York. Land clearing remained the most difficult task and the ax the most common farm tool in the North as well as the South. At best, a farmer could clear about 10 acres of timber per year without hired labor, but his fields might be pocked with stumps for more than a generation. Hogs remained the most common livestock and corn the universal grain. Although the plantations symbolized southern agriculture, the planters were merely farmers who operated on a larger scale than most. Although the planters' labor supply was based on slaves rather than free white workers and their lifestyle differed considerably from that of the yeomen in the North, they experienced the same cash-flow and credit problems of farmers in other areas and enjoyed the gains from escalating land prices and profits earned from the sale of staple crops. Bountiful production in the Midwest, however, forced other northern farmers to specialize for local and regional markets.

Although the American farmer suffered from the economic depressions, known as the panics of 1819, 1837, and 1857, most experienced general economic improvement, even though those gains cannot be viewed as entirely progressive. Rather, American agriculture remained dynamic. Although it changed rapidly, not all of those changes were for the better. The Civil War era would compound the successes and problems of the farmer and change American agriculture for all time.

SUGGESTED READINGS

Atack, Jeremy, and Fred Bateman. *To Their Own Soil: Agriculture in the Antebellum North.* Ames: Iowa State University Press, 1987.

Atherton, Lewis E. *The Frontier Merchant in Mid-America.* Columbia: University of Missouri Press, 1971.

______. *The Southern Country Store.* Baton Rogue: Louisiana State University Press, 1949.

Bidwell, Percy Wells, and John I. Falconer. *History of Agriculture in the Northern United States, 1620–1860.* New York: Peter Smith, 1941.

Blassingame, John W. *The Slave Community: Plantation Life in the Antebellum South.* New York: Oxford University Press, 1979.

Bogue, Allan G. *From Prairie to Corn Belt: Farming on the Illinois and Iowa Prairies in the Nineteenth Century.* Chicago: University of Chicago Press, 1963.

Broehl, Wayne G., Jr. *John Deere's Company: A History of Deere & Company and Its Times.* New York: Doubleday, 1984.

Craven, Avery O. *Soil Exhaustion as a Factor in the Agricultural History of Virginia and Maryland, 1660–1860.* University of Illinois Studies in the Social Sciences, vol. 13, no. 1, 1925.

Crockett, Norman L. *The Woolen Industry of the Midwest.* Lexington: University of Kentucky Press, 1970.

Danhoff, Clarence H. *Change in Agriculture: The Northern United States, 1820–1870.* Cambridge Mass.: Harvard University Press, 1969.

Degler, Carl N. *Place Over Time: The Continuity of Southern Distinctiveness.* Baton Rouge: Louisiana State University Press, 1977.

Eaton, Clement. *The Growth of Southern Civilization.* New York: Harper and Row, 1961.

Ellis, David M. *Landlords and Farmers in the Hudson-Mohawk Region, 1790–1850.* Ithaca, N.Y.: Cornell University Press, 1946.

Fox-Genovese, Elizabeth. *Within the Plantation Household: Black and White Women in the Old South.* Chapel Hill: North Carolina University Press, 1988.

______. "Women in Agriculture during the Nineteenth Century". In *Agriculture and National Development: Views on the Nineteenth Century,* ed. Lou Ferleger, 267–301. Ames: Iowa State University Press, 1990.

Gates, Paul W. *The Farmer's Age: Agriculture, 1815–1860.* New York: Holt, Rinehart and Winston, 1960.

______. *History of Public Land Law Development.* Washington, D.C.: Government Printing Office, 1968.

Genovese, Eugene. *The Political Economy of Slavery: Studies in the Economy & Society of the Slave South.* 2d ed. Middletown, Conn.: Wesleyan University Press, 1989.

Gibson, James R. *Farming on the Frontier: The Agricultural Opening of the Oregon Country, 1786–1846.* Seattle: University of Washington Press, 1985.

Gray, Lewis Cecil. *History of Agriculture in the Southern United States to 1860,* vol. 2. Gloucester, Mass.: Peter Smith, 1958.

Hardeman, Nicholas P. *Shucks, Shocks, and Hominy Blocks: Corn as a Way of Life in Pioneer America.* Urbana: University of Illinois Press, 1975.

Heitmann, John Alfred. *The Modernization of the Louisiana Sugar Industry, 1830–1910.* Baton Rouge: Louisiana State University Press, 1987.

Hibbard, Benjamin Horace. *A History of the Public Land Policies.* Madison: University of Wisconsin Press, 1965.

Hurt, R. Douglas. *Agriculture and Slavery in Missouri's Little Dixie.* Columbia: University of Missouri Press, 1992.

______. *American Farm Tools: From Hand Power to Steam Power.* Manhattan, Kans.: Sunflower University Press, 1982.

______. "Out of the Cradle: The Reaper Revolution." *Timeline* 3 (October–November 1986): 38–51.

Jensen, Joan M. *Loosening the Bonds: Mid-Atlantic Farm Women, 1750–1850.* New Haven, Conn.: Yale University Press, 1986.

Jones, Donald P. *The Economic and Social Transformation of Rural Rhode Island, 1780–1850.* Boston: Northeastern University Press, 1992.

Larkin, Jack. *The Reshaping of Everyday Life, 1790–1840.* New York: Harper and Row, 1988.

McDonald, Forrest, and Grady McWhiney. "The Antebellum Southern Herdsman: A Reinterpretation." *Journal of Southern History* 41 (May 1975): 147–66.

McMillen, Sally G. *Black and White Women in the Old South.* Arlington Heights, Ill.: Harlan Davidson, 1992.

McNall, Neil A. *An Agricultural History of the Genesee Valley, 1790–1860.* Philadelphia: University of Pennsylvania Press, 1952.

Moore, John Hebron. *The Emergence of the Cotton Kingdom in the Old Southwest: Mississippi, 1770–1860.* Baton Rouge: Louisiana State University Press, 1988.

Neely, Wayne Caldwell. *The Agricultural Fair.* New York: Columbia University Press, 1935.

Oakes, James. *The Ruling Race: A History of American Slaveholders.* New York: Knopf, 1982.

Oberly, James W. *Sixty Million Acres: American Veterans and the Public Lands before the Civil War.* Kent, Ohio: Kent State University Press, 1990.

Opie, John. *The Law of the Land.* Lincoln: University of Nebraska Press, 1987.

Rikoon, J. Sanford. *Threshing in the Midwest, 1820–1940: A Study of Traditional Culture and Technological Change.* Bloomington: Indiana University Press, 1988.

Rogin, Leo. *The Introduction of Farm Machinery in Its Relation to The Productivity of Labor in the Agriculture of the United States During the Nineteenth Century.* Berkeley: University of California Press, 1931.

Russell, Howard S. *A Long, Deep Furrow: Three Centuries of Farming in New England.* Hanover, N.H.: University Press of New England, 1982.

Schlebecker, John T. *Whereby We Thrive: A History of American Farming, 1607–1972.* Ames: Iowa State University Press, 1975.

Schob, David E. *Hired Hands and Plow Boys: Farm Labor in the Midwest, 1815–60.* Urbana: University of Illinois Press, 1975.

Scott, Anne Firor. *The Southern Lady: From Pedestal to Politics, 1830–1930.* Chicago: University of Chicago Press, 1970.

Sitterson, J. Carlyle. *Sugar Country: The Sugar Cane Industry in the South, 1753–1950.* Lexington: University of Kentucky Press, 1953.

Stampp, Kenneth M. *The Peculiar Institution: Slavery in the Antebellum South.* New York: Knopf, 1956.

Tadman, Michael. *Speculators and Slaves: Masters, Traders, and Slaves in the Old South.* Madison: University of Wisconsin Press, 1989.

Van Deburg, William L. *The Slave Drivers: Black Agricultural Labor Supervisors in the Antebellum South.* Westport, Conn.: Greenwood Press, 1979.

Van Wagenen, Jared, Jr. *The Golden Age of Homespun.* Ithaca, N.Y.: Cornell University Press, 1953.

Wilson, Harold Fisher. *The Hill Country of Northern New England: Its Social and Economic History, 1790–1930.* New York: Columbia University Press, 1936.

Woodman, Harold D. *King Cotton & His Retainers: Financing & Marketing the Cotton Crop of the South, 1800–1925.* Lexington: University of Kentucky Press, 1968.

THE CIVIL WAR

When the guns opened fire on Fort Sumter on April 12, 1861, more than half of the 31.4 million people in the North and South lived on farms, and agriculture remained the most important component of the national economy, with three-fourths of the country's exports by value derived from farming. At that time, 75 percent of the nation's farm families owned their land.

Quickly, the relative prosperity of American agriculture changed dramatically as both Union and Confederate armies demanded large quantities of food supplies. Compared to the North, southern farmers could not meet the military needs of the Confederacy because of inadequate diversification, poor transportation, and slave-labor problems. Some southerners in some areas who had relied on northern agricultural commodities such as pork and flour prior to the war suffered hardship when their sources of supply ended with the closure of the Mississippi River. Moreover, Confederate and Union troops usually took what they needed from southern farmers, and northern soldiers also destroyed buildings, machinery, and fences. When transportation problems from west to east were resolved early in the war, midwestern farmers found a ready market and high prices with Union contractors.

The Confederacy needed its farmers to produce food for both the army and citizenry, but draft policy severely restricted their ability, and army requisitions sometimes reduced their desire to aid the government. Southern leaders urged farmers to diversify by raising more corn, wheat, and livestock and to reduce their reliance on cotton. Early in 1862, the editor of the Columbus (Georgia), *Sun* told farmers to "Plant Corn and Be Free, or plant cotton and be whipped." Soldiers needed food, not cotton.

The Confederate Congress even considered banning the cultivation of cotton during 1862.

With the Union navy blockading the coast, little cotton could reach Europe via blockade runners, and northern textile factories had adequate quantities of fiber remaining from the record crop in 1859. Southern farmers and planters responded by substituting corn on a large scale for cotton, tobacco, and sugar cane. Texas proved an exception. There, farmers continued to emphasize cotton and to trade across enemy lines. At the same time, the Confederate government called on southern forges and foundries to make guns rather than the plows, reapers, and corn shellers that farmers needed to aid the process of diversification and food production. Agricultural implements became expensive, and most farmers got by with the tools they already owned. As agricultural implements in the South deteriorated, so did agricultural efficiency.

Southern agriculture also suffered from an inadequate labor supply. With most men between the ages of 18 and 50 subject to the draft, few remained to till the soil. Only those who held public office or managed plantations with at least twenty slaves were exempt from compulsory military service. The slaves became even less efficient with freedom in sight and were often taken from the farms and plantations to build fortifications. In addition to the loss of human labor, southern farmers also suffered great losses of horses and mules from impressment by the Confederate army or capture by Union troops.

Despite these problems, southern farmers were able to feed most of the Confederate population through 1862. Thereafter, war and the Union blockade destroyed or interrupted transportation and food distribution systems and caused food problems despite increased acreages planted to corn, wheat, and potatoes. Food shortages and "extortionate prices" in Virginia, created by Robert E. Lee's Army of Northern Virginia draining supplies, caused a "bread riot" in Richmond during the spring of 1863. In April, a mob of sixty women shouting "Bread or Blood" began plundering stores for food and flour. Soon they were joined by a host of other women as well as men. Additional food riots on a smaller scale occurred in other Confederate cities.

By 1863, Union troops prevented many small-scale

farms and plantations from meeting the food needs of southerners by confiscating livestock, grain, and forage for their own use or destroying it and farm implements. The destruction of the limited railroad system in the region also contributed to hunger in the South. Moreover, with the cotton market destroyed, planters could not meet their debts and taxes. Only a host of "stay laws," laws that delayed payment of debts without retribution, prevented the loss of their property.

Despite these problems, southern farmers tried to profit from the war as much as possible, often asking from 200 to 400 percent more for their commodities than they received before secession. Jefferson Davis, president of the Confederacy, complained about farmers and other war profiteers: "The passion of speculation has become a gigantic evil. It has seemed to take possession of the whole country, and has seduced citizens of all classes from a determined prosecution of the war to a sordid effort to amass money. It destroys enthusiasm and weakens public confidence." But agricultural profits were not guaranteed, and many farmers hid livestock and other foodstuffs to prevent impressment by the Confederate army, which, they contended, offered below-market prices. Some farmers even refused to accept the increasingly worthless Confederate paper money for agricultural commodities, particularly if Union troops threatened to occupy the area. One Virginia soldier complained that southern farmers had an "infernal lust for property" and a "hellish greed for gold."

In the absence of men, farm women worked the fields, tended livestock, and marketed produce. In 1864, one Confederate agent reported, "It is a common thing to see the women and children industriously engaged on their farms and . . . the people as a general thing are doing all they can to supply themselves." Production, however, decreased from that achieved when husbands and sons were home to conduct the heavy work of plowing and harvesting corn and other grains. In 1863, however, the *Charleston Courier* reported that much of the binding and stacking of the South Carolina wheat crop was done by "brave and faithful women who are worthy to be sisters of the soldiers of our army."

In the Confederate West, Texas cattlemen profited at

first from the war. On the New Orleans cattle market, beef brought $30 to $50 per head, double the prewar prices. Texas stockmen also drove cattle to Natchez, Memphis, and Shreveport. In January 1862, Marcellus Turner, a stockman in Lavaca County, Texas, reported that stockmen in his neighborhood had "gone crazy driving–anyone who had a hundred head expected to drive." Contractors for the Confederate army also paid high prices for Texas beef. This cattle trade peaked in the summer of 1862 when the beef markets in New Orleans and Memphis came under federal control and Union gunboats prevented most Texas cattle from reaching the east bank of the Mississippi River. West of the river, however, Texas cattlemen continued to supply the Confederate army in Missouri, Arkansas, and the Indian Territory, although they could not meet the demand.

As the Confederate currency depreciated, however, Texas stockmen became increasingly reluctant to sell their cattle. In 1864, John Maton, a cattleman in Refugio County, Texas, complained that "the beef is no account and the money worse." Confederate purchasing agents had so much difficulty buying Texas beef that the army began the impressment of cattle for $25 per head. Some pro-Confederacy citizens chided the cattlemen for a lack of patriotism. One group of Texas stockmen responded by threatening to hang any Confederate commissary agent who seized their cattle. Profits, not patriotism, mattered the most to these cattlemen, whose social and political attachments were not as strong as those of other southerners and were not dependent on slavery for their livelihood. Animosity characterized the relationship between the Texas cattlemen and the Confederacy during most of the war, even though many served in the Confederate army.

By the end of the war in 1865, southern farmers had experienced great hardships and faced severe problems that would take years to solve. In some areas such as Virginia, Georgia, Tennessee, Mississippi, Louisiana, and the coast of the Carolinas, houses, barns, and fences lay in ruin, fields abandoned, livestock for food and draft power killed or captured, and farm equipment worn out or destroyed. One Georgia planter wrote: "I had the misfortune to be in the line of Sherman's march, and lost everything—Devon cows, Merino sheep, Chester hogs, Shanghai chickens, and in fact everything but my land, my wife and children and the clothing we had at that time."

Confederate money was worthless, and southern credit institutions no longer existed. The slaves had been emancipated, and farmers would not be paid for the commodities that the Confederate army impressed. Most important, 260,000 Confederate soldiers, mostly farmers and rural people, had been killed. Those who survived faced a gigantic task restoring southern agriculture.

In the North, farmers did not confront the same problems. Northern farmers did not experience any invasion of consequence or destruction of their property. At most, they endured temporary shortages of labor, but these difficulties were not serious. Union enlistments and the draft generally did not take many men from the fields, although Iowa, Minnesota, and Wisconsin were exceptions. Those who were overage or who could not pass their physical examinations returned home because the Union army had an extensive manpower pool to draw upon. In contrast, the Confederate army was hard pressed to find enough men. In addition, immigrants, discharged veterans, and farm women made up many losses in the northern work force.

Northern farmers also increasingly adopted new agricultural implements such as reapers, threshing machines, and mowers. In 1862, Henry F. French, who later became the president of Massachusetts Agricultural College, reported that almost every New England farmer owned a mower because of a scarcity of labor for haying. He wrote: "While the war has prevented extensive and permanent improvements upon the farm, it has . . . greatly hastened their introduction of useful machines, and of useful processes in farm labor." This observation applied to all areas of the Union, including the Midwest. Women operated much of this new equipment, and it enabled agriculturists to expand production, save time, and reduce labor expenses. Northern farmers became increasingly dependent on specialized production and a market economy.

The federal government also called on northern farmers to increase production of food and fiber. They responded by expanding their use of dual-purpose cattle for meat and milk, particularly the Shorthorn and Devon breeds. Gail Borden's newly developed process for condensing milk, removing three-fourths of the water, provided a new food product that the military used regularly in the field, at sea, and in hospitals, and it increased the demand for milk. The extension of railroad lines into the Midwest and the development of new farms also brought more grain and livestock to market. Crop failure due to bad weather in 1863 damaged the corn crop and caused a decline in cattle, and hog cholera limited swine production, but the war ended before these difficulties caused food problems for the army and civilian population. Great demand for flour by the Union army and insufficient grain crops in Great Britain and France absorbed large wheat crops in the North and kept prices high.

Banking and other credit sources remained strong throughout the war. In addition, "greenbacks," paper money that the federal government issued to help finance the war, helped farmers to pay off debts with an inflated currency that did not always reflect the value of their debts. Moreover, high prices and the demand for increased production encouraged farmers to use more fertilizer, although this improvement was not universal.

For northern farmers, then, the war paid, and they took advantage of commercial markets both civilian and military. Many farmers used wartime profits to pay debts and improve land with fences, buildings, and drain tile. These improvements, together with growing demands for agricultural commodities, often doubled land values. In Iowa, for example, the average value per acre rose from $12 in 1860 to $25 by 1870. Farm values also increased during that decade. New York farm valuations rose from $4,452 to $5,668, and the average price of Illinois farms increased from $2,774 to $4,480. The Civil War also caused financial changes that gave farmers increased credit and borrowing power to improve their operations. For most northern farmers, the Civil War brought prosperity.

---

5 • THE GILDED AGE

IN 1873, MARK TWAIN AND CHARLES DUDLEY WARNER PUBLISHED A satire of the Grant presidency, *The Gilded Age*. Their book captured the hard-driving, speculative, and sometimes unethical nature of the American people as they worked to make or improve their fortunes. The book quickly became a best seller and left its imprint as a sobriquet for the entire late nineteenth century. It showed clearly that wealth can be little more than a veneer over problems and poverty. Twain's Gilded Age particularly applied to American agriculture. Although some farmers profited from agriculture, social and political problems and poverty dulled the luster of rural life.

Between the Civil War and the turn of the twentieth century, American agriculture underwent revolutionary change, not all for the better. The Civil War not only revolutionized the labor system in the South by abolishing the institution of slavery but stimulated the mechanization of agriculture as northern farmers rapidly shifted from handpowered to horsepowered implements. During the war years, agriculture rapidly expanded into the Great Plains and Far West. Farmers increasingly specialized, often with a single crop. They gave less attention to self-sufficiency and more to producing a commodity for sale in a market economy. This specialization often tied farmers to bankers, railroads, and businessmen in a way that often caused bitterness and encouraged organization and protest.

In the South, farmers remained reliant on cotton, but thousands became trapped in the sharecropping system—a new combination of land, labor, and credit that held many farmers in a state of peonage. Although the federal government gave increasing attention to agricultural aid and education, the benefits of that support often depended on the slow development of science. In contrast, the agricultural regulatory activities of the federal government expanded to serve both rural and urban communities. By the end of the century, the best agricultural land had been claimed from the Atlantic to the Pacific. But while not all farmers prospered during the Gilded Age and many did not believe that agriculture paid, most remained convinced that farming provided the best way of life.

THE SOUTH

The rural South of the late nineteenth century was considerably different from the antebellum region. The Civil War proved devastating in many areas. In May 1865, Eliza F. Andrews, a Georgia woman, wrote: "A settled gloom, deep and heavy, hangs over the whole land." Black farmers who wanted land could not afford it, and the Southern Homestead Act of 1866 provided only 2.9 million acres for 27,800 black families. Much of this land was too poor to support independent farms for the newly freed men and women, and many eventually lost their homesteads. Although the Civil War destroyed slavery, the plantations remained, often under absentee and sometimes northern ownership. Henry Grady, editor of the *Atlanta Constitution,* put it best when he noted that after the war the planters were "still lords of acres, though not of slaves."

The war did not result in the redistribution of slaveowners' property to the blacks. There would be no "40 acres and a mule" as rumor had led them to believe. Although Confederate farmers and planters either abandoned or lost 850,000 acres to the federal government by 1865, private property was too sacred to take from one group and give to another, no matter how repressed, and few white northerners or southerners wanted to challenge the tenets of racism. Almost all confiscated land would be restored to former owners, which caused one freedman in Virginia to complain: "We has a right to the land where we are located. . . . Our wives, our children, our husbands, has been sold over and over again to purchase the lands we now locates upon; for that reason we have a divine right to the land."

Rather than redistribute land, the Republican party, which controlled Reconstruction, introduced a contract labor system that essentially kept former slaves bound to the land and poverty stricken until the sharecropping system replaced it during the late 1860s. Yet without land or financial support from the federal government, the blacks could not gain the economic independence necessary to preserve their newly gained civil and political rights, let alone adequately care for their families. Moreover, they could not purchase the seed, implements, and livestock necessary to begin farming, even if they had been given land, because they lacked financing.

Although the planters still controlled the bulk of the land, they encountered serious labor problems now that slavery had been destroyed. Many whites believed working for wages was beneath their dignity, particularly if planters required them to work alongside the

AFTER THE CIVIL WAR, black farmers usually worked as sharecroppers, often living in the same houses and cultivating the same plantation lands that they had known during slavery. In this photograph, ca. 1895, a group of black women and children pose before a log cabin home, which had probably been slave quarters before emancipation. *Georgia Department of Archives and History.*

former slaves. Planters also expressed dissatisfaction with the white labor available, in part because of their long custom of using black workers. In 1874, one white planter complained: "The attempt to introduce hired white men from any quarter to carry out the exclusive cotton system is, from various causes, preposterous. In the southern climate, with the habits of the people and the wages they can afford to pay, no laborer can be a substitute for the negro."

The 4 million freed blacks in the South depressed wages. When landlords hired white workers, they demanded higher pay than the black field hands, who had little alternative but to work for lower wages than most whites would accept. In 1866, Thomas Jackson of Princess Ann County, Virginia, reported that landowners found white labor too expensive and were "anxious to make engagement with Freedmen." Soon white farmers praised black labor, because in the words of one Arkansas farmer, they were "easier to get on with and less apt to grumble and find fault than native whites." Most important, these farmers had little or no legal protection in the South, and landowners could treat the blacks as they pleased.

Moreover, the banking system lay in ruins, and credit remained inadequate. In the absence of labor and money or credit, yet with an abundance of land ready for renewed cultivation, the planters devised a new agricultural system to bring land, labor, and capital together—the sharecropping system. Sharecropping involved the subdivision of large-scale farms or plantations into small-scale farms. This system enabled black and white farmers who could not afford to purchase their own land or pay rent as tenants to cultivate 20 to 50 acres and pay the landlord with a portion of the crops. Or, more appropriately, the landlords let them keep a portion of the crop as a wage for working their land.

Although sharecropping arrangements varied from state to state and region to region, the average sharecropper could expect to pay one-third to one-half of his crop, usually cotton, to the landlord. If the landlord furnished seed, feed, fertilizer, implements, and a mule instead of just land and a house, the sharecropper might be required to pay two-thirds of his crop in rent. Sharecroppers moved frequently, often annually, in pursuit of landlords who would offer a better arrangement. Sharecropping also enabled white and black farmers to separate themselves by families in the fields, in contrast to contract or day work, which required both races to work together.

Invariably, sharecroppers suffered excruciating poverty and perpetual debt. They became virtual peons in the rural South. Without money to purchase needed items for daily living and farming such as coffee, calico, and tools, sharecroppers became bound to the local merchant. These entrepreneurs provided the basic necessities of life on credit. They became known as furnishing merchants, but they also took a lien on the farmer's crop; sometimes the landlord also provided the "furnish." When the cotton harvest came, the furnishing merchant presented his bill for the year. Invariably, it tallied more than the farmer earned from his crop after paying the landlord his share. The furnishing merchant took the remainder of the crop and required the farmer to pledge his next year's crop to pay off the debt.

This procedure became a continuing contract that locked sharecroppers under the control of the landlord if he served as the furnishing merchant. Moreover, furnishing merchants required their debtors to raise staple crops, particularly cotton. The fiber stored easily, and a market always existed for cotton even though overproduction kept the price low. With interest rates for credit reaching usurious levels and higher prices prevailing for purchases on credit, these southern farmers had no chance to work their way out of poverty, nor could they control their own lives. One merchant-landlord showed his control over his sharecroppers when he said, "If I say 'plant cotton' they plant cotton."

THESE AFRICAN-AMERICAN men, women, and children picked cotton in Thomas County, Georgia, about 1895. Note the sack tied around the waist of the boy in the front and the basket on the shoulder of the worker to the rear. *Georgia Department of Archives and History.*

Although cotton prices remained high for a decade after the Civil War (overproduction ultimately caused prices to fall from $0.43 to less than $0.05 per pound between 1866 and 1898), the cost of needed items did not decline as sharply or as rapidly. In 1898, a record crop of 11.5 million bales dropped the price below $0.05 per pound when farmers needed $0.08 per pound to earn a profit. In the absence of adequate markets and transportation for other agricultural commodities, southern farmers continued to specialize in cotton. By the turn of the twentieth century, more than 1 million farms or 54 percent of the farms in nine states emphasized cotton production because it remained the best and most reliable cash crop. Still, most farmers could not save enough to buy land or get out of debt. Poverty would characterize the southern countryside until New Deal agricultural policies changed the credit system and World War II took surplus labor from the farms.

THE NORTH

Northern farmers in New England and the Midwest also experienced great change during the late nineteenth century. Although western lands and improved railroad transportation affected New England farmers before the Civil War, by 1870 both began to undermine the northern agricultural economy. New England farmers, for example, were commercial agriculturists by the end of the Civil War, when sheep raising was their most profitable endeavor. After the war, however, northern flockmasters were unable to gain adequate tariff protection against wool shipped from Australia and South America; confronted with increasing land values, they abandoned sheep raising. The sheepmen who survived resided in the Far West where they grazed their animals on cheap public land and used the railroads to ship large quantities of wool to eastern markets at prices below the competitive level for New England farmers. By 1870, agriculture in New England, particularly for sheep raising, had become increasingly unprofitable, and thousands of farmers who could not change their agricultural operations left the land for better-paying jobs in the cities or moved west in search of cheap, productive land. As the rural population rapidly declined, abandoned farms grew in number. Not all the population loss from the countryside was bad, however, because many farms had been founded on marginal or submarginal lands far removed from adequate transportation facilities.

As agricultural profitability declined precipitously in New England, land values quickly dropped. Those farmers who remained on the land necessarily adopted new agricultural practices to survive. Many New England farmers who remained on the land emphasized dairying. The cool climate, rocky terrain, and lush grasses of New England, together with the proximity of many urban markets, made dairying ideal there. Dairying provided more profit than sheep raising and enabled New England farmers to maintain profitable and competitive operations. By the late nineteenth century, improved refrigeration technology and increased urban demands caused dairy farmers to emphasize the production of fluid milk, which brought even higher profits than the traditional sale of that commodity to secondary markets for the manufacture of butter and cheese.

IN NEW ENGLAND, tobacco farming remained a moderately profitable form of agriculture through the 1890s. This tobacco farm near Westchester, Connecticut, typified the small-scale, labor-intensive nature of the crop in the Northeast. After this farmer cut his tobacco, a helper tied the leaves into hands for transport via oxcart to the barn for curing. *Connecticut Historical Society, Hartford, Connecticut.*

By 1900, New England farmers had shifted from general diversified agriculture to intensive specialized production to survive. Northern New England, for example, had changed from a meat-wool-grain-producing region to a dairy-fruit-truck-crop area during an age of rapid scientific, technological, and economic change. High land values, small farms, and poor soils prevented the New England agriculturists from competing with the abundant production of meat and cereal grain in the Midwest and Great Plains. Relatively low transportation rates from west to east made matters worse, and labor costs became prohibitive for New England farmers. While a midwestern agriculturist spent $9 for labor to raise an acre of corn at the turn of the twentieth century, a New England farmer paid $17 for that production.

South of New England, the number of small farms devoted to specialty production, such as poultry and vegetables, increased. Most of the agricultural change in New Jersey, however, involved emphasis because the major field crops remained. Farmers merely intensified the cultivation of some crops, in part with the aid of improved varieties. Forage crop production, for example, increased to feed the numerous horses that farmers had acquired to pull their implements as well as streetcars in urban areas. Fruit production also expanded, particularly apples, to meet the demands of the urban markets. Farmers who experienced economic stress in the Garden State, in contrast to those in New York and New England, responded not by selling out and moving west or abandoning their lands but by farming more intensively, efficiently, and scientifically. Confronted with rising labor costs after the Civil War, New Jersey farmers also turned to technology to increase production. By adopting reapers, mowers, threshing machines, binders, hay rakes, and fodder choppers, they boosted production and capitalized on commercial markets. Technology not only helped New Jersey farmers solve problems caused by inefficient labor but enabled them to emphasize production for commercial sale.

Not all the changes in northern agriculture were positive. Increased capital costs for machinery slowed the rise from tenancy to full ownership. In addition, investments in technology and science such as commercial fertilizers and improved seeds made farmers more dependent on bankers and other moneylenders. Increased production also required farmers to use middlemen to arrange transportation and sale of their crops. Efficient production placed agriculturists at the mercy of others. Once they became commercial farmers, they were no longer masters of their destinities.

New economic conditions and rapid technological change, then, enabled major readjustments in northern agriculture. Not all of these adjustments, however, saved the family farm. Lost homes, broken families, and economic hard times became realities for many farmers. After the panic of 1873, for example, new economic conditions in American agriculture kept the attrition rate high among farmers in Pennsylvania. Although many remained on the land, the majority who left their farms were simply surplus population. Yet despite the potential for greater agricultural wealth in the West, emigration depended on birth order and ties to parents, relatives, and friends. Older sons who had little prospect of immediately inheriting land moved west to acquire land and begin farming, and the younger sons stayed behind and eventually gained control of the family farm. Still, if the technology and capital were available, northern farmers quickly adjusted their mode of operation and emphasis, remained on the land, and produced for a profit.

THE GREAT PLAINS

The open-range cattle industry continued to dominate the agriculture of the southern Plains after the Civil War. An estimated 100,000 cattle, primarily in the diamond tip of Texas, grazed the range and stockmen still trailed them to New Orleans markets. Supply, however, exceeded demand, and cattle remained unprofitable until the stockyards were built in Chicago in 1865 and railroads carried beef to eastern markets. With longhorns worth only $3 per head in Texas, cattlemen were anxious to capitalize on the high prices in Chicago that brought $30 or more per head, and they began driving them north. By 1866, however, farmers in Missouri and Kansas blocked the Shawnee Trail to Sedalia, where cattlemen could ship their livestock to market on the Missouri Pacific Railroad, because of Texas fever. No one understood that ticks transmitted this disease, but they knew that their cattle became sick and died whenever Texas livestock came into their vicinity, and they kept drovers and their cattle away from agricultural settlements by force of arms.

In 1867, Joseph G. McCoy, a cattle dealer in Springfield, Illinois, convinced the Kansas Pacific Railroad to build a line west of Kansas City about 140 miles where he would establish stockyards for the collection of Texas cattle that drovers brought north on their long drives. This settlement would be far removed from the cattle of farmers. By negotiating reasonable shipping rates, a portion of which were returned to him for the services he rendered, and extensive advertising and recruiting, McCoy created a major cattle town almost overnight. In August 1867, 35,000 head arrived from Texas, although 1869 was the first big year of the long drive. As the railroad extended to the west, other towns like Newton, Wichita, Ellsworth, and Dodge City became cattle towns. In 1871, when the long drive peaked, 600,000 head had been driven from Texas to the railheads in Kansas during that year alone.

The cattlemen who sent their beefs to market via the Chisholm and Shawnee trails and to Montana over the Goodnight-Loving Trail continued the Spanish tradition of the roundup to collect and brand their cattle. The spring roundup enabled cattlemen to locate newborn calves for branding, and they used the fall roundup to separate steers for market and brand late arrivals and calves that had been missed in the spring. Cattlemen preferred this technique particularly where several ranchers grazed cattle on the same grasslands of the public domain. Roundups were efficient, communal activities, and with the formal organization of cattle growers' associations in the 1870s, they became even more important than before.

During the late 1870s after the Indians had been forced onto the

reservations, the northern Great Plains opened for the range-cattle industry. Many cattlemen in the southern Plains drove cattle north to take advantage of lush grasslands. With cattle fattened on northern ranges bringing profits of between 25 and 40 percent, the open-range cattle industry boomed between 1880 and 1885. Many individual cattlemen, however, could not compete for the range because a host of cattle investment companies organized and forced them out of business.

Cattle companies usually consisted of foreign or eastern investors and absentee owners. The Spur Land and Cattle Company was organized in London, and the Matador Land and Cattle Company was formed in Scotland. These companies did not always function efficiently or with the best management, but some earned great profits for investors. Only large-scale cattlemen could compete efficiently with the cattle companies. Charles Goodnight joined with John Adair, an English investor, to purchase more than 700,000 acres, and they grazed 40,000 head in the Lone Star State by 1888. C. C. Slaughter owned more than 1 million acres for cattle grazing in Texas by 1890.

The XIT ranch, however, overshadowed all others on the open

THE OPEN-RANGE cattle industry dominated the Great Plains after removal of the Indians to the reservations. Spring and autumn roundups enabled the collection and branding of the cattle owned by the stockmen who used the grasslands. The cattle selected for sale were driven to market or to a nearby railroad for shipment to meat-packers. In 1887, these cowboys pause for a midday meal around the chuck wagons on the range in the Dakota Territory. *South Dakota State Historical Society.*

range. Created in 1885 from 3 million acres in the Texas panhandle, which the state gave to the London-Chicago–based Capitol Freehold Land and Investment Company in exchange for building the capitol in Austin, the ranch began with a $125 million capital base and headquarters at Buffalo Springs, Texas. The XIT served as the major market for small-scale cattlemen in West Texas and grazed as many as 150,000 head and branded 35,000 calves annually. The XIT managers used Hereford and polled Angus bulls to upbreed their herds and improve the quality of their beef.

The development of large-scale cattle-ranching operations by investment companies changed the use of the range in the Great Plains. Before the mid-1880s, few ranchers owned the grasslands their cattle grazed, freely using the public domain. In time, they considered those lands their own, despite legitimate claims of homesteaders. With pressure from major cattle companies and later farmers to use or claim the open range, individual cattlemen began to purchase land from the federal and state governments and railroads for exclusive use.

The cattlemen also turned to collective action by forming cattle growers' associations for mutual aid. On January 19, 1872, the Colorado Stock Growers' Association formed, in part to prevent cattle on Colorado's ranges from being swept northward by drovers who trailed herds to the northern ranges. On November 29, 1873, the Laramie County Stock Growers' Association organized in Wyoming. In 1879, it changed its name to the Wyoming Stock Growers' Association and became the most politically powerful cattlemen's organization in the West.

The stock growers' associations all had the same basic objective—to aid the economic fortune of their members by providing organization and rational practices for the cattle-ranching business. These associations helped prevent rustling by creating brand books in which the members registered their unique marks and brands. The associations set roundup dates and created the rules for working the range at branding time. These groups circulated blacklists of cowboys suspected of rustling cattle and in the case of the Wyoming Stock Growers' Association, placed inspectors at shipping points along the railroad and in the major market cities to determine whether association cattle reached the meat-packers illegally through rustling.

Stock growers' associations also worked to improve cattle by controlled breeding practices with purebred bulls and the use of barbed wire. Cattlemen found that barbed wire, patented by J. F.

ALTHOUGH THE DAYS of the open range had passed by the late nineteenth century, the work of branding cattle and managing the herds remained basic tasks on the cattle ranches that survived on the Great Plains. In 1891, these cowboys in South Dakota were at work branding cattle. The glass windows and neat shutters suggest a relative affluence for this ranch. *South Dakota State Historical Society.*

BAYLOR COUNTY

	PAIR OF SPURS G. F. Boone Seymour, Texas		ANCHOR W. F. Robertson Seymour, Texas
	BRIDLE BIT S. A. Clark Seymour, Texas		HASH KNIFE E. C. Sterling & Sons Seymour, Texas
	7H HALF CIRCLE Bud Holt Seymour, Texas		DOUBLE ANCHOR Lou Stout Bomarton, Texas
	PAIR OF BOOTS H. R. Martin Comanche, Texas		SCISSORS R. A. Talley & Co. Seymour, Texas
NO	KEY NO Cablero Mathis Seymour, Texas		A HORN R. W. Talley Seymour, Texas
	3 BLOCKS W. H. Portwood Seymour, Texas		C LAZY T Carter Taylor Seymour, Texas

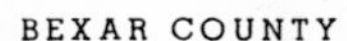

BEXAR COUNTY

	RECTANGLE LAZY R J. A. Ackerman San Antonio, Texas		RECTANGLE L J. M. Hawkins San Antonio, Texas
	CA CONNECTED A. J. Cichon San Antonio, Texas	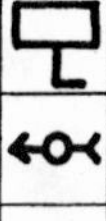	CIRCLE ARROW J. L. Hensley San Antonio, Texas
IOU	IOU Ruth P. Davidson San Antonio, Texas	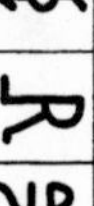	LAZY LR Leon Ramzinski San Antonio, Texas
	HALF CIRCLE LONG D Michael Delany San Antonio, Texas		R DOWN & UP HALF CIRCLE Rosa Reisinger San Antonio, Texas
	INVERTED 2 DJ Juan Delgads San Antonio, Texas		BR CONNECTED Benito Rodriquez San Antonio, Texas
	RAFTER DOT J. M. Dobie San Antonio, Texas		SNAKE S Albert Schmitt San Antonio, Texas
$	DOLLAR MARK Ernest Eastwood San Antonio, Texas		WA CONNECTED George Washington San Antonio, Texas
	LAZY ANCHOR M. B. Farin San Antonio, Texas	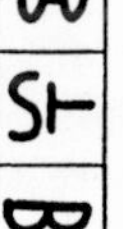	S LAZY T Charles Weir San Antonio, Texas
	RF CONNECTED Alfred Friesenbohm San Antonio, Texas		LAZY B Bud Wood San Antonio, Texas

BRANDS ENABLED cattle raisers to identify their livestock. Usually they registered their brands and earmarks with a county clerk or livestock association. These brands were registered in Baylor and Bexar counties, Texas. Abstract designs show a Spanish heritage; letter and numerical brands indicate an Anglo influence. *A Century of Texas Cattle Brands.* Fort Worth: Fair, 1936.

Glidden in Dekalb, Illinois, in 1874, kept aggressive Longhorn bulls from their heifers. Although barbed wire was expensive at first, the Great Plains did not have wood for fencing, and smooth wire proved useless because cattle pushed through it. By 1897, however, the price had dropped to affordable levels for most farmers and cattlemen.

Much of the early fencing on the Great Plains was illegal because the cattlemen did not have the right to enclose the public domain. They used barbed wire not only to keep their cattle in but also to keep homesteaders out of range lands they considered their private property. By so doing, they formed barbed-wire kingdoms in the open-range cattle country. The result was violence and acrimony between cattlemen and settlers. By the early 1880s, the western states made fence cutting illegal and required cattlemen to place gates at frequent locations. In 1885, Congress passed legislation prohibiting the fencing of the open range.

By the late 1890s, barbed wire had ended the open range, but the harsh environment of the Great Plains had already destroyed the large-scale open-range cattle industry. Because the cattlemen customarily let the livestock run loose without supplemental feeding, adequate shelter, or special care, their cattle often survived the winter in poor condition. During the early 1880s, cattlemen also overstocked the northern ranges, and a series of frigid winters and scorching summers placed the industry in a precarious position where disaster proved inevitable.

Drought followed the hard winter of 1884. The winter of 1885–86 was even worse, and another drought followed. By the summer of 1886, range cattle in the northern and central Great Plains were in bad condition, and the cattle companies and ranchers began to sell their livestock. Oversupply and the flooding of the cattle market caused prices to fall from $40 to less than $10 per head. Then the winter of 1886–87 delivered the economic death blow. Heavy snow began in November and continued into January. The northern ranges filled with snow; the wind remained nearly constant, and temperatures dropped to 40 below zero. Cattle could not reach the grass. Many froze in upright positions; others drifted with the wind until they hit a fenceline and smothered. In the spring, the gullies, canyons, and river valleys were littered with cattle carcasses. Losses ranged from 30 to 85 percent. By 1900, many of the land and cattle companies had become agricultural colonization companies. The sale of their vast acreages to small-scale farmers proved more profitable than raising cattle.

Before the open-range cattle industry collapsed, however,

farmers began their rapid settlement of the Great Plains during the late 1870s. By 1880, they had crossed the 100th meridian. Although drought and crop failure slowed their migration, a "boomer" psychology prevailed in Kansas, Nebraska, and the Dakotas. In March 1884, one observer reported that "for more than four weeks almost every train from the east has been loaded with large parties of excursionists . . . seeking homes in various portions of Kansas." The idea that the Great Plains was the "Great American Desert" began to fade. Sod and frame houses began to dot the landscape. By late 1887, settlers had occupied most of the Dakota Territory east of the Missouri River. During the boom years from 1870 to 1890, the population of Kansas, Nebraska, and the Dakota Territory grew from less than 500,000 to nearly 3 million.

Some believed that settlement would increase precipitation, giving heed to the axiom that rain followed the plow. When these settlers came, they invariably tried to extend their eastern agricultural practices to the Plains. Instead of planting wheat, they seeded corn. When a decade-long drought began in 1887, those practices failed. The farmers who remained adjusted their agricultural practices by planting more wheat and drought-resistant grain sorghums, letting their lands lie fallow (resting their fields every third year to enable the soil to rebuild moisture), and diversifying with more livestock.

In addition to the problem of drought, many farmers failed on the central and northern Great Plains because they did not have sufficient capital to purchase labor-saving equipment that would enable them to plant more acres and raise a bigger crop to compensate for low prices. Although some critics have charged that small-scale farmers often failed on the Great Plains because 160 acres was insufficient to provide a living in this semiarid region, inadequate acreage was not the most significant problem. Most farmers could not cultivate that much land because they needed money to purchase seed, equipment, fertilizer, food, and other domestic items and to pay the mortgage. Low prices combined with bad weather and poor management skills prevented many settlers from earning a profit and keeping their farms. Until normal precipitation returned during the late 1890s and prices increased, many settlers lost their farms and moved away.

To the south, settlement lagged in Texas until after 1875 with the defeat of the Comanche. In addition, the federal government reserved the present state of Oklahoma for Native American groups being moved to reservations. The settlers who first occupied the Texas plains attempted to grow corn, but they experienced the same problems of crop failure from drought as the homesteaders to the north. During the 1870s, however, they introduced the cotton culture to the region, and it became a common crop as far west and north as Abilene, Texas, by the late 1880s. Even so, large-scale cattle ranching continued to dominate the landscape of West Texas until the twentieth century.

AFTER THE FEDERAL government opened much of the Indian Territory for white settlement, this portion of the southern Plains became important for cotton farming. Near the turn of the twentieth century, East Main Street in Norman, Oklahoma, was crowded with wagons during the autumn ginning season. Farmers made their way to the gin with wagonloads of loose cotton and returned with tightly bound bales. *Western History Collections, University of Oklahoma Library.*

As settlers claimed the best lands on the Great Plains, others began to covet land in the Indian Territory. White squatters began illegally to settle the Indian Territory during the early 1870s, and the U.S. Army could not keep them out. Eventually, the political pressure became so great to open those lands for settlement that Congress transferred most of the area to the public domain on March 2, 1889. Little more than a month later, President Benjamin Harrison proclaimed those lands open for settlement under the Homestead Act, and the federal government supervised a great land rush to give settlers an equal opportunity to claim land in the Oklahoma District. Some cheated, however, and entered the district early to stake out their claim. These settlers became known as "Sooners," a sobriquet that applies to Oklahomans today.

Some settlers came to the Great Plains with the aid of ethnoreligious colonization companies, such as the Irish Catholic Colonization Society, the Swedish Agricultural Company, and the Hebrew Union Agricultural Society. These companies worked closely with railroads, steamship companies, and state boards of immigration to locate land for their clients, but the Homestead Act also provided a great lure. Lutheran, Catholic, Mennonite, and Evangelical churches also provided social and cultural support for immigrants and encouraged their settlement in cultural or religious groups. Still, government and railroad land policies prevented amassing large blocks of land by a group or individual to replicate the peasant agricultural villages of Europe. In contrast to many native-born Americans, however, who considered land a means to wealth rather than an end in itself, foreign immigrants seldom sold their land for a profit and relocated. They stayed and maintained their ethnic identity until, with the exception of their church, American culture overwhelmed their institutions by the early twentieth century.

Germans were the most numerous of the foreign immigrants to the Great Plains, with the highest concentration in Nebraska. The German-Russians from the Volga region and Bessarabia settled in Kansas, Nebraska, Oklahoma, and the Dakotas. German Mennonites from Russia revolutionized wheat production on the Great Plains by introducing Turkey Red wheat, a hard red winter wheat that could be planted in the autumn yet survive the harsh winters in the Plains for harvest in the early summer. Immigrants from Great Britain and Scandinavia also settled the Great Plains in large numbers.

THE OKLAHOMA LAND RUN of 1889 was followed by an even greater land rush on September 16, 1893, when some 100,000 settlers established 40,000 claims in the Cherokee Outlet of the newly created territory. When the gun sounded at the starting line, prospective settlers made a mad dash to claim 160-acre homesteads. In the dozen years following 1889, settlers claimed most of the Oklahoma Territory. No other portion of the Great Plains was settled so rapidly. *Western History Collections, University of Oklahoma Library.*

THE FAR WEST

Farmers moved into the Pacific Northwest over the Oregon Trail as early as 1841. Those who made the long, dangerous trek westward found a climate well suited for wheat in present eastern Washington and for fruit in the Willamette Valley around Puget Sound and in the Walla Walla area. Long distances from settled areas in the East and the lethargy of the federal government to survey public domain in the Pacific Northwest slowed the settlement process, and grain and livestock farmers remained isolated west of the Cascades until the 1870s. Until the early 1880s, livestock raising remained the most important activity. Those settlers who migrated to Washington and Oregon during the late nineteenth century came primarily from the Old Northwest, Iowa, and Missouri. In 1887, the Oregon Board of Immigration urged settlers to have $500 to $800 after traveling expenses to meet the costs of building a house and buying seed, implements, and livestock. If they did not settle on public land, they would need even more money to buy land.

Because California with its large population had first priority for getting a transcontinental railroad, commercial agriculture in the Pacific Northwest languished. Still, nearly eighty ships transshipped grain raised on the Columbia plateau from Portland to California, East Coast, and British markets during the late 1870s. When the Northern Pacific and Union Pacific built lines into the Pacific Northwest in the 1880s, commercial agriculture developed rapidly, particularly for grain and fruit.

In California, the gold rush of 1849 changed the nature of agriculture in that territory. The rapid influx of population created a great demand for beef that benefited the rancheros for a brief time. Many miners quickly discovered that they had a better opportunity to improve their economic fortunes by becoming ranchers and grazing cattle on the open range or cultivating wheat, grapes, or other fruit than searching for gold. By the early 1870s, California had become a major wheat state, and the cultivation of fruit rapidly gained importance. During the 1860s, middlemen in San Francisco began exporting wheat to Great Britain, Australia, China, and the East Coast. Wheat farms in California quickly became large-scale or bonanza enterprises characterized by absentee ownership, financing based on extensive credit, and mechanization, particularly the combine, which was well suited for the flat terrain and dry climate.

The hard-drinking miners also stimulated production of wine. When the mining boom ended, California's grape industry had been well established, particularly near San Francisco Bay and in the Sacramento and San Joaquin valleys. By 1870, horticulturists had planted extensive orchards for all deciduous fruits except apples as well as oranges and lemons. The moderate climate and rich soils gave California's fruit growers distinct advantages over horticulturists in other sections of the country. The development of the refrigerated railway car enabled these growers to ship their crops to eastern markets at prices that were competitive or lower than those of eastern growers. In southern California, the navel orange became an important crop by 1889, and lemons became a major crop during next decade.

Without irrigation, however, specialty-crop agriculture would have failed in California, and during the 1870s, irrigation developed rapidly. Little land remained for dryland farming, and irrigation boosted land prices to $70 per acre. During the late 1870s, private canal companies irrigated nearly 293,000 acres. A decade later, 1 million acres, 26 percent of California's farms, were irrigated, more irrigated land than in any other state. In addition, land values had climbed to $1,500 per acre, far beyond the ability of most farmers to purchase it, and California increasingly became a place for large-scale capital-intensive agriculture.

Conflicts between English common law and agricultural needs in this arid region forced adjustments that created a new form of water law during the late nineteenth century. In California and the West, *riparian rights,* which enabled a landowner to use the water that bordered his property, did not meet the needs of farmers who settled far from a stream. The new settlers developed the concept of *prior appropriation*—first in time, first in right. Simply put, water could be diverted for use miles beyond a stream, and it could be claimed and used based on the date that a farmer appropriated (claimed) it for use.

Although California courts recognized both water doctrines, the state legislature partially solved the legal difficulties created by the acceptance of two forms of water law when it passed the Wright Irrigation Act in 1887. This legislation authorized the creation of irrigation districts on a county level whenever fifty landowners petitioned for it. The members of the irrigation districts, usually large-scale landowners, could then exercise the right of eminent domain, issue bonds, and impose taxes for the construction of irrigation works.

But the development of extensive irrigation was only one contributing factor to the development of large-scale agriculture in California. Cheap labor was another. During the late nineteenth century, farmers in the Central, Imperial, and Coachella valleys also began emphasizing vegetable production and used railroads to reach eastern markets and steamship companies to carry their produce to Central and South America. These growers needed a large number of workers at low cost, and they used Chinese immigrants who sought work after they helped complete the transcontinental railroad in 1869. California farmers preferred to hire Chinese laborers because they were readily available and worked harder for lower wages than Anglo-Americans or other ethnic groups.

Cheap Chinese labor helped offset high shipping costs to distant markets and enabled growers to cultivate larger acreages. By 1882, Chinese workers composed more than 50 percent of the agricultural labor force in California and as much as 75 percent of the specialty-crop workers. Chinese workers planted, cultivated, and harvested most of the acreage in the vineyards, vegetable fields, and orchards. Intensive agriculture like this required extensive "squat labor," and although few whites wanted to do it, they pressed for federal legislation to exclude the Chinese from the United States, especially California.

Congress responded by periodically passing exclusionary legis-

lation against the Chinese, and by the turn of the twentieth century, their numbers in the fields dwindled. For a time thereafter, the large-scale growers hired white laborers to avoid the hostility of the anti-Chinese movement in the state. But the growers' need for extensive cheap labor continued to create serious labor problems during the twentieth century. Moreover, as agriculture transformed from family farm to an agribusiness, California became the symbol of western farming.

LAND POLICY

In 1862, after secession of the southern states that had opposed a lenient policy for the disposal of the federal lands, Congress provided free public domain with the Homestead Act, which President Lincoln signed on May 20. This act enabled any head of a household or any male or female 21 or older who was a citizen or who had filed for citizenship to claim 160 acres (a quarter section) of the public domain without cost. The homesteader had to live on the land for five years and make improvements such as plowing, building fences, and constructing buildings. Then the homesteader received title to

THE HOMESTEAD ACT of 1862 provided free land on the Great Plains. In the absence of trees, the early settlers who claimed homesteads had to use sod to build their houses. The homesteaders cut the sod into blocks 12 to 18 inches wide, 4 inches thick, and 3 feet long. Then they laid up the sodlike bricks. Fortunate homesteaders acquired wood for a roof that they also covered with sod. This family proudly poses with their most-prized possessions—a workhorse, table, and a few potted plants. The glass window and sturdy cellar behind signify hard work and success on this homestead in Finney County, Kansas, during the 1890s. *Kansas State Historical Society.*

the land. If homesteaders wanted to pay for this land and receive title quickly, they could do so after six months' residence and payment of a commutation fee of $1.25 per acre. The federal government permitted only one homestead per family.

The Homestead Act was one of the most important laws in American history. Between the claiming of the first homestead on January 1, 1863, near Beatrice, Nebraska, and 1880, homesteaders established approximately 57 percent of the farms on the frontier. Homesteading boomed during the 1870s and 1880s in the central and northern Great Plains. Although speculators gained considerable acreage under false pretenses by using dummy entrymen (those who made a claim, then turned it over to a speculator), the Homestead Act enabled many settlers to acquire land, gain a propertied stake in society, and farm for profit. Many immigrants settled lands in the Great Plains under the provisions of the Homestead Act.

The transcontinental railroads such as the Northern Pacific and Union Pacific also stimulated agricultural development and western expansion. Between 1862 and 1871, the major transcontinental railroads received nearly 110 million acres of public land to support railroad construction. The railroads received alternating sections along a proposed route for as many as ten miles on either side of the track. Railroad land often sold for $10 per acre, but settlers preferred to purchase those lands because the railroads would conveniently link them to commercial markets. Moreover, the railroad companies provided credit, helped farmers locate suitable farms, and provided transportation and other aid during the settlement process. Alternating sections of federal land within the railroad strips could be homesteaded, but a settler could claim only 80 acres, with a commutation fee of $2.50 per acre.

In 1873, the federal government provided another method to purchase land in the transmississippi West with the Timber Culture Act. This statute enabled a farmer to claim 160 acres of public domain, provided he planted 40 acres of trees on this land. Congress intended the Timber Culture Act to promote tree planting and reduce wind velocities. The harsh semiarid environment of the Great Plains, however, proved too difficult for farmers to raise large acreages of trees, and in 1878 Congress amended the statute to require only 10 acres of trees. After a farmer kept his trees, in the form of shelterbelts, alive for ten years, he received title to his claim. (*Shelterbelts* are long lines of trees planted in several rows that form a windbreak and provide protection for fields, buildings, and homes.) Farmers took advantage of the Timber Culture Act primarily in Kansas, Nebraska, and the Dakota Territory, claiming about 9.7 million acres by the early twentieth century.

In the more arid West, neither the Homestead nor Timber Culture act was sufficient to enable economic survival. As a result, in 1877, Congress passed the Desert Land Act, enabling a settler to claim 640 acres (a section) of land for $1.25 per acre. Unfortunately, Congress also required the buyer to irrigate that land within three years. Few settlers had the capital or the technical ability to practice irrigated agriculture on such a vast scale, particularly with inadequate resources. In 1891, Congress reduced the irrigation requirements to 80 acres, but speculation and fraud plagued land acquisition under the Desert Land Act. Although it applied to every western state except Colorado, settlers received title to only 8.6 million acres, even though they had tried to claim nearly 33 million acres under this legislation.

The Timber and Stone Act of 1878 was the last major public-land legislation of the late nineteenth century. This statute enabled an individual to purchase 160 acres for $2.50 per acre to cut the timber and quarry the stone on land unfit for agriculture. The Timber and Stone Act applied to land primarily in California, Oregon, Nevada, and Washington Territory. Anyone who claimed land under this act had to pledge that he was not a speculator, but fraud characterized land acquisitions under this legislation. Even so, free or cheap public land stimulated the rapid settlement and exploitation of the trans-mississippi West, and cattlemen and farmers were among those who profited from this legislation.

AGRICULTURAL AID AND EDUCATION

Before the Civil War, the southern states had blocked any governmental aid to farmers because of economic and political ideology. But when the southern states seceded from the Union in 1861, the way was opened for the creation of a Department of Agriculture. Abraham Lincoln and the Republican party supported the creation of a Department of Agriculture, and on December 3, 1861, the president recommended in his first annual message to Congress that such a department be established. Congress moved slowly on Lincoln's proposal but ultimately approved legislation, which Lincoln signed on May 15, 1862, authorizing a department to "acquire and to diffuse among the people of the United States useful information on subjects connected with agriculture in the most general and comprehensive sense of the word." Specifically, the bill authorized the department to acquire, test, and distribute new and valuable seeds and plants; to conduct "practical and scientific experiments"; to collect agricultural statistics and other information; and to publish annual and other reports that could improve agriculture.

Few congressmen opposed the Department of Agriculture in principle because their constituents favored such an agency and they believed it would provide a useful public service. However, they disagreed about the organizational structure of the department and the scope of its authority. Some wanted an independent department led by a secretary with cabinet status. Others sought to create a lower-level bureau, fearing department and cabinet status would promote too much centralized authority at the expense of state governments. Some congressmen wanted to give the Department of Agriculture broad powers; others wanted its authority strictly limited.

The final bill provided for a department directed by a commissioner the president appointed rather than a secretary with cabinet status. Congress, however, had created the Department of Agriculture in perilous times. With the Civil War raging, few people paid much attention to the news on July 1, 1862, that Isaac Newton had taken the oath of office as the first commissioner of agriculture. Under Newton, the department stressed research and education to help farmers improve their operations. In 1862, for example, the department began publishing reports on crop conditions, the weather, acreage under cultivation, yields per acre, and numbers of livestock on the nation's farms. Annual reports also presented a broad statistical survey of national agricultural conditions and affairs.

Until the late nineteenth century, the scientists, technicians, and officials of the department were primarily dedicated to making two blades of grass grow where only one grew before—that is, making farms more productive. Departmental officials based this policy on the assumption that increased productivity would enable farmers to sell larger quantities of agricultural commodities and earn more money. As farmers increased their income and improved their standard of living, the public would also enjoy an abundant supply of food and fiber. During the late 1860s and early 1870s, the department expanded its scientific work, particularly in plant and animal disease and nutrition. Scientists in the department also began experimenting with chemicals to control or prevent insect damage to crops.

In the 1880s, departmental policy began to change as farmers and consumers became increasingly hostile to food processors who manufactured oleomargarine from the by-products of slaughtered cattle and hogs. The Division of Chemistry, under Harvey W. Wiley, began to examine food for adulteration. In 1884, the department also created the Bureau of Animal Industry, marking an important new direction for the Department of Agriculture. Not only was it the first bureau in the department with a special mission, but it also wielded regulatory powers that placed the department in a stronger position to aid the American farmer. The bureau enforced regulations to prevent the importation of diseased livestock and began investigations to determine the cause of Texas fever.

TEXAS FEVER plagued the range cattle industry during the late nineteenth century. Eventually, USDA scientists discovered that ticks caused the fever, which spread rapidly as cattle from southern ranges were driven to northern markets or rangelands where they mingled with other livestock. Cattle raisers used oil and arsenic dips. Here, cattle are being driven through a chute while cowhands push them under to ensure that the oil solution covers the entire animal. *Western History Collections, University of Oklahoma Library.*

Although the department continued to seek new ways to improve agricultural production and farming practices, the regulatory activity of the 1880s was a major departure from previous policy that stressed research and education to improve agricultural productivity. Thereafter, the department continued to expand its regulatory powers to aid farmers by helping eliminate crop and livestock diseases and unethical business practices such as selling oleomargarine as butter. During the early 1890s, Congress passed legislation that authorized the department to inspect meat and live cattle and hogs shipped in interstate commerce. The department also worked to improve roads in the countryside to help farmers reach their markets and began an enduring policy of seeking new and expanded foreign markets for American farm products. By the turn of the twentieth century, the scientific, educational, and regulatory responsibilities of the Department of Agriculture made it one of the largest and most important agencies of the federal government.

As departmental responsibilities expanded during the late nineteenth century, farm groups across the nation urged Congress to elevate the department to cabinet status to increase the prestige of the agency so that farmers would have a stronger voice in governmental affairs. After lengthy debate, Congress approved this measure, and on February 9, 1889, President Grover Cleveland signed the legislation and appointed Norman J. Colman as secretary of agriculture. But Colman had little opportunity to influence the department's new level of importance because the Cleveland administration came to an end in early March. Jeremiah McLain Rusk became the first really full-time secretary of agriculture. Appointed by President Benjamin Harrison, Rusk took office on March 6, 1889. Under Secretary Rusk, the department worked to make its practical scientific discoveries available to farmers across the nation.

In 1862, Congress passed, and President Lincoln signed, the Morrill Land-Grant College Act. This legislation provided for the creation of state colleges with federal support to teach "agriculture and mechanic arts." Although many had advocated agricultural education at the college level since the early 1850s, more than a decade passed before the idea reached fruition. In 1857, Justin S. Morrill, congressman from Vermont, introduced a land-grant college bill in the House of Representatives. Although it passed both houses, President James Buchanan vetoed it because southerners, who had supported him for the presidency, feared it would enlarge the powers of the federal government at the expense of the states. With the election of Abraham Lincoln and the secession of the South, Congress passed the bill, and Lincoln signed the act on July 2, 1862. Iowa became the first state to authorize a land-grant college when it accepted the terms of the Morrill Act on September 11, 1862.

The Morrill Act authorized the federal government to provide each state with 30,000 acres of public land for each member of the Senate and the House. Those states that had no public land within their boundaries were issued scrip that could be used to claim land in states and territories that still had large areas of public domain for sale. Congress authorized the states without public land to sell their scrip and invest the money to support a land-grant college. The states without public land could also give the scrip to a college, which could claim public land in another state to support that institution. Usually the states sold their scrip to land dealers at low prices varying from $0.50 to $1.00 per acre. Then these states used that money to establish an agricultural and mechanical college.

At first the agricultural colleges suffered from a lack of trained faculty to teach agriculture and an insufficient base of knowledge. Critics legitimately charged that these new agriculture colleges did not meet the needs of farmers. Young men who wanted to become farmers preferred to learn their trade at home, and with abundant rich lands available in the West, they could see no reason to go to college and learn agriculture. Few sons and daughters of farmers, then, enrolled in the land-grant colleges, and those who attended seldom returned to the farm.

Congress eventually helped resolve these problems with the creation of agricultural experiment stations. Although Connecticut established the first agricultural experiment station in 1875, primarily to test chemical fertilizers to ensure that farmers were not cheated by manufacturers, the federal government did not establish an agricultural experiment station system until Congress passed the Hatch Act, which became law on March 2, 1887. Introduced to Congress by William Hatch of Missouri and J. Z. George of Mississippi, the Hatch Act provided annual federal support for agricultural research, both pure and applied, in every state. The state experiment stations were closely linked to the agricultural colleges, and the stations issued research bulletins to help farmers learn about useful scientific discoveries. Most farmers did not read this literature, and the federal government was still unable to communicate its agricultural discoveries to farmers. It would not solve this problem until the development of the agricultural extension system in the early twentieth century.

While the federal government attempted to use the land-grant colleges and the agricultural experiment stations to increase and diffuse agricultural knowledge, it also made amends for the Morrill Act, which failed to provide support for the education of African-Americans, most of whom were slaves in 1862. In 1890, Congress passed the Second Morrill Act, which became law on August 30, prohibiting racial discrimination at land-grant colleges but permitting states to establish separate institutions of higher education for African-Americans, provided federal funds for support were "equitably" distributed between the black and white institutions. Eventually, seventeen states either created separate black land-grant colleges or provided the funds to existing African-American colleges. These land-grant colleges, however, suffered from chronic underfunding and discrimination. Instead of becoming major institutions for the training of African-American men and women in agricultural practices and science, they served as little more than preparatory or high schools well into the twentieth century.

During the late nineteenth century, state agricultural colleges organized agricultural institutes to help farmers learn about better agricultural practices. Colleges or experiment stations sent lecturers and scientists to rural communities to meet with groups of farmers for two or three days. They would conduct meetings to discuss subjects such as crop rotation, soil conservation, and plant and animal diseases. These educational efforts failed to achieve noticeable improvements in local agricultural practices. Most farmers still considered the dissemination of agricultural information in this form little better than "book farming," whose practicality they rejected out of hand. They wanted to see the results of such advice for themselves. Although Seaman H. Knapp began providing practical agricultural guidance to Louisiana rice planters during the 1880s, large-scale federally supported extension work under Knapp's leadership did not occur until the early twentieth century.

AGRICULTURAL IMPROVEMENTS

During the 1860s, implement manufacturers primarily continued to produce plows with wrought-iron moldboards because the cost of high-quality steel remained prohibitively expensive and steel could not yet be uniformly tempered and shaped. In 1868, however, John Lane improved plow technology when he developed a process for making "soft-center" steel. Lane welded a soft bar of cast iron between two bars of steel, then rolled the block into a thin plate for tempering and shaping into moldboards. Virtually unbreakable moldboard plows could now be fashioned that maintained their scouring ability. A year later, James Oliver patented a technique for hardening cast iron so that the moldboard would wear longer and scour better than regular cast-iron plows. This process involved passing a stream of warm water over hot cast iron. Oliver called the result "chilled iron" because it cooled rapidly and became exceptionally strong. In 1870, Oliver produced his first plow for commercial sale. Soon his plows became popular nationwide because they were light and durable and required less draft than other iron models.

By the last quarter of the nineteenth century, a variety of moldboards reached the market that scoured and turned a furrow under many soil conditions. Plows with iron beams gave the implement superior strength. Wooden beams, however, were lighter and less likely to be sprung, although they might break if the share hit a rock that did not give way. Furthermore, plow makers abandoned cast iron for chilled iron and soft-center steel. Additional plow improvements would come only with the introduction of gasoline tractors that increased draft power, during the twentieth century.

In the meantime, farmers moved beyond the midwestern prairies to the Great Plains. Agriculture now became more extensive than ever before. Large-scale "bonanza" farms developed in the Red River Valley between North Dakota and Minnesota and in California. Farmers now found single-moldboard walking plows, sometimes called "swing" plows or "foot burners," too slow and began using sulky and gang plows to speed their work. Sulky plows had a seat so that the operator could ride. Sometimes these plows had two shares. Farmers could now take advantage of the extra power of their horses, cover more acreage during the course of the day, and ride at the same time. With a two-bottom sulky plow, a farmer could turn from 5 to 7 acres per day. The second moldboard made it a gang plow; however, it substantially increased the draft required, and farmers called these plows "horse killers."

By the mid-1860s, farmers commonly used the grain drill in the mid-Atlantic states, Pennsylvania, and Ohio. A decade later, farmers drilled about half the winter wheat planted in the Mississippi Valley. During the 1880s, Great Plains farmers also began using grain drills. By the 1890s, a 16-foot grain drill pulled by four horses easily planted 15 to 20 acres per day. However, the field had to be smoothly prepared for these implements to work efficiently; even then, many grain drills did not measure the seed accurately. But generally grain drills produced greater yields and planted less seed in a field than broadcast seeders, thereby saving money.

During the 1860s, corn farmers also began using mechanical

planters on a wide scale. These implements had two runners that opened furrows in which a tube deposited the seed whenever the rider, usually a boy seated in the front of the machine, moved the lever that released the seed at the proper moment. Farmers usually marked their fields with a two-runner wooden sled in a checkerboard fashion, then dropped the seeds at the intersections of the marks so that the corn would grow in straight rows for easy cultivation both ways across the field. With the hand-operated corn planter, a farmer could seed from 12 to 20 acres per day, approximately twenty times more than he could plant with a hoe. During the 1870s, the check-row corn planter became popular. This implement used a knotted wire that was strung across the fields to mark the spots where the seed would be dropped. As a knot passed through a mechanism on the planter, it triggered the release of the seed into the tube behind the furrow opener. This check-row planter became the standard corn-planting implement during the 1870s.

During the Civil War, grain farmers rapidly adopted the reaper. Although farmers harvested an estimated 70 percent of the wheat in the West with the reaper, wartime labor shortages further encouraged grain farmers to adopt this machine. At that time, reapers required a crew of eight to ten men—one to drive the horse, a second to rake the platform, and six to nine for binding and shocking. This crew could harvest from 10 to 12 acres per day. By 1864, two-thirds of the reapers produced by McCormick were self-rake models. Labor shortages during the war years made the self-rake reaper extremely popular, and the increased price of wheat made it more affordable than before.

In 1872, Syvanus D. Locke produced the first commercially successful wire binder. This implement cut the grain, wrapped a strand of wire around the sheaf, and tied it. By the late 1870s, the major implement companies were manufacturing wire binders. Although this implement eliminated the need for rakers and binders and required only a driver to operate it, the wire binder created new problems. The wire often caught in the machine's moving parts, and once the sheaves had been broken for threshing, disposal became a nuisance because the wire did not rot or burn. Cattlemen and millers also complained that bits of wire harmed their livestock or damaged their grinders.

In 1880, the Deering Company solved the problems of the wire binder by marketing a twine binder. With the invention of a device that would tie twine around the sheaf, the disposal problem was eliminated. During the mid-1880s, competition among farm machin-

DURING THE 1870s, grain binders further eliminated the number of workers needed at harvesttime for small grains such as wheat. Binders cut the grain and bound it into sheaves. Workers called shockers followed behind and placed the sheaves in shocks. The shockers would gather the sheaves later for threshing. Harvesting rates averaged between 10 and 20 acres per day, depending on field conditions and the crop stand. *National Archives.*

ery companies caused the cost of twine binders to decline, and the price of hemp for twine also fell, so that many farmers could now afford to purchase this machine. By the end of the decade, farmers completed almost the entire wheat harvest with the twine binder. Twine binders averaged about 12 acres per day, although 20 acres per day was possible depending on field conditions and the thickness of the crop stand.

Between the adoption of the self-rake reaper and the binder, horses and even steam engines could not provide the speed necessary to increase substantially the total number of acres that could be harvested in one day. An increase in daily harvested acreage would not come until the twentieth century when gasoline tractors provided the draft power necessary to expand the binder's daily cutting capacity. Still, these implements made harvesting substantially easier for grain farmers.

While the grain binders clattered across the wheat fields in ever-increasing numbers, other farmers wanted to apply the same kinds of technological change to the corn crop, and during the 1880s inventors began designing mechanical corn harvesters to help lessen the labor of cutting the stalks by hand. By the mid-1890s, most farmers used the corn binder to harvest the crop. These machines cut the stalks and bound them into sheaves with twine. The sheaves then dropped to the ground where workers placed them into shocks. Corn binders harvested about 7 to 9 acres per day. Although the cost of cutting corn with a binder was about the same as cutting it by hand, farmers preferred the binder because it made the corn harvest easier. By the turn of the twentieth century, binders were standard implements in the corn belt.

During the last decade of the nineteenth century, inventors gave increased attention to mechanical corn threshing and shelling. The implements tested at that time, however, so pulverized the stalks that inventors soon realized a mechanical shredder would have great appeal to farmers who used the stalks for fodder. The finely

ON THE GREAT PLAINS, the header served as a useful grain-harvesting implement. A four-horse team pushed the header. The cut grain fell onto a platform before being elevated into a wagon drawn alongside. Headers eliminated the work of binding, shocking, and hauling the sheaves to the threshing site. With a 12-foot cutter bar, a header could harvest from 15 to 25 acres per day. *Smithsonian Institution.*

ground stalks were more palatable to cattle, and livestock wasted less fodder if it was cut into fine pieces. This discovery led to the invention of the combined husker-shredder, which removed the ears from the stalks, stripped the husks, and ground the stalks into fodder. During the 1890s, the husker-shredder gained popularity. By the turn of the twentieth century, implement manufacturers produced husker-shredders that husked from 100 to 1,000 bushels per day. Horses and steam engines provided the power to operate these implements and shellers that processed 2,500 bushels per day.

During the nineteenth century, farmers who threshed their own wheat commonly owned small portable machines. The larger and more efficient implements were usually owned by an entrepreneur who sent a thresher with an itinerant crew from farm to farm. These operators contracted their work each season. Although contract threshing imposed an immediate labor expense on the farmer, it freed him from the capital investment necessary to purchase a large threshing machine. By 1860, grain farmers commonly hired itinerant crews with threshing machines to do their work quickly and efficiently. Large threshing machines saved time and enabled many farmers to get their crops to market before prices fell. By the turn of the twentieth century, steam engines had replaced horses for powering threshing machines.

During the 1860s, several inventors continued to devise an implement that would combine the harvesting and threshing processes for small grains. In California, large-scale wheat farmers anticipated a combined harvester-thresher that would make the harvest more efficient, easier, and profitable. A gearing problem, however, slowed development of a commercially viable combine until 1884 when the newly organized Stockton Combined Harvester and Agricultural Works solved this difficulty. By 1886, combines cut and threshed from 25 to 35 acres per day. Although ten to forty horses and mules pulled these giant machines, the combines often used a steam or gasoline engine to power the moving parts of the machine in the early twentieth century. These combines were suitable only for flatland. Wheat fields planted on hilly ground still had to be cut with reapers or binders because the gigantic combines toppled over when used for hillside harvesting or the long cutter bars dug into the ground. By 1900, the combine harvester cut approximately two-thirds of California's wheat crop.

The combine increased productivity and reduced labor costs because it wasted less grain than reapers and threshers and fewer hired hands were needed at harvesttime. Indeed, for farmers who could afford the machine, the combine made them almost independent of hired help. Not only did the combine remove the crop at once and clear the fields for immediate plowing, but it also scattered the cut straw across the ground to help build soil humus. By the turn of the twentieth century, large-scale farmers used the combine pri-

marily along the West Coast. Most of the implements east of the Rocky Mountains before World War I operated on the Great Plains where the dry climate made wheat suitable for combining. These combines, however, remained too large and expensive to merit adoption by small-scale grain farmers in the Midwest.

Technological change in the wheat harvest had another important effect besides saving the farmers time and money. As the development of the grain harvesters and threshing machines progressed, fewer hired hands were needed. This change lightened the cooking burden on the farmer's wife, who had the responsibility of preparing three hearty meals and frequently an afternoon lunch for the harvest and threshing hands from the time the cutting began until the last sheaf was in the shock or the grain threshed and hauled to the bin or elevator.

During the late nineteenth century, important changes in plowing, planting, and harvesting technology increased grain production. With the expansion in production, additional changes were needed to enlarge the threshing machine's capacity and increase its operating speed. As manufacturers built larger threshing machines, more power was needed to operate them. Horsepowered treadmills and sweeps could not provide sufficient power to operate the machines.

BETWEEN THE MID-1880s and the adoption of the gasoline tractor in the early twentieth century, combines powered by as many as forty horses became popular on the large-scale wheat farms of Washington and California. The gearing of this combine was linked to a wheel that depended on the forward speed of the horses to power the implement. Note the driver sitting on the seat above while workers bag the grain on the platform in front. This combine harvested the wheat crop in the vicinity of Davenport, Washington, near the turn of the twentieth century. *Washington State Historical Society.*

Moreover, horses tired and required frequent rest or changing. Because of these problems, by the late 1860s, an increasing number of grain farmers were beginning to use portable steam engines to power their threshing machines. By 1870, steam engines were an important aspect of the threshing machine business. During that decade, many farmers began applying steam power to their wheat, corn, and rice threshing and cotton-ginning operations.

Although the portable steam engine increased the speed and capability of the threshing machine, it had one serious flaw—it was not self-propelled. Lacking traction, steam engines could neither plow nor move under their own power. Consequently, their use was limited to belt work on threshing, grinding, milling, and ginning machines. To move the portable steam engine, the operator had to hitch a team of horses to it and pull it from place to place. Some steam engines had a seat or platform on the front or the rear from which the operator steered the horses with reins. On other models, the farmer simply held reins as he walked alongside the engine.

These problems were resolved in 1873 when the Battle Creek, Michigan, firm of Merritt and Kellogg offered the first self-propelled model for sale. By the late 1870s, the C. and G. Cooper Company of Mount Vernon, Ohio, had won the reputation of being the first company to manufacture steam engines in quantity and to market them nationwide. Because of the steam tractor's great weight, most of the power was used to propel it across the field or down the road or to operate implements. Until inventors solved the power-weight relationship, most farmers preferred to use horsepower for plowing rather than invest in an expensive "elephantine" monster. By the late 1870s, however, manufacturers were making great strides toward the development of a steam tractor that could pull a plow and operate a threshing machine. By the 1890s, steam-traction engines easily plowed from 35 to 45 acres per day in the Great Plains and Far West. Although the production of steam engines increased rapidly during the last quarter of the nineteenth century, the average farmer remained hesitant to make the large investment required to purchase a steam tractor. Consequently, itinerant or custom threshermen owned most of the steam engines used to power threshing machines. Because steam-traction engines were not only expensive but awkward to maneuver on small-scale farms, these iron monsters soon became dinosaurs when tractors with internal-combustion gasoline engines replaced them during the early twentieth century.

Scientists also contributed to agricultural improvement. William O. Atwater at Wesleyan University described the transfer of nitrogen to the soil by leguminous plants. Professor Eugene W. Hilgard developed procedures for soil analysis. Francis Storer at Harvard University conducted research in agricultural chemistry. William G. Farlow, also at Harvard, made pioneering contributions to mycology, the study of funguses. American agricultural scientists also gained an international reputation in the newly developed field of plant pathology. In 1889, Charles V. Riley, the Department of Agriculture's entomologist, discovered that the Australian ladybug was a natural enemy of a scale insect that plagued the citrus industry in California. He introduced the ladybug to American agriculture, and it has thrived for the benefit of both farmers and gardeners. In 1890, scientists in the Bureau of Animal Industry proved that the cattle tick spread Texas fever. This discovery was important because it established that a disease-carrying organism could be spread from one animal to another. Soon scientists applied this discovery to the study of yellow fever and other insect-borne diseases. In time, all of these scientific discoveries benefited the commercial development of American agriculture.

THE AGRARIAN REVOLT

After the Civil War, farmers in the Great Plains and South experienced problems created by surplus production, high railroad rates, and mortgage indebtedness. Although farmers had never been able to organize and cooperate to solve common problems for any length of time or over a large area, many agriculturists, particularly in the Midwest, Great Plains, and South, now became part of the movement to gain economic, social, and political change. The first organization that enabled farmers to address economic problems through collective action, however, did not form for that purpose. The Patrons of Husbandry, also known as the Grange, organized in December 1867 to improve educational and social opportunities for farmers.

In 1866, Oliver Hudson Kelley, a Minnesota farmer and clerk who worked for the U.S. Department of Agriculture (USDA) in Washington, D.C., toured the South to survey the agricultural conditions in that war-torn region. Kelley began his tour in January 1866 and returned to the capital city in late April convinced that the economically depressed conditions of southern farmers could not await the whims of politicians in Congress or state legislatures for relief. Kelley also believed that northern and southern farmers had little understanding of one another's needs and ended his tour convinced that many rural problems stemmed from inadequate educational and social opportunities for farm families. To band farmers together nationwide, help restore order, and improve the educational, social, and cultural level of the agrarian class, Kelley developed plans for a secret agricultural organization modeled after the Masonic Lodge. By November 1867, his plans were nearly complete, and he sent out more than three hundred circulars proposing such an organization. A month later, on December 4, Kelley and six associates founded the National Grange of the Patrons of Husbandry. Almost immediately they organized the first subordinate grange, Potomac No. 1, to serve as a model and school of instruction. The next year, the first permanent subordinate grange was organized in Minnesota, and in February 1869 Minnesota patrons created the first state grange.

Kelley expected the members of the subordinate granges to hold meetings and listen to lectures about ways to improve farming practices, share agricultural publications, and socialize. Kelley, however, had largely misjudged the needs of the farmers in the Midwest, Great Plains, and South. They wanted an organization that would defend them politically and protect them economically. All other concerns were of secondary importance.

Quickly the state granges became interested in cooperative business ventures to solve their economic problems. During the 1870s, the cooperative movement dominated grange activities. Grangers on the local and state levels organized cooperative stores to save money by purchasing plows, sewing machines, clothing, food, and other items. They believed prices were too high at retail stores because middlemen and businessmen gouged them. They also used the Grange to market cattle, wool, and grain cooperatively to eliminate the middlemen and increase their profits. Later Kelley observed: "The education and social features of our Order offer inducements to some to join, but the majority desire pecuniary benefits—advantages in purchase of machinery, and sales of produce."

During the 1870s, the Grange expanded rapidly, primarily because most members misunderstood the purpose of the organization. They believed the sole intent of the Grange was to withhold business from local merchants and to save money through collective action based on the English Rochdale Plan whereby members bought stock in the organization for the creation of cooperative stores stocked at reduced prices, sold at retail prices, and divided the dividends periodically. This collective action would force price reductions for needed farm goods at neighborhood stores.

Although patrons expected "glorious results," almost immediately their cooperative ventures experienced difficulties. Farmers could not estimate their needs in a timely fashion to place orders, and they often did not provide sufficient working capital for the cooperative stores to operate efficiently. Some manufacturers, such as the McCormick Company, refused to deal with them, and local businesses tried to destroy the organization by reducing prices sufficiently to encourage disloyalty among the members, and with much success. Grangers often bought at prices that were too expensive and sold for prices that were too low. Good business practices in the absence of experience were often wanting. While the cooperative Grange stores existed, however, they saved farmers money through price reductions on a host of items.

Some Grangers also wanted railroad regulation. They demanded that rates be based on the cost of service rather than bonded indebtedness and complained that railroads charged more for a short haul than a long one when competitive lines were not available. Although the Grange was not a political party, some members advocated nonpartisan political action to solve their economic problems. Although a number of "Granger laws" that provided for the regulation of railroads emerged from midwestern legislatures, the Grange was not the primary instrument for gaining that legislation. Rather, businessmen, who also suffered from railroad discriminatory practices, had greater political influence in the state legislatures and with Grange support molded much regulatory railroad legislation that helped reduce rates.

The Grange advocated a host of progressive reforms such as the direct election of U.S. Senators, woman suffrage, the initiative and referendum, inflation of the currency with greenbacks and silver coins, prohibition of pooling by railroads and grain elevators, rural free mail delivery, and government control of the railroads and telegraph systems, but the organization most successfully achieved its educational and social goals. It advocated free textbooks and lengthening the school year, sought the improvement of education and the elimination of gender barriers at the land-grant colleges, and supported adult education through a series of farmers' institutes sponsored by the agricultural colleges and the creation of state departments of agriculture.

By 1875 the Grange had reached its zenith. Rapid decline followed. Many farmers had joined but lacked the interest or commitment to keep the organization active. Poor business ventures destroyed the organization in some areas, and jealousies and suspicion weakened the internal structure and order of the organization. Unwilling and unable to become a political party, the Grange failed to organize sufficiently to ensure economic success through a cooperative movement. During the 1880s and 1890s, many Grangers joined the Farmers' Alliance and the People's party. Although the Grange became essentially dormant during these years, at the turn of the twentieth century it reemerged in some areas as the social and educational organization that Kelley had sought more than thirty years previously. Before that happened, however, farmers in the Great Plains and the South had moved into the political arena with considerable force.

While the Grange peaked during the mid-1870s, another agricultural organization developed that soon overshadowed all agrarian organizations in the South and Great Plains. Known as the Farmers' Alliance, this group organized in Lampassas County, Texas, during the mid-1870s when a group of cattlemen and farmers joined to fight cattle rustling through a secret order. Although this vigilante association soon disappeared, these farmers in central Texas founded another organization in 1877, the Knights of Reliance. Organized on the same principles of secrecy and ritual as the Grange, the Knights helped locate stray cattle, provided social activities for its members, and supported the Greenback party in 1878. Two years later, it reorganized as the Farmers' State Alliance and sought better financial terms from cotton-furnishing merchants, opposed land speculators who charged usurious interest rates, and advocated interregional agricultural cooperation.

Like the Grange before it, the Texas Alliance considered all agricultural problems economic and advocated cooperative marketing and selling to ensure the financial well-being of farmers. The members of the Alliance were more militant and political than the Patrons of Husbandry and soon overwhelmed that organization. Although the Alliance was not engaged in party politics, it supported political education to help members know the issues and the politicians that advocated their cause. In this way, the Alliance could engage in political activities for the benefit of farmers but avoid becoming a third political party.

By the late 1880s, farmers across the South looked to the Alliance cooperative system to break the bondage of the furnishing merchants. By providing credit based on a farmer's pledge to market his cotton crop through the Alliance, the organization intended to eliminate the furnishing merchants, who in symbol and fact controlled their lives. One Alabama woman said, "The furnishing merchant was the boss, pure and simple . . . his word was law. If there were differences in a community or a family they were settled by 'The Man.'" In 1887, the Texas Alliance began spreading its gospel of economic cooperation and organization to other southern states. In North Carolina, Leonidas L. Polk, editor of the *Progressive Farmer,* became a leading supporter of the organization. Like Polk, many farmers had faith in the Alliance, and it spread rapidly, especially among small-scale owners and stable tenant farmers. Cooperative stores, warehouses, gins, and elevators followed the organizers, but these operations were usually poorly capitalized and quickly failed. Democratic politicians soon became aware of the political potential of the Farmers' Alliance and joined the organization for their own purposes.

The Farmers' Alliance also served as a social institution. Membership was limited to rural whites, and the organization did not distinguish between planters, tenants, farm workers, or small-scale owners, provided everyone was of "good moral character" and believed in a supreme being. Country mechanics, merchants, schoolteachers, physicians, and ministers could also join. However, few wealthy farmers became members, and the Alliance did not recruit impoverished tenants because they did not have the financial reserves to participate in the cooperative movement.

Women composed about 25 percent of the membership, but unlike their sisters in the Grange, few held office. Women were primarily relegated to servers at Alliance-sponsored dinners, picnics, rallies, camp meetings, and other social engagements. While the role of women was primarily social, they were at least included. African-American farmers were excluded. Racial prejudice prevented unity to achieve common goals of economic cooperation and political solidarity. In March 1888, however, a Colored Farmers' National Alliance organized in Houston County, Texas. It had little contact with the white Alliance until 1890, by which time southern agrarians realized that it did not threaten the political and social status quo concerning race relations. Like the white Alliance, the Colored Farmers' National Alliance spread quickly across the South, eventually claiming approximately 1.2 million members, about the same strength as the white Alliance.

In 1888, the Farmers' Alliance began organizational activities in the Great Plains. There, it appealed to wheat farmers suffering from overproduction and low prices. This Southern Farmers' Alliance, however, experienced some competition with the National Farmers' Alliance, also known as the Northern or Northwestern Farmers' Alliance. In 1880, Milton George had founded the National Farmers' Alliance in Chicago to oppose monopolies. The Northern Alliance, however, refrained from the aggressive cooperative movement and nonpartisan political activity of the Southern Alliance, which appealed to many farmers in the Great Plains.

Organizational growth and economic hard times increased demands by some Alliance members for political action. In response to a growing demand for political influence, a variety of reform organizations met in St. Louis in December 1889. The Northern Alliance and the Southern Alliance, the latter now the National Farmers' Alliance and Industrial Union, failed to merge. The Northern Alliance opposed the secrecy of the Southern Alliance and advocated the admission of African-American farmers to the organization. In addition, the Northern Alliance opposed the use of cottonseed oil for the manufacture of oleomargarine as a butter substitute.

Temporarily foiled, the advocates of agrarian-based political activity carried the Southern Alliance into politics in December 1890 when the organization held its annual convention in Ocala, Florida. The delegates devised a platform that advocated government ownership of the railroads, abolition of national banks, free and unlimited coinage of silver, and a subtreasury. This platform thrust the Alliance into politics, and the subtreasury became a new panacea for solving the problems of inadequate farm credit and an insufficient monetary supply.

The subtreasury plan involved the creation of a government system in which farmers could turn over a staple crop like wheat or cotton to a federal warehouse and receive a loan for 80 percent of the value of the crop in negotiable certificates at 2 percent interest per month. If prices rose above the loan value during the following year, the farmer could sell his crop and pay off the debt. But if prices remained lower than the loan value, he could forfeit his crop to the government and wipe out his obligation. This plan would provide the financing that southern farmers needed to break the credit system of the furnishing merchants and the crop-lien system. Politicians who supported the subtreasury measured up to the "Alliance yardstick" and merited the votes of the membership on election day. Most southerners, however, were unprepared to support the creation of a large, expensive federal bureaucracy and government centralization that would intimately affect their lives and agricultural operations.

The Alliance achieved greater success in the Great Plains states where its members moved beyond nonpartisan pressure-group influence into the world of third-party politics. In June 1890, Kansas farmers founded the People's party based on the Southern Alliance platform. South Dakota followed a month later. With the Alliance unable to win political support in the South, the cooperative stores bordering on economic ruin, and the organization in the Great Plains states moving rapidly to create third parties, the Southern Farmers' Alliance suffered a swift decline from which it never recovered.

In May 1891, at the National Union Conference in Cincinnati, western Alliancemen reasserted a call for the creation of a national People's party. Later, in February 1892 at the Conference for Industrial Organizations in St. Louis, agrarian radicals prepared a political platform for a national third party based on Southern Alliance goals and called for a national nominating convention to be held in Omaha, Nebraska, on July 4, 1892. Leonidas L. Polk, the leading candidate for the presidential nomination by the People's party, died before the convention met, and James B. Weaver, an Iowan who had been the Greenback party's presidential nominee in 1880, received the nomination. With the creation of the People's party, which co-opted the program of its parent organization, the Farmers' Alliance essentially died.

Weaver won only 22 electoral and little more than 1 million popular votes and carried only four states in the West, however this was the best showing for a third-party candidate since John C. Frémont had run for the presidency in 1856 on the Republican ticket. The Populists were confident that they could win the presidential election in 1896 and began to work hard to gain political support by reiterating the problems of farmers in the Great Plains, South, and Far West. The movement, however, was the strongest in the Great Plains, where farmers complained about unfair railroad practices, mortgage rates, and an inadequate monetary supply.

Essentially, the Populists sought economic change by political means, in part by advocating public ownership of railroads, direct election of senators, the initiative and referendum, a subtreasury, a progressive income tax, postal savings banks, and antitrust regulation. The Populists particularly urged the federal government to intervene in the economic system and guarantee, in the words of Minnesota Populist Ignatius Donnelly, that the "public good is paramount to private interests." Based on that premise, the Populists launched a crusade against the social Darwinists, those who believed in survival of the fittest in the economic world.

By advocating the greatest good to the greatest number, the

Populist movement became the closest approximation of nationwide utilitarianism in American history. To the Populists, government existed to serve mankind, not just to protect property. Specifically, the Populists argued that "special privilege" in the form of monopolies oppressed farmers and demanded equal economic rights, especially a fair return for their labors. Within this premise, the Populists championed an active federal government to solve their economic difficulties. Operating on the philosophy that "self-interest rules," the Populists attempted to unite white and black voters for political purposes, but this effort failed because white Southerners could not break the bonds of racism and their allegiance to the Democratic party.

MARY ELIZABETH LEASE

Kansas State Historical Society.

Known as Mary Ellen to her friends and branded Mary "Yellin" by her enemies, Mary Elizabeth Lease became one of the most vocal leaders of the Populist movement. Born on September 11, 1853, in Ridgeway, Pennsylvania, she moved to Kansas at age 17 in 1870. After teaching school and raising four children as a farm wife, she studied law at home and gained admission to the bar in 1885. Although she supported the Republican party, she left it for the reform-oriented Union Labor party in 1888. With the organization of the Kansas People's party in 1890, she became an ardent Populist and enjoyed considerable success on the lecture circuit by attacking the railroads and loan companies.

Newspaperman William Allen White reported that she had a "golden voice—a deep, rich contralto, a singing voice that had hypnotic qualities." Lease projected her voice powerfully, and she had a great sense of cadence and timing that in the words of White could "set a crowd hooting and hurrahing at her will," even over the most boring economic or monetary policies. Standing nearly 6 feet tall, she had a presence and speaking ability that captivated audiences. She did not mince words when she attacked the monopolists, institutions, and policies that trampled the "will" of the people, particularly farmers, whom she called the "rough-handed, kingly-hearted, sons of toil—chieftains fit to guard the ark and covenant of liberty." She had remarkable oratorical skills. Her opponents, however, considered her "foul of tongue" with a mind that scattered like a shotgun for a fee of $10 per night.

Lease toured the South for the People's party during the congressional campaign of 1890 and the presidential campaign two years later. There, conservative males reacted to her much as they did in Kansas. One editor wrote that "the sight of a woman traveling around the country making political speeches [was] . . . simply disgusting." Another observer noted that she had a "voice like a cat fight and a face that [was] rank poison to the naked eye." Political perspectives, of course, color everyone's vision of the world, and during the Populist revolt it prejudiced the public's view of Lease for both good and ill. Whether she won friends for the People's party or made enemies, no one could doubt that she was an untiring campaigner for the party. She is best known for allegedly telling farmers to "raise less corn and more hell." Although she denied ever making that statement, she essentially did just that in her campaign speeches, and the Populists loved every word.

After the failure of the People's party during the election of 1896, Lease moved to New York and worked as a political correspondent for the New York *World*. She supported the Republican party but jumped to Theodore Roosevelt's Bull Moose party in 1912. Lease left public life in 1921 and died on her farm in Sullivan County, New York, on October 29, 1933.

The Populists' enthusiasm, however, quickly waned when their candidates fared poorly in the election of 1894. Many Populists now abandoned their far-reaching reform program for a panacea or quick fix of the economic system by advocating monetary inflation, commonly known as the free and unlimited coinage of silver at the ratio to gold of 16:1. This plan did not mean minting 16 silver dollars to 1 gold but referred to the weight of the metal in each. Populists believed that this formula and its expansion of the silver coinage would adequately increase the circulating currency, boost prices, and help farmers get out of debt. The doctrinaire party members (known as the middle-of-the-road Populists) advocated the maintenance of a broad reform platform, but those who favored monetary inflation above all else succeeded in nominating William Jennings Bryan for the presidential ticket in 1896.

Unfortunately for the Populists, who waited until after the Democratic and Republican nominating conventions assuming both would select a candidate who supported the gold standard, the Democratic party nominated Bryan. The Populists, now unable to nominate a free-silver candidate that would stand alone, offered to "fuse" with the Democrats and nominate Bryan if he would replace his vice-presidential nominee, Arthur Sewell, a banker from Maine, with agrarian Thomas E. Watson of Georgia. But Bryan would not forsake the Democrats for the Populists. Ultimately the Populists "fused" with the Democratic party on the national level and in the South with the Republican party in some states to win offices through the strength of combined numbers.

With the Democratic party stealing a major plank from the Populist platform on the national level, southerners unwilling to leave the party of white supremacy on the state level, and many farmers in the Midwest and Northeast relatively secure economically because of diversified farming operations, the Populists were unable to muster sufficient strength to elect Bryan. Republican William McKinley won the election, and with an increase in foreign demand for agricultural products soon after the election, a return of normal precipitation in the Great Plains, and new discoveries of gold in Alaska and Australia for coinage, agricultural prices increased, and the Populist movement essentially disintegrated. Tom Watson put it best for the Populists when he said, "Our party, as a party, does not exist anymore. Fusion has well nigh killed it." Watson's analysis was an oversimplification. Populism failed because it could not muster sufficient political support. White southerners betrayed it, and New England and midwestern farmers did not need it. Great Plains and Far West farmers and miners did not have enough votes to carry the party to victory.

Although the People's party died, the agrarian reform movement remained, but in different institutional form. During the twentieth century, most of the reforms that the Patrons of Husbandry, Farmers' Alliance, and People's party had advocated gained the support of other constituencies and Congress. Primarily a political party of landowning cotton and wheat farmers who suffered severe economic hardship, the Populists pioneered the use of politics to achieve substantive economic, social, and political reform. That is the greatest legacy of the late-nineteenth-century agrarian revolt.

RURAL LIFE

Great diversity characterized rural life during the late nineteenth century. While farm men and women often lived in brick or frame houses in the North and Midwest, clapboard shacks typified the homes of sharecroppers, and settlers on the Great Plains built sod houses. Many farmers still lived isolated lives, particularly on the Great Plains and far western frontier. Outdoor toilets prevailed; indoor plumbing and hot running water lay in the distant future. Wood and coal stoves kept the kitchen hot in the summer and made it a gathering place in the winter. Saturday trips to town provided a wonderful escape from the routine and drudgery of most farm life. No matter the region, however, the typical farm family did not enjoy the convenience of a telephone, paved roads, electricity, or internal-combustion engines. Most rural children still attended one-room country schools where a teacher instructed a variety of age groups in the basics of reading, writing, and arithmetic. Field work, whether plowing, planting, or harvesting, often took children from school, and most rural children were fortunate to have three or four years of education. Low incomes and tax bases, especially in the South, perpetuated poor education and poverty.

Towns and villages continued to serve as the primary service centers for farmers, who frequently made trips for supplies or to conduct business on Saturdays. Small-town businessmen were dependent on the success of farmers and in the Great Plains and Far West the railroads. Many merchants invested in land and farmed on the side or rented to tenants, especially in the South, Great Plains, and Midwest. With so many people reliant on agriculture for their prosperity, weather and crop conditions became standard topics of conversation.

By the mid-1880s, the federal government had provided post offices in most rural areas, and by the mid-1890s it experimented with mail delivery directly to the farm, known as rural free delivery.

Approximately 2 million people enjoyed this service by the turn of the twentieth century. Mail-order businesses began to thrive on the rural trade. In 1872, Montgomery Ward, commonly called "Monkey Ward," used the railroads and the mail service to provide farmers with access to a wide variety of goods that they could not purchase locally or at a reasonable price.

The farm family continued to characterize American agriculture. It remained a kinship network, agent of socialization and education, and commercial producer. In the South, a "good name" improved

NEIGHBORS OFTEN SHARED the work of big projects. During the 1890s, a group of farmers joined to build a barn somewhere in northeastern Ohio. Note the women on the left. The accepted division of labor relegated the heavy work of lifting the timbers to the men, and the women prepared the food for on-site meals. *Ohio Historical Society.*

the credit rating of a family. Although many families sought self-sufficiency in food for security, commercial production for profit governed their activities whenever markets developed. Moreover, in the Great Plains and Far West, subsistence agriculture was usually impossible, and in other regions it was not practical. Farm families always served as a profit-oriented economic unit on the land. In every region, though, farm families were divided by class based on wealth, race, and culture.

Farm women were still locked in the combined roles of housewife, mother, and helpmate. Women milked cows, churned butter, hoed cotton, picked worms off tobacco plants, and raised children. Butter and egg money gave women some economic independence from their husbands, contributed to their sense of dignity and self-worth, and helped the family get by and sometimes get ahead. One farm woman complained that she had become "slightly demoralized" by the demanding work of cooking and washing for the hired plowboys, wheat stackers, threshers, and haying men. She wrote: "I have been going all spring and summer like a well-regulated clock, am set running every morning at half-past four o'clock, and run all day, often until half-past eleven P.M." If farm women were fortunate, they could hire a neighbor's daughter or a girl from town, particularly during the summer months when their work was often the heaviest. Most could not, and farm life remained tiring labor with few material rewards. Edith Duncan Pitts, who settled on the Llano Estacado (Staked Plains) of Texas, spoke for the most optimistic rural women during the late nineteenth century, however, when she said that "life is so beautiful after all; for memory gathers only the bright spots from our lives." For the most part, those good memories involved social activities such as neighborly visits, church services, occasional trips to town, and raising their children.

American agriculture underwent revolutionary change between 1865 and 1900. The Civil War not only ended slavery in the South but caused fundamental economic and social readjustment in the North. High wartime prices and labor shortages encouraged farmers to purchase a host of labor-saving implements. Improved plows, reapers, and threshing machines helped northern farmers increase production and improve their profits. The availability of free western lands under the Homestead Act, the completion of a transcontinental railroad system, and the development of extensive, one-crop agriculture in the Great Plains soon brought even more substantive changes to northern agriculture.

The great regional diversity of American agriculture during the late nineteenth century makes it impossible to characterize any particular farm as typical in size or operation. But farmers customarily rejected any lingering attachment to self-sufficient agriculture. One Populist spoke for all farmers when he wrote that self-sufficiency "requires the farmers to step out of the line of progress, to refuse to avail themselves of the industrial improvements of the nineteenth century, turn back the wheels of civilization three thousand years, become a hermit and have nothing to do with the outside world." A

North Carolinian agreed: "True wisdom does not sanction a retrograde movement . . . to 'hog and hominy.'"

Although farmers sought commercial production to make a profit and improve their standard of living, many were locked in poverty and suffered poor health due to inadequate diets, especially in the South. In that region, many people suffered from pellagra, the result of diets mainly of pork, cornmeal, and molasses, and hookworm, the result of unsanitary practices. Periodic illnesses and an absence of doctors or money to pay a medical bill made life worrisome for farmers. One Mississippi mother reflected, "A farmer only gets money twice a year, and if the children get sick between seasons they have to get along." Home remedies for illnesses prevailed throughout rural America.

In the absence of applied science and technology, the small-scale southern farms remained unproductive and unprofitable, and sharecropping continued to encourage soil exhaustion. Black farmers owned only 13.5 million acres at the turn of the twentieth century, compared to more than 100 million acres owned by white farmers. They remained the poorest farmers in the South at a time when most sharecroppers and tenant farmers usually earned about $100 annually. Agricultural income like this fostered and perpetuated insufficient education, inadequate diet, and poor health. Families, rather than slaves, bound by contracts to the landlords and furnishing merchants, worked the land.

Farmers in the Midwest and Great Plains also often experienced hard times, but they usually had some capital, credit, or diversified operation to make their lives easier than the sharecroppers who lived without land or hope. In the Far West, bonanza wheat farming and specialty fruit and vegetable cultivation became profitable new forms of agriculture. Environment and culture determined the nature of agricultural and rural life, particularly on the frontier. There, the environment set the parameters, but people adapted based on their traditional ways of living.

Although American agriculture became heavily capitalized during the Gilded Age, the late nineteenth century was a period of both progress and poverty. At the turn of the twentieth century, one northerner who visited the South reported that in the countryside "the wretchedness is pathetic and the poverty colossal." Many farmers in every region quit or saw their children move to town. New technology often improved production, but surpluses frequently reduced prices. Extensive one-crop agriculture in the Midwest and Great Plains provided a living, if not always comfort, for grain farmers, but it forced agriculturists in New England and along the Atlantic seaboard to adjust their farming operations to practices that would not compete with the Westerners. Although economic hard times drove many farmers to organize the Patrons of Husbandry, Farmers' Alliance, and People's party to achieve economic ends by political means, their interests were too diverse to maintain long-term solidarity. But although many problems remained, particularly in the South, the twentieth century dawned bright with promise for many American farmers.

SUGGESTED READINGS

Atherton, Lewis. *The Cattle Kings.* Bloomington: Indiana University Press, 1961.

Ayers, Edward L. *The Promise of the New South: Life After Reconstruction.* New York: Oxford University Press, 1992.

Baker, Gladys, et al. *Century of Service: The First Hundred Years of the United States Department of Agriculture.* Washington, D.C.: U.S. Department of Agriculture, 1963.

Barron, Hal S. *Those Who Stayed Behind: Rural Society in Nineteenth-Century New England.* Cambridge, England: Cambridge University Press, 1984.

Black, John Donald. *The Rural Economy of New England: A Regional Study.* Cambridge, Mass.: Harvard University Press, 1950.

Bogue, Allan G. *From Prairie to Cornbelt: Farming on the Illinois and Iowa Prairies in the Nineteenth Century.* Chicago: University of Chicago Press, 1963.

Buck, Solon Justice. *The Granger Movement: A Study of Agricultural Organization and Its Political, Economic, and Social Manifestations, 1870–1880.* Cambridge, Mass.: Harvard University Press, 1913.

Chan, Sucheng. *This Bittersweet Soil: The Chinese in California Agriculture, 1860–1910.* Berkeley: University of California Press, 1986.

Clanton, O. Gene. *Populism: The Human Preference in America, 1890–1900.* Boston: Twayne, 1991.

Cronon, William. *Nature's Metropolis: Chicago and the Great West.* New York: Norton, 1991.

Dale, Edward Everett. *The Range Cattle Industry: Ranching on the Great Plains from 1865 to 1925.* Norman: University of Oklahoma Press, 1930.

Danhof, Clarence. *Change in Agriculture: The Northern United States, 1820–1870.* Cambridge, Mass.: Harvard University Press, 1969.

Daniel, Cletus E. *Bitter Harvest: A History of California Farmworkers, 1870–1941.* Ithaca, N.Y.: Cornell University Press, 1981.

Daniel, Pete. *Breaking the Land: The Transformation of Cotton, Tobacco and Rice Cultures since 1880.* Urbana: University of Illinois Press, 1985.

Dunbar, Robert G. *Forging New Rights in Western Waters.* Lincoln: University of Nebraska Press, 1983.

Fahey, John. *The Inland Empire: Unfolding Years, 1879–1929.* Seattle: University of Washington Press, 1986.

Fite, Gilbert C. *Cotton Fields No More: Southern Agriculture, 1865–1980.* Lexington: University of Kentucky Press, 1984.

———. *The Farmers' Frontier, 1865–1900.* New York: Holt, Rinehart and Winston, 1966.

Gates, Paul W. *Agriculture and the Civil War.* New York: Knopf, 1965.

———. *History of Public Land Law Development.* Washington, D.C.: Government Printing Office, 1968.

Goodwyn, Lawrence. *The Populist Moment: A Short History of the Agrarian Revolt in America.* New York: Oxford University Press, 1978.

Hahn, Steven. *The Roots of Southern Populism: Yeoman Farmers and the Transformation of the Georgia Upcountry, 1850–1890.* New York: Oxford University Press, 1983.

Hibbard, Benjamin Horace. *A History of the Public Land Policies.* Madison: University of Wisconsin Press, 1965.

Hicks, John D. *The Populist Revolt.* Minneapolis: University of Minnesota Press, 1931.

Hurt, R. Douglas. *American Farm Tools: From Hand Power to Steam Power.* Manhattan, Kans.: Sunflower University Press, 1982.

Jelinek, Lawrence J. *Harvest Empire: A History of California Agriculture.* 2nd ed. San Francisco: Boyd and Fraser, 1982.

Lanza, Michael L. *Agrarianism and Reconstruction Politics: The Southern Homestead Act.* Baton Rouge: Louisiana State University Press, 1990.

Larson, Robert W. *Populism in the Mountain West.* Albuquerque: University of New Mexico Press, 1986.

McMath, Robert C. *American Populism: A Social History, 1877–1898.* New York: Hill and Wang, 1993.

———. *Populist Vanguard: A History of the Southern Farmers' Alliance.* Chapel Hill: University of North Carolina Press, 1975.

Malin, James C. *Winter Wheat in the Golden Belt of Kansas: A Study in Adaptation to Subhumid Geographical Environment.* Lawrence: University Press of Kansas, 1944; repr., New York: Octagon, 1973.

Marcus, Alan I. *Agricultural Science and the Quest for Legitimacy.* Ames: Iowa State University Press, 1985.

Marti, Donald B. *Women of the Grange: Mutuality and Sisterhood in Rural America, 1866–1920.* Westport, Conn.: Greenwood Press, 1991.

Miller, George H. *Railroads and the Granger Laws.* Madison: University of Wisconsin Press, 1971.

Nordin, D. Sven. *Rich Harvest: A History of the Grange, 1867–1900.* Jackson: University of Mississippi Press, 1974.

Osgood, Ernest Staples. *The Day of the Cattleman.* Minneapolis: University of Minnesota Press, 1929.

Pisani, Donald J. *From the Family Farm to Agribusiness: The Irrigation Cru-*

sade in California and the West, 1850–1931. Berkeley: University of California Press, 1984.

______. *To Reclaim a Divided West: Water, Law, and Public Policy, 1848–1902*. Albuquerque: University of New Mexico Press, 1992.

Pollack, Norman. *The Just Polity: Populism, Law, and Human Welfare*. Urbana: University of Illinois Press, 1987.

______. *The Populist Response to Industrial America*. Cambridge, Mass.: Harvard University Press, 1962.

Rikoon, J. Sanford. *Threshing in the Midwest, 1820–1940: A Study of Traditional Culture and Technological Change*. Bloomington: Indiana University Press, 1988.

Russell, Howard W. *The Long Deep Furrow: Three Centuries of Farming in New England*. Hanover, N.H.: University Press of New England, 1982.

Saloutos, Theodore. *Farmer Movements in the South, 1865–1933*. Berkeley: University of California Press, 1960.

Skaggs, Jimmy M. *The Cattle Trailing Industry: Between Supply and Demand, 1866–1890*. Lawrence: University Press of Kansas, 1973.

Sutherland, Daniel E. *The Expansion of Everyday Life, 1860–1876*. New York: Harper and Row, 1989.

Webb, Walter Prescott. *The Great Plains*. New York: Grosset and Dunlap, 1931.

Woods, Thomas A. *Knights of the Plow: Oliver H. Kelley and the Origins of the Grange in Republican Ideology*. Ames: Iowa State University Press, 1991.

Woodward, C. Vann. *Origins of the New South, 1877–1913*. Baton Rogue: Louisiana State University Press, 1951.

6 • THE AGE OF PROSPERITY

AMERICAN FARMERS ENJOYED CONSIDERABLE ECONOMIC GROWTH and prosperity during the first two decades of the twentieth century. Agricultural prices generally increased more than farm costs, particularly from August 1909 to July 1914, a period that has been called the golden age of American agriculture. During that time, agricultural prices gave farmers a purchasing power that equaled or exceeded that of other workers in the national economy. During the 1920s when the agricultural economy experienced recession, farmers, agricultural economists, and policymakers would look back to that period as a time of agricultural "parity." Soon the concept of parity prices and income became the hallmark of American agricultural policy.

World War I caused farm prices to rise dramatically. The federal government urged farmers to plant as much grain, feed as many livestock, and raise as much cotton as possible to aid the war effort. Farmers broke new lands, some acquired through favorable though unethical leases on Indian reservations. Some of the land in the Great Plains was submarginal; that is, given current agricultural prices it did not merit cultivation. Although submarginal lands were productive in years of above-normal precipitation, they suffered severe wind-erosion problems when the rains stopped falling. Wartime prices, however, did not remain high for long. By 1920, European recovery; international competition from Australia, Argentina, and Canada for the beef and wheat trade; and high American surplus production caused prices to fall precipitously. Many farmers who had used their wartime profits to buy more land and purchase new equipment on credit could not meet their obligations and lost their farms and implements at foreclosure sales. Although some farmers, such as cotton producers, saw modest recovery before the end of the 1920s, most agriculturists experienced hard times before the agricultural economy totally collapsed after the stock market crash in 1929 and the onset of the Great Depression.

Despite both boom and bust in agriculture during the early twentieth century, farmers experienced an age of great technological and educational change. Although not all farmers could take advantage of the opportunities to mechanize or learn new techniques between 1900 and 1932, farmers laid the foundation for a new commercial agriculture based on the application of science and technology and a reliance on federal governmental policies and efficient business practices. The age was neither entirely golden nor totally depressed. It was a time when many farmers enjoyed a relatively good life, with the exception of sharecroppers and tenants in the South and farmers with debts in the Midwest and Great Plains.

THE SOUTH

During the early twentieth century, the South remained rural and poor. Nearly half of the region's farmers were tenants, and African-American farmers remained at the bottom of the agricultural ladder. More than 75 percent of all farmers were sharecroppers and tenants. Few white or black farmers had any hope of purchasing land, producing a surplus crop for profit, and becoming economically independent of the furnishing-merchant system. Landowners still required sharecroppers and tenants to raise cotton, which perpetuated soil depletion from one-crop agriculture. Most farmers in the South remained locked in an economic and social position that bordered on peonage. Insufficient funds to purchase and improve breeding stock, along with Texas fever, prevented the improvement of southern livestock. Moreover, the farms, averaging between 20 and 50 acres, were too small to permit diversification with wheat and cattle.

The absence of markets for commodities other than cotton, tobacco, and rice also prevented diversification. Cotton was not perishable. It always had a market, and tenants could not eat it. Thus, southern farmers without capital or credit despaired of changing the land-tenure and cropping systems. Few had enough money to buy land or pay a cash rent. Most did not have funds to purchase fertilizer, implements, and cattle. Most important, with production at subsistence levels after the furnishing merchant and landlord took their share of the crop, sharecroppers had too little to pay their debts or improve their farming practices.

Although most cotton farmers accepted their fate, they felt powerless to change. In contrast, tobacco farmers in the upper South took a militant approach to low prices. Under the direction of James B. Duke, the American Tobacco Company crushed its competitors, manipulated prices, and often paid farmers less than their cost of production. In September 1904, the tobacco farmers in the dark-leaf areas of Kentucky and Tennessee rebelled against these monopolistic practices and organized the Planters' Protective Association. Planters and sharecroppers provided 30,000 members in those states alone. The burly-tobacco growers of central Kentucky also organized, and other tobacco farmers provided additional members to these and similar organizations, all designed to improve the tobacco farmer's economic position.

Unable to gain concessions from the American Tobacco Company and with the federal government unwilling to provide relief, these tobacco growers took the law into their own hands. Between 1906 and 1908, night riders burned tobacco warehouses and destroyed crops in the "Black Patch War." They whipped and murdered farmers who sold to the tobacco trust and intimidated the courts. Not since the wealthy tobacco planters in colonial Maryland had used force to reduce surplus production during the 1730s had such violence occurred to achieve economic gain. This vigilante action quickly spread from Kentucky and Tennessee to Arkansas, Mississippi, Alabama, and Louisiana, and it reached a viciousness unparalleled since the lawlessness of the Ku Klux Klan during the late nineteenth century. Although the tobacco planters achieved higher prices for a while, military force finally suppressed their activities.

Before World War I, then, sharecropping and tenancy, one-crop agriculture, tradition, and resistance to learning new methods at farmers' institutes or land-grant colleges or state agricultural experiment stations, together with insufficient markets, capital, and long-term credit and the control of the furnishing merchant, prevented the development of efficient and profitable agriculture in the South. During World War I, however, the economic position of cotton and tobacco farmers improved as international demand increased. Cotton prices reached nearly $0.28 per pound, the highest price since

the Civil War, and southern farmers enjoyed this new prosperity. Tobacco and rice prices also rose dramatically, and farmers increased production to earn as much profit as possible.

At the same time that these farmers were improving their standard of living, wartime industries in the North lured thousands of African-American farmers from the South. They left one-crop agriculture and the peonage of sharecropping behind, never to return. These farmers would prosper from the war, but they would do so in the industrial cities of the North rather than on the small farms in the South. The flight of black farmers, however, threatened to create

SHARECROPPERS received a portion of their crop, usually half, as pay from the landlord. Consequently, they planted as much of their allotted land as possible because the largest possible crop meant more money for the family. In 1939, this sharecropper in Chatham County, North Carolina, planted his tobacco crop nearly to the house. *Southern Historical Collection, The University of North Carolina at Chapel Hill.*

a farm labor shortage, and wages substantially increased across the region.

Despite wartime prosperity, the structure of southern agriculture remained unchanged. By 1920, nearly half the farms contained less than 50 acres. Small farms like these prevented diversification and perpetuated low incomes. Tenancy remained at 50 percent or more for the white farmers, and tenancy rates for African-American farmers reached as high as 90 percent in some areas. In the mountain region, farmers earned only a "scant living," and poverty remained the prevailing agricultural condition across the South. Few southern farmers owned telephones, trucks, or tractors or lighted their homes with electricity. High wartime prices collapsed in 1920 and made the economic problem worse for southern farmers because operating costs remained high.

During the early twentieth century, southern farmers also fought and endured the boll weevil. In 1892, the boll weevil entered Texas from Mexico. By 1901, it had become a serious problem for cotton growers in the Lone Star State and across the South. This insect bores into the cotton bolls and lays eggs; upon hatching, the young weevils devour the lint. The boll weevil did not wreak uniform havoc in all areas, but it was particularly destructive in the Black Belt of Alabama and Georgia. The destructiveness of the boll weevil and the economic hardship that it left behind caused Ned Cobb, a black sharecropper, to reflect that "all God's dangers ain't a white man."

In response to the boll weevil, southern farmers welcomed federal agents and the advice of the USDA for the first time. They fought the boll weevil by cultural methods such as fall plowing and burning the stalks after the harvest and with the application of calcium arsenate to the plants. In 1904, W. J. Spillman, with the USDA, observed that "the boll weevil is really a blessing in disguise" because it began to destroy the one-crop system of the South. Significant diversification, however, did not occur until after 1910. Even so, cotton remained king with about 37 million acres planted in 1914, approximately 10 million more than a decade earlier. Prices dropped to $0.065 per pound on some local markets. If cotton could not pay debts, one Georgia farmer said, creditors "take everything the farmer has."

Wartime demands and voluntary acreage reductions temporarily improved prices, which caused a reporter for the New York *Republic* to write that "the white fiber still holds the imagination of the South." When prices fell to $0.10 per pound in 1926, cotton farmers suffered. Near the end of the decade, one agricultural editor wrote: "When southern cotton prices drop, every man feels the blow; when cotton prices advance, every industry thrives with vigor." By 1929, cotton farmers were beaten men and women.

Although some southern editors, agricultural leaders, and bankers advocated limiting acreage and production and even urged farmers to "burn-a-bale" to help reduce the price-depressing cotton surplus, most responded to economic hard times in their traditional fashion. They increased cotton acreage, especially in Oklahoma and Texas where mechanization enabled farmers to reduce labor costs. But credit, cultivated acreage, transportation, markets, capital, and mechanization remained inadequate to enable most southern farmers to improve their economic well-being. Rural schools also remained isolated and without sufficient resources to improve educational and economic development. On the eve of the Great Depression, then, most southern farmers lived in desperation and without alternatives to improve their economic fate. They were locked in perpetual poverty and without hope. Little did they know that agricultural conditions and rural life would soon get worse or that southern agriculture was on the threshold of revolutionary change.

BEFORE FARM CHILDREN began riding the familiar yellow bus to school during the mid–twentieth century, some took a schoolwagon. These children used a wagon to reach their school near Popular Springs, Georgia, about 1912. Isolation, distance, and poor roads often made school attendance difficult for farm children nationwide. *Southern Historical Collection, The University of North Carolina at Chapel Hill.*

RURAL CHILDREN often did not have the opportunity to gain more than a basic education in country schools. Students of all grade levels usually attended the same one-room school where the teacher taught the basics of reading, writing, and arithmetic and perhaps some geography and history. Note the pile of coal by the stove. These students in Randolph County, Georgia, pose in their Sunday-best clothes sometime during the early 1900s. This spartan classroom was typical of country schools, where attendance depended on the weather and the work to be done on the farm. *Georgia Department of Archives and History.*

THE NORTH

During the early twentieth century, farmers in New England continued to adjust their agricultural practices by abandoning unproductive lands. The number of farms declined, and more rural people moved for better-paying jobs and an improved standard of living. Those who stayed often consolidated their holdings by purchasing neighboring lands and emphasized speciality production. They also abandoned crops and livestock that could not be produced on a profitable scale because of competition from the Midwest.

With urban populations burgeoning, many of these farmers stressed dairying for the fluid-milk market rather than the secondary markets for the manufacture of butter and cheese. Fluid milk brought higher prices and greater profits, and farmers far removed from urban markets could not compete with them for the production of this perishable commodity. As a result, the production of cereal grains declined, and farmers raised more hay. One New England farmer clearly recognized the importance of adjusting agricultural operations to new economic conditions: "Sheep are a frontier crop. The frontier in that sense is gone. . . . I can make more money from something else—dairying."

Dairy farmers who could afford to used the Babcock tester to chart the butterfat productivity of their cows and improve their herds through selection and better breeding practices. Silos for the storage of fodder, which permitted winter feeding and year-round milk production, became the symbol of good dairying. But testing for butterfat and tuberculosis and improved breeding and feeding practices required more labor and expense. Often small-scale farmers with little capital failed and left the land. By 1920, however, the Boston milkshed reached north into southern Maine and northeastern New York. Branch railroad lines and trucks enabled distant farmers to ship milk to market relatively easily, if not always cheaply, and milk dealers handled the distribution and sales. By 1930, the majority of the farmers in New Hampshire, Vermont, and Maine depended on the monthly milk check for their primary income.

Farmers in New Jersey and Delaware also increasingly specialized in the production of eggs, poultry, and truck crops. Agriculturists in Connecticut and New York devoted more attention to fruit production, particularly apples and peaches, for the eastern cities. Farmers in Maine's Aroostook Valley gained a reputation for producing seed potatoes. With the help of science and tariff protection, Connecticut and Housatonic Valley farmers began to produce high-

CHILDREN have always been actively involved in the work required to make the farm produce food and fiber. Most people considered farm work a healthy activity for children, and child labor in agriculture did not receive even modest regulation until the Fair Labor Standards Act of 1938. Although many children willingly helped their parents on the farm without damage to their health, precious childhood years could easily be lost in the drudgery of farm-related tasks. In 1917, these girls strung tobacco near Hazardville, Connecticut. *Connecticut Historical Society, Hartford, Connecticut.*

grade tobacco suitable for cigar wrappers. By the mid-1920s, however, overproduction and technological change in the cigar-manufacturing industry had decreased the need for Connecticut binders and wrappers, and tobacco raising largely passed from the domain of the small-scale farmer. Farmers who raised sweet corn, onions, and cucumbers usually sold their crops under contract to local processors, avoiding some financial risk and receiving a guaranteed price, but they lost the economic flexibility and independence that had long characterized farm life.

In the Midwest, farmers continued to raise corn, which they used to fatten hogs and cattle, but they also produced large quantities of milk in Wisconsin, Minnesota, and Michigan. Farmers in the cornbelt of the Midwest produced most of the high-protein foods in the form of meat and dairy products for family tables across the nation. By 1930, midwestern farms averaged 130 acres, large enough to justify investment in new technology. During the 1920s, these farmers doubled their use of tractors. By the end of the decade, 30 percent of the farmers in Illinois and Iowa owned one.

Midwestern agriculturists also increased their acreage in soybeans during the 1920s. Introduced from China in 1804, soybeans were used primarily for a legume-hay, which they cut with a mower and binder. Although soybeans were first harvested by combine in Illinois in 1924, the beans easily shattered from the pods, and the loss proved substantial. During the mid-1920s, combine tests in Illinois reduced seed losses to about 9 percent. By 1930, floating cutter bars clipped the vines close to the soil, but a satisfactory pickup reel had still not been developed to lift the vines over the sickle and onto the canvas aprons that led to the combine's threshing cylinder. Until agricultural engineers solved that problem, losses from shattering could not be further reduced. Even so, the combine enabled midwestern farmers to expand production of soybeans, which brought high prices for processing into livestock feed and oil.

THE GREAT PLAINS

Homesteading, the breakup of large-scale land and cattle companies, and extensive wheat cultivation, often on submarginal lands, characterized early-twentieth-century agriculture on the Great Plains. In this new age of tractor power, homesteaders still plowed with horses unless they could afford the services of a plowman with a steam or gasoline tractor. Many farmers in the Great Plains lived with one foot in the past and the other stretching onto the tenuous soil of the future. In the "west river country" of South Dakota, for example, 100,000 settlers arrived between 1900 and 1915 to build a new life. Some settlers, such as Mary Bartels, moved to South Dakota because "homesteading was the spirit of the times—a big adventure." Lindsay Denison, an eastern reporter for the *American Magazine,* wrote that the last homesteading on the Plains proved that "the best that was in our fathers . . . is with us yet." Optimism characterized the spirit of twentieth-century homesteaders.

With better railroad transportation services available, homesteaders now built tar-papered shacks from lumber rather than sod houses. Like those who came to the Great Plains before them, they experienced the harsh realities of drought, grasshoppers, and the incessant wind, together with temporary isolation, crop failure, and unending toil. Even so, promoters billed unsettled areas such as western South Dakota as the "last great frontier" and lured settlers who yearned for the economic independence that only the land could provide. Many new settlers claimed land that the federal government had recently acquired from the reservation Indians or acreage purchased from railroad companies.

During the early twentieth century, the homesteaders on the

northern Great Plains found that their lives were as tightly circumscribed by the annual cycle of plowing, planting, cultivating, and harvesting as those of the settlers who had come to the Plains during the nineteenth century. They planted spring and winter wheat and oats, and prayed for rain during the growing season and blue skies at harvesttime. Occasionally the harsh environment defeated them, but they could master their own destiny by renouncing much of the agricultural knowledge they brought with them from the Midwest, at least in relation to planting corn. The grasslands of the western Great Plains were unsuitable for corn without irrigation. Water was the lifeblood of the West then as it is now, and those who depended only on the sky to provide it for their corn usually experienced crop failure. Many lost their farms and moved away, but others stayed. Like the homesteaders before them, these twentieth-century settlers helped make the Great Plains the best "next-year" country in the nation.

Those who remained adjusted their agricultural practices to the dictates of the Great Plains environment. They planted more spring wheat, grain sorghum, and alfalfa and let some fields lie fallow every other year to build soil moisture. They also raised more cattle to diversify and spread their risks. High prices enabled profits, which permitted the accumulation of capital and the purchase of implements and more land. As northern Great Plains farmers adopted tractors and combines after World War I, they solved many of their most expensive and technical problems such as planting and harvesting at the right time or when the weather mandated haste. Yet they never solved the social problems of isolation or insufficient wealth to support adequately rural schools, churches, and organizations. Still, those who survived financially became "stickers."

At the turn of the twentieth century, many of the great land and cattle companies on the southern Great Plains were turning from raising livestock to financing land sales for settlers, which returned greater profits. Many large ranches with more than 50,000 acres remained. Still, by 1910, wheat farmers had settled all but the most arid portions of the southern Great Plains. Few farmers practiced soil conservation. As a result, the cultivation of wheat, cotton, and corn invariably left the soil exposed to wind and water erosion. As the plow pulverized sandy soils, the land became more susceptible to blowing. Although livestock raising remained important, farmers emphasized wheat, but that one-crop economy meant financial ruin during severe drought. Moreover, when the wheat crop failed, more land laid exposed to wind erosion.

THIS COUNTRY SCHOOL was more spartan than most. In the autumn of 1914, Eva Deem taught these rural children in Chouteau County, Montana. Deem paid for the school with subscriptions she received from parents to support their children's education. The school measured about 14 feet by 14 feet. *Montana Historical Society.*

Shortly before World War I, Great Plains farmers increasingly used tractors, gang plows, and headers to expand their wheat acreage. With the outbreak of war, wheat prices soared to $2 per bushel. When the United States entered the conflict in 1917, Great Plains farmers responded to the "wheat will win the war" campaign by planting still more. Annual precipitation remained sufficient to allow that expansion. After the war, wheat expansion continued, and grazing land appraised at $10 an acre increased ten times in value when farmers planted it in wheat. The proceeds of a good crop year might equal the profits received from a decade of stock raising. When wartime prices collapsed in the early 1920s, plainsmen broke more sod with the newly adopted one-way disk plows to plant still more wheat and offset their economic losses.

FARMING BECAME a patriotic endeavor during World War I. The Department of Agriculture reminded farmers that wheat would win the war and urged maximum production to help feed American soldiers in France. This cartoon appeared in the *Farm Implement News* on May 16, 1918. *Author's collection.*

Competent Threshing Here Will Hasten the Thrashing Over There

New technology, war, and depressed prices stimulated Great Plains farmers to break 32 million acres of sod between 1909 and 1929 for new wheat lands. In the southern Plains, wheat acreage expanded 200 percent between 1925 and 1931; in some counties, this expansion varied from 400 to 1000 percent. Lawrence Svobida, a farmer in Meade County, Kansas, reflected: "My tractor roared day and night, and I was turning eighty acres every twenty-four hours, only stopping for servicing once every six hours." During the years of adequate precipitation, almost all of these acres were harvested, but in dry years nearly complete abandonment prevailed.

In the Great Plains, where crops need every drop of moisture that falls, a deficiency of only a few inches can mean the difference between a bountiful harvest and economic disaster. When drought struck the region during the early 1930s, it caused crop failure and encouraged farmers to abandon land unprotected by vegetative cover. The prevailing winds, averaging 10 to 12 miles per hour throughout the region, increased evaporation, weathered the soil, and brought dust storms to the southern Great Plains in January 1932.

THE FAR WEST

California epitomized commercial agriculture of the Far West during the early twentieth century. With major trading arrangements established with Great Britain, a growing population needing bread, and large-scale mechanized farms already operating, the Bear Flag State became an area for bonanza wheat farms. Low world prices for wheat, soil depletion, increased demand by settlers for land, and higher profits to be earned from irrigated fruit and vegetable production, however, caused these extensive wheat farms to fail in the first decade of the twentieth century. Growers in southern California and the Central Valley planted more fruit trees, especially citrus, and took advantage of refrigerated railroad cars and cooperative marketing to dominate the eastern markets. Improved transportation enabled vegetable growers to send perishable crops to distant markets in Central America, Canada, and the eastern United States. The Central, Imperial, and Coachella valleys became the most important areas for vegetable production and canning in the first decade of the twentieth century.

The growers continued to use cooperative exchanges to facilitate efficient and rational marketing of their crops. In 1905, citrus growers organized the California Fruit Growers' Exchange. This exchange enabled producers to collect their crops for distribution to specific markets based on the concepts of supply and demand. The Exchange avoided flooding a market with more fruit than could be sold, kept prices profitable, and reduced the costs and services of the middlemen who had traditionally brokered the crop. This control and systematic marketing enabled the growers to earn greater profits than ever before. The growers, both large and small, quickly saw the economic advantages of the Exchange. By the early 1920s, the Exchange represented more than two-thirds of California citrus growers in the marketplace. To create consumer loyalty, the Exchange adopted the brand name Sunkist to convey the message of quality and reliability, and it soon became a household word. The deciduous growers, unable to develop a similar marketing exchange because of the greater diversity of their fruit crops, created individual commodity associations, which gave them a competitive advantage over other producers who were far less organized for marketing peaches, apricots, and plums.

By 1910, then, California had developed large-scale commercial agriculture, particularly for fruit and vegetable production. This farming practice was based on extensive lands, irrigation, railroad and steamer transportation, and strong marketing cooperatives. Specialization rather than diversification became the key to economic success. But specialized commercial agriculture had little place for the small-scale family farmer who had little capital, few acres, and no irrigation or access to a marketing cooperative. As California's agriculture became more industrialized, more farmers could not compete. Economic gain drove the agricultural economy at a rapid pace of development. Landholdings became increasingly concentrated among a few wealthy growers. At the same time, large-scale growers and corporations began the process of vertical integration—controlling the crop from the fields to the consumer.

California's growers based their speciality-crop agriculture not only on the ownership of extensive fertile lands and access to cheap water but also on immigrant labor. Although racism and xenophobia in California had driven many Chinese workers from the fields and

IRRIGATION made the Central and Imperial valleys of California among the most productive in the world. Law and politics in California, however, determined that irrigation water would benefit the farmers with the largest landholdings and the most capital. State water policy also made irrigation water relatively cheap for farmers but expensive for urbanites. Fruit growers became particularly dependent on a bountiful supply of cheap water. *California Section, California State Library.*

IN CALIFORNIA, the fruit growers organized cooperative marketing associations to enable systematic shipment and sales. By the early twentieth century, California had become the leading producer of citrus fruits. The growers usually preferred Anglo workers, based on the prejudice that orchard work required greater skill than laborers of other heritages could provide. These workers picked grapefruit about 1930. *California State Archives.*

into the sweatshops of the cities during the late nineteenth century, the growers still needed a large work force. After the turn of the twentieth century, they recruited Japanese, Filipino, and Mexican workers. Japanese immigrants, who were particularly willing to labor for lower wages and accept living conditions other workers rejected, pooled their earnings and used that capital to become tenants. Japanese laborers were efficient workers, and the growers in the Sacramento, San Joaquin, and Imperial valleys preferred them to others. But because the Japanese worked hard, saved their money, and bought land, the nativists who had been responsible for driving out Chinese labor began to attack the Japanese workers, figuratively and literally. By 1920, the Japanese no longer contributed a significant labor force to the vegetable fields, vineyards, and orchards of California. Their fate as landowners would worsen because of restrictive legislation and war.

With the prejudice against Asian workers too powerful and ingrained to overcome, the growers turned to Mexican and Mexican-American workers, particularly after 1915. During World War I, Mexicans became the major work force in the citrus and cotton areas of the Imperial and San Joaquin valleys. By the early 1920s, Mexican workers had become essential to the agricultural production of the Central Valley. As higher-paying industrial jobs lured whites to the cities, Mexican and Mexican-American workers increasingly assumed the stoop work in the vegetable fields. By 1930, this work force provided as much as 80 percent of the fruit and vegetable harvest labor in California. Growers favored Mexican and Mexican-American workers because they were docile; collective action, particularly unionization, did not appeal to them, and they left the area upon completion of their tasks. As the nation moved closer to economic collapse, California became a land not only of opportunity (if a farmer could meet the financial qualifications) but of haves and have-nots.

LAND POLICY

In 1902, Congress passed the most important land legislation since the Homestead Act. Known as the Reclamation Act or Newlands Act, after its sponsor, Congressman Francis G. Newlands of Nevada, this legislation enabled the creation of major irrigation projects in the arid West on federal land. Enacted on June 17, 1902, with President Theodore Roosevelt's support, the Reclamation Act authorized the secretary of the interior to construct irrigation projects in sixteen western states. These projects would be financed by the sale of public land in those states. Public land in the project areas would be settled under the terms of the Homestead Act. Those settlers who used impounded water for irrigation would pay for the preparation of their land for irrigation over a ten-year period.

Congress created the Reclamation Service, which administrated the provisions of the Newlands Act under the jurisdiction of the U.S. Geological Survey. By 1907, this agency had authorized twenty-four projects in fifteen western states. Although these projects were engineering triumphs, economic and social problems soon plagued the Reclamation Service because its officials made many of the mistakes of their late-nineteenth-century predecessors who tried to bring irrigation agriculture to the West. Most important, they underestimated construction costs and the ability of newly settled farmers to pay for the preparation of the fields for irrigation within a decade. Moreover, the service did not screen applicants, and many settlers did not have the financial reserves to meet their indebtedness or the experience necessary to become successful irrigation farmers. In August 1914, Congress responded to pressure from the settlers to allow more time to pay their irrigation expenses by extending the repayment period to twenty years. Ten years later, Congress again lengthened the payment period, to forty years. By 1924, the newly designated Bureau of Reclamation had provided water to irrigate 1.2 million acres on more than 34,000 western farms. Project dams also provided flood control and generated hydroelectric power for the farmers and nearby cities. Complaints abounded, however, that land speculators, who had claimed acreage in the project areas before passage of the Reclamation Act, benefited the most from this land policy.

Congress also gave attention to the inadequacy of 160-acre homesteads in the semiarid and arid regions of the West. No farmer could earn a living on such small farms without irrigation in an area where one cow might need at least 20 acres for grazing. In 1904, Moses P. Kinkaid, congressman from Nebraska, introduced a bill that became the Kinkaid Act. This legislation provided 640-acre homesteads in the western two-thirds of Nebraska for settlers who would live there for five years and make improvements. In contrast to the original homestead law, Congress did not permit commutation. Cattlemen primarily benefited from this statute, but Congress had intended them to do so via a land policy that encouraged the establishment of small-scale cattle-raising homesteads. Overall, however, the large-scale ranchers benefited by purchasing these homesteads from the settlers who found they could not raise sufficient fodder or other crops to support their cattle-raising endeavors. The Kinkaid Act enabled economic improvement for some cattlemen in Nebraska, but they were the stockmen who already enjoyed financial success. Settlers who hoped to begin a mixed agricultural operation of crops and livestock on a 640-acre homestead in Nebraska usually failed because of inadequate acreage, insufficient capital, and a hostile environment.

In 1909, Congress also attempted to meet the needs of settlers in the West by providing more land for homesteaders. Under the Enlarged Homestead Act of that year, a settler could claim 320 acres of nonirrigable and nonmineral land in nine western states. By 1915, five other states had agreed to participate in this federal land policy. Once again, a settler had to live on the land and cultivate it for five years. To prevent speculators from quickly purchasing the acreage of discouraged homesteaders, Congress prohibited commutation. Under the Enlarged Homestead Act, many settlers opened new lands in the Great Plains and Far West. Yet 320-acre farms beyond the 100th meridian were still too small for profitable operation. Moreover, the best lands had already been claimed. As a result, many of these homesteads were consolidated under a single owner in years to come.

During the early twentieth century, western cattlemen also urged Congress to provide 640-acre homesteads for them. In 1916, they succeeded when Congressman Edward T. Taylor of Colorado won approval of the Stock Raising Homestead Act. With the best agricultural lands essentially taken, Taylor argued, the nonirrigable public lands remaining in the West were suitable only for grazing cattle and sheep. He contended that a homestead act for stock raisers would mean "more people, more homes, and more property on the tax rolls." The economic fortunes of the settlers and the states would both improve. Congress agreed and authorized 640-acre homesteads on public lands in the West that were "chiefly valuable for grazing and raising forage crops." This acreage could not be irrigable or contain commercial timber. Homesteaders had to improve their lands with fences and wells. A host of homesteaders quickly filed claims, and entries continued to be made under this act until western land policy changed in 1934 with the Taylor Grazing Act.

Although Congress designed the land policy of the early twentieth century to put more public land under private ownership for development, taxation, and profit, the semiarid and arid climate of the West often negated federal land policy. The allotted acreages remained too small for viable crop or livestock operations, and large-scale cattlemen with investment and operating capital often manipulated federal land policy to their advantage or bought out those who failed. In the end, federal land policy profited those who needed it least.

AGRICULTURAL IMPROVEMENTS

By the dawn of the twentieth century, great technological change had occurred in American agriculture. Most of that change evolved throughout the nineteenth century as farmers increasingly adopted horsepowered machinery to replace handpowered implements. During the early twentieth century, they began to replace their horsepowered implements with traction steam engines and internal-combustion gasoline tractors. Steam-traction engines easily plowed from 35 to 45 acres per day and pulled grain drills at the same time. These giant implements, weighing from 10 to 25 tons, burned coal, wood, oil, and straw. By the beginning of World War I, steam engines provided power equivalent to 7 million horses and mules. Still, most farmers could not afford to own a steam engine. From 1908 to 1915, during the peak of the steam engine's popularity, only one farmer in twenty owned a steam tractor. Where steam power was used for threshing and plowing, most farmers hired custom operators who owned their own implements to do the work.

Farmers still needed a more consistent power source than horses and mules. They needed power that would not tire, power that would drive the mechanisms of large implements more effectively than gearing to a ground wheel. In addition, farmers needed a more suitable, cheaper, and more maneuverable source of draft power than steam engines. At the turn of the twentieth century, inventors and mechanical engineers were on the verge of developing that new source of farm power—the internal-combustion tractor.

In 1905, the Hart-Parr Company began using the term *tractor* to replace the expression *gasoline traction engine* to identify the powerful traction internal-combustion implements becoming available. Farmers, implement dealers, and manufacturers quickly adopted the term, and the word has been used since that time. Still, farmers did not rush to purchase this new technology. Because these early tractors were unreliable, expensive, and gigantic, farmers took a wait-and-see attitude. In the meantime, they used their horses to provide steady, reliable, and inexpensive service.

As late as 1912, farmers remained hesitant to purchase tractors

FARMERS PRIMARILY used steam engines to power threshing machines during the late nineteenth and early twentieth centuries. Steam engines provided the consistent power to operate large-capacity threshing machines. This crew threshed wheat near Russell, Kansas, about 1908. The long belt that linked the steam engine to the threshing machine enabled the operator to keep the engine as far away from the haystacks as possible. A spark from the chimney could set the stacks ablaze. Note the coal wagon to the rear and water tender to the right. *Smithsonian Institution.*

because they were impractical, too large and unwieldy for most farms. Farmers primarily used tractors for drawbar work—for pulling, especially for plowing or breaking new ground. On the eve of World War I, however, several manufacturers like the Avery and Chase companies tried to build lighter, cheaper, more efficient and maneuverable tractors for work on the average farm. By 1915, implement companies manufactured tractors that weighed between 3,000 and 5,000 pounds and pulled two plows. These tractors each replaced three horses in the field at plowing time and approached practicality for small-scale farmers. Tractor production increased from an estimated 5 in 1900 to more than 17,000 annually by 1917. Demand for tractors also increased during World War I when farmers in the Midwest and West increasingly used this technology to alleviate labor shortages. When the war ended, however, many farmers still could not afford a tractor or did not want one because these implements could be used only for plowing or powering stationary machinery such as threshing machines or corn shellers. Tractors remained unsuited for general-purpose farm work or row-crop cultivation on small-scale farms. By 1920, farmers still preferred horses; only 3.6 percent owned tractors.

While various inventors and entrepreneurs tried to develop an efficient, affordable tractor suitable for the small-scale farmer, Henry Ford decided to enter the field. He left a lasting mark on the technological history of American agriculture. Between 1906 and 1908, Ford built his first tractor from a variety of parts used in the automobile industry. This tractor did not perform adequately, and Ford did not succeed in building a lightweight, low-cost, two-plow tractor until 1917. This tractor, known as the Fordson, achieved immediate success because of postwar shortages of men and horses for farm work and because Ford used his automobile dealers to sell tractors. Moreover, many farmers could afford the price, $750.

Ford's innovation had several serious problems, even though it was the most practical and affordable tractor on the market. First, the Fordson was too light and had insufficient power to meet the needs of the farmers who tilled large acreages. Those farmers needed a tractor that could pull at least three to four plows instead of two. Nevertheless, the Fordson proved that small, versatile tractors could be produced at an affordable price and that such tractors could meet the power needs of many small-scale farmers. Moreover, the Fordson's design became standard for other tractor makers. With the Fordson, mass production entered the tractor industry, and this

THE EARLY gasoline-powered tractors looked much like steam-traction engines, primarily because the manufacturers of steam engines tried to convert to gasoline-tractor production without making the necessary structural changes. The result was a giant iron monster like the Mogul, Type C, manufactured by the International Harvester Company between 1909 and 1914. *Smithsonian Institution.*

implement began the second revolution in American agriculture.

By the early 1920s, the advantages of internal-combustion tractors compared to horses, mules, and steam-traction engines were readily apparent. Horses were relatively expensive to purchase and costly to maintain. Farmers needed about 5 acres to supply the oats, hay, and fodder each horse required for an entire year. In contrast, if a tractor did not work, it did not require fuel. Land that had been used for raising animal feed could now be seeded with more profitable crops for human consumption. In addition, tractors could be operated day and night, while horses and mules had to be rested or

THE TRANSITION from horse to tractor power proved difficult and expensive for many farmers. Tractors did not begin to replace horses on a wide basis until Ford, International Harvester, John Deere, and other companies began building tractors designed for the small-scale grain farmer during the 1920s. These farmers in Gallatin County, near Bozeman, Montana, use a tractor to pull one grain binder, but five horsepowered binders followed. As soon as farmers could afford to, they replaced their horses with tractors. *Montana Historical Society.*

changed. Compared to horses, tractors did not require as much daily care. Moreover, tractors were immune to the effects of hot weather and insects. Gasoline tractors also saved the time needed to put a steam engine into operation. Gasoline tractors started relatively easily, worked efficiently, and shut off quickly. Tractors did not emit sparks from the exhaust and were safer in the field than steam engines. Furthermore, gasoline tractors were easier to maneuver than most steam-traction engines.

Still, problems remained. Only large-scale farmers had sufficient capital or enough land adaptable to tractors to merit owning one. Small-scale row-crop farmers found tractors too cumbersome to maneuver in their smaller fields and too expensive to purchase. As a

result, tractors remained almost entirely restricted to the western Middle West, the Great Plains, and the Far West, where farmers used them for plowing and powering threshing machines. In this respect, gasoline tractors remained about as useful and practical as steam-traction engines. Moreover, row crops such as corn and cotton required cultivation, and tractors like the Fordson could not be driven down the rows without damaging the plants. Consequently, most farmers delayed the purchase of a tractor until an all-purpose implement could pull both plows and cultivators and could be purchased for a reasonable price.

In 1924, the International Harvester Company (IHC) contributed to the great technological change under way in American agriculture by introducing the Farmall tractor, which soon made the Fordson obsolete. The Farmall was the first low-priced, tricycle-designed tractor built for row-crop farming. In front the Farmall had two closely spaced wheels designed to travel between the crop rows. In the rear it had a high axle to straddle the growing crop during cultivation, and a brake on each rear wheel enabled sharp turns in small fields. It had three forward speeds and a reverse. Although the Farmall was too light for work on grain farms, it was ideal for truck and dairy farms. By 1925, small tractors such as the Farmall were in wide use on dairy, vegetable, and fruit farms from New York to the Pacific Coast.

While some agricultural engineers and entrepreneurs such as Henry Ford worked to develop a reliable internal-combustion trac-

GASOLINE-POWERED TRACTORS helped revolutionize American agriculture after World War I. Tractors like this John Deere were readily adaptable to the small-scale farm and enabled farmers to complete a host of tasks with less labor, more speed, and greater efficiency. In 1928, this farmer used a tractor to power a feed grinder and elevator. *Deere & Company Archives.*

BEFORE the widespread adoption of the tractor after World War I, farmers seeded their corn crop with horse-drawn planters. The seat at the rear was a safety device to prevent the driver from falling beneath the implement if he lost his balance. The disks in front cut through the stalks from the previous harvest and aided the uniform planting of the seed, which dropped down tubes behind. The box at the top held the seed. *Special Collections, William Robert and Ellen Sorge Parks Library, Iowa State University.*

tor, others devoted their attention to the most important regional problem in twentieth-century agricultural technology—mechanizing the cotton harvest. During the first half of the twentieth century, the cotton harvest differed little from that of antiquity. Technological change came slowly to this aspect of American agriculture, and nearly a century would pass from the patent of the first cotton picker in 1850 until International Harvester began quantity production of a mechanical picker in 1948. To develop a cotton picker, agricultural engineers faced a multiplicity of problems: diverse soil and climate conditions, various plant types and ripening times, technical and economic difficulties, and an abundance of cheap farm labor. A marketable cotton picker had to be reasonably priced and easy to operate and repair; it had to be maneuverable on moderately rough terrain and in muddy fields; and it had to pick a high percentage of trash-free cotton without damaging the lint.

The inherent characteristics of the cotton plant posed a significant problem for the development of a universally acceptable mechanical picker. Because the cotton bolls did not ripen uniformly and early-opened cotton deteriorated rapidly if it was left unpicked, a farmer could not wait until all the bolls opened to harvest his crop. The height of the plants varied in the fields and among geographical regions, requiring machine adaptability to plants ranging in height from a few inches to several feet. During the late 1920s, a cotton plant that matured early and produced bolls in clusters on short stems, a development that would facilitate mechanical harvesting, had reached only the experimental stage. Not until scientists and agricultural engineers mastered the biological and technical problems of the cotton harvest would a mechanical picker have beltwide appeal. Ultimately, the cumulative contributions of mechanical engineers, plant breeders, and chemists made the development of the cotton picker possible by the early 1940s.

The solutions to these problems were not easy. Inventors patented more than 1,800 cotton pickers between 1850 and 1942 when International Harvester tested the first reliable implement. These numerous inventions included threshing machines that cut the entire plant and separated the cotton from the bolls in a process similar to a grain combine; pneumatic pickers that removed the cotton from the bolls with suction or blasts of air; electrical machines that used static electricity to extract the cotton; strippers that pulled the entire boll from the stalk; and spindles that used fingers and prongs to remove the lint.

John and Mack Rust were the most successful inventors to work on a cotton picker that led to the eventual development of a reliable implement. As early as 1924, John Rust experimented with a barbed or serrated spindle designed to twist the lint from the boll. Although the barbs caught the cotton, the Rusts could not devise a method to remove the lint from the spindles. Then, three years later, John remembered that dew caused the cotton to stick to his fingers when he had picked it in boyhood and that his grandmother moistened the spindle of her spinning wheel to get the cotton to adhere properly. Rust redesigned his implement to include an endless belt to which he attached vertical rows of dampened smooth wire spindles that rotated against the cotton plants. Steel ribbons stripped the cotton from the spindles, and the machine then conveyed the lint to a storage basket. Rust filed patent papers for his cotton picker in 1928 and completed building his first test model that year.

In 1931, near Waco, Texas, Rust's machine harvested 1 bale of cotton during the test day. Two years later, he had improved it to pick 5 bales per day at the Delta Experiment Station at Stoneville, Mississippi. Rust's cotton picker, however, collected a large amount of trash with the lint, and the machine left much of the cotton in the bolls. Even so, this implement proved that mechanical harvesting was possible with a few improvements because it picked forty to fifty times faster than hand workers. Financing became a problem for the Rust brothers, particularly during the Great Depression, and as late as 1941 their picker remained in the developmental stage with no prospect for commercial production.

Still, a cotton picker's potential savings in labor were clearly apparent. Cotton farming now hovered on the threshold of major technological change. The harvest remained the great bottleneck in production since cotton still had to be picked by hand. Since the early twentieth century, cotton farmers had become increasingly interested in mechanization to offset rising costs and an uncertain labor supply. During World War I, higher-paying jobs lured southern black workers to northern cities, and schools sometimes closed to provide an adequate number of picking hands. Farmers therefore began anticipating the development of a cotton picker that would solve their harvesting problems.

In January 1924, International Harvester began work on a cotton picker and tested a spindle picker that autumn. Over the next few years, IHC tested both self-propelled and tractor-drawn harvesters. Invariably, these one-row machines brushed too much cotton onto the ground, damaged the plants, clogged, and bogged down in muddy fields. From 1926 to 1930, IHC simplified and improved the spindle picker by developing short tapered barbed spindles and applying Rust's principle of moistened spindles. These adjustments ultimately proved successful.

But although the scientific and technical problems of harvesting cotton by machine would soon be resolved, the adoption of cotton pickers to southern agriculture also depended on the development of a new credit system and cropping pattern. These changes would not come until Franklin D. Roosevelt initiated a host of agricultural programs that provided direct economic aid and suggested changes in land use under the New Deal. This relief program enabled the eventual adoption of cotton harvesters. In the meantime, cotton farmers picked their crop by hand.

While the gasoline tractor became the most important development in agricultural technology nationwide during the early twentieth century and the development of a reliable cotton picker was on the verge of becoming the most significant regional agricultural technology, combines designed for the small-scale grain farmer also increased productivity and made the harvest easier. Until the turn of the twentieth century, farmers primarily used the combine in the Central Valley and the plains of eastern Washington. Soon thereafter, implement manufacturers began to develop small horse-drawn combines that cut a 7–10-foot swath. World War I stimulated these efforts and encouraged farmers to adopt combines by raising grain prices and reducing the labor supply. About 1917, Great Plains farmers began purchasing 12–16-foot combines powered by an auxiliary gasoline engine mounted on the harvester. Tractors or horses still pulled these combines, but the auxiliary power source, which drove the cutting and threshing mechanisms of the machine, provided the consistent and uniform power that could not be achieved from the ground wheel characteristic of previous combine harvesters. The collapse of the farm economy in 1920 prevented immediate wide-scale adoption of this new harvesting implement, but farmers began using the combine on an expanded scale during the mid-1920s.

Grain farmers adopted the combine in part because this implement reduced their labor costs. Indeed, during the early twentieth century, labor costs required major expenditures at harvesttime. Prior to the adoption of the combine, wheat farmers customarily doubled their labor force for the harvest. A timely and profitable harvest depended on a wheat farmer's ability to secure adequate help at a reasonable cost. The annual scarcity of harvest and threshing hands drove labor prices up and forced farmers to rely on itinerant workers; rural neighborhoods and nearby small towns usually could not furnish an adequate number of hands.

With a tractor and combine, a farmer no longer required hired hands to gather the binder's sheaves into shocks or a crew to operate the threshing machine. At most, the farmer needed one man to drive the tractor; another to ride the combine, operate the cutter bar, and monitor the gasoline engine; and one to haul the threshed grain to a storage bin or grain elevator. In addition to the savings in labor costs, combines sped the harvest. Binders averaged approximately 15 to 20 acres per day; a 15-foot combine harvested 35 to 40 acres per day. Speed like this reduced the risk of bad weather at harvest time and encouraged farmers to expand their production because they could harvest more acres than ever before. By 1929, Great Plains farmers harvested approximately 75 percent of their winter wheat with combines.

Meanwhile, scientists were on the verge of revolutionizing crop production with the development of hybrid corn. Although Native American farmers had produced a form of hybridization long before European contact, successful experiments designed to improve the productive capacity of corn by using "hybrid vigor" did not occur until the early twentieth century. Corn breeders had relied on seed selection based on the physical appearance of the ears and kernels to improve production. Essentially, they believed that good-looking corn would produce excellent progeny. Soon after the turn of the twentieth century, however, George Harrison Shull, working at the Carnegie Institution's laboratory at Cold Spring Harbor, Long Island, showed that genetic characteristics determined productivity and that those characteristics could be isolated or controlled. Shull inbred corn varieties to produce relatively pure strains. He then crossed two pure varieties, those with uniform genetic characteristics, to produce a "single-cross" plant superior to the parents. Shull termed the improved productivity or "hybrid vigor" of this plant "heterosis." In 1917 Donald Jones, an agricultural scientist at the Connecticut Agricultural Experiment Station, refined Shull's cross-breeding technique by breeding two "single-cross" parents to produce a "double-cross" plant. In other words, Jones used four purebred lines and two generations to produce hybrid seed. This double-cross technique produced seed that would enable farmers to raise more corn on fewer acres.

GEORGE WASHINGTON CARVER

Special Collections, William Robert and Ellen Sorge Parks Library, Iowa State University.

George Washington Carver (c. 1864–January 5, 1943) was born to a slave woman owned by Moses Carver sometime during the Civil War on a farm near Diamond Grove, Missouri. Before the war ended, bushwhackers kidnapped Carver and his mother. His owner rescued George and raised him, and Carver took the last name of his foster parent. As a boy, Carver had a keen interest in plants and animals, but racial discrimination and inadequate schools limited his formal education. Carver eventually finished high school during his late twenties while working as a farmhand in Kansas.

In 1890, Carver enrolled at Simpson College in Iowa, but he transferred to the Iowa Agricultural College (now Iowa State University) in 1891. Carver studied primarily horticulture and mycology and received a bachelor's degree in agriculture in 1894. After completing his undergraduate work, Carver accepted a faculty position in botany with the responsibility of supervising the greenhouse, and he taught biology. Two years later, he earned a master's of science. His thesis was entitled "Plants as Modified by Man." In the autumn of 1896, Booker T. Washington invited Carver to become the head of the newly created agriculture department at the Tuskegee Institute in Alabama, and Carver accepted.

At Tuskegee, Carver conducted research and led innovative extension programs to improve southern farming. He worked particularly to disseminate practical agricultural knowledge to poor black and white farmers and to develop new food and industrial products from plants. In this latter area, he developed more than three hundred products from the peanut and discovered more than one hundred uses for the sweet potato. His work with peanuts and soybeans eventually helped southern farmers end their dependence on cotton and improve soil fertility.

In 1902, Carver wrote, "It is not unusual to see so-called farmers drive to town weekly with their wagons empty and return with them full of various kinds of produce that should have been raised on the farm." Carver could not single-handedly change the bondage of the sharecropping system, although he urged farmers to diversify to reduce their reliance on the furnishing merchant. Moreover, although his experiments in the field of applied agricultural science did not lead to the development of a new scientific theory or the discovery of the immediate commercial feasibility of new products, neither had been his goal. Instead, Carver sought to aid farmers with practical agricultural research and advice. He believed agricultural education was "the key to unlock the golden door of freedom to our people." Most black students at Tuskegee, like white students at other institutions of higher education, however, saw education as a means to escape the farm.

Soon after World War I, hybrid-corn-breeding programs developed at many agricultural experiment stations and private seed companies. By 1933, hybrid corn producers like the Hi-Bred Corn Company in Iowa and DeKalb Agricultural Association in Illinois operated on a substantial commercial scale. These and other companies not only produced hybrid corn that produced heavy yields but bred the plant to resist disease and insects, facilitate mechanical harvesting, and adapt to specific soils and climate. Yet despite the benefits of hybrid seed, a danger of losing the open-pollinated genetic base from which all hybrid corn derives soon appeared because so few farmers continued to raise traditional varieties. This

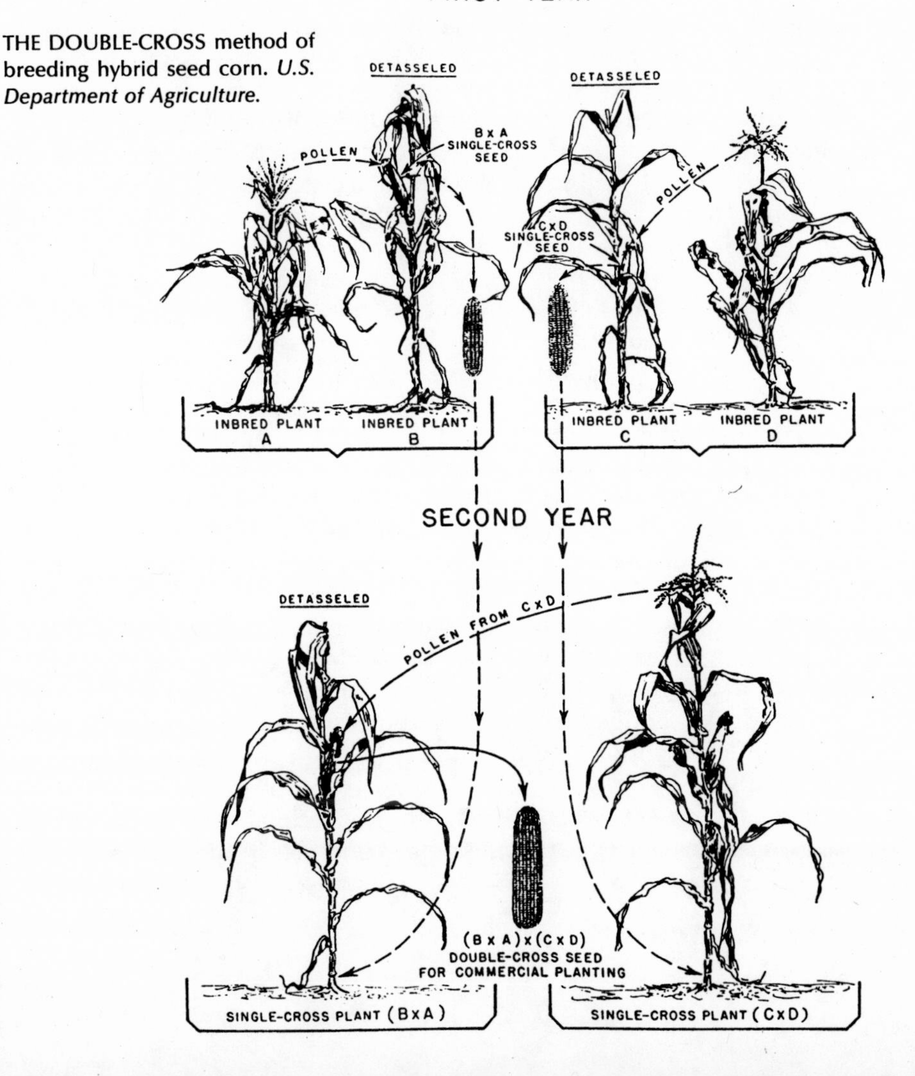

THE DOUBLE-CROSS method of breeding hybrid seed corn. *U.S. Department of Agriculture.*

problem had not been completely resolved by the late twentieth century.

Agricultural scientists, particularly those in the USDA, contributed to improvements in livestock breeding and feeding. In 1903, Marion Dorset discovered that a microorganism caused hog cholera. Three years later, he developed a serum that prevented and cured this devastating disease. Wheat and sorghum breeders used Gregor Mendel's genetic principles of plant breeding and produced new varieties such as Marquis, a hard-red spring wheat well suited to the harsh environment of the northern Great Plains. Plant breeders also used Turkey, a hard-red winter wheat, for crossbreeding experiments to improve winter hardiness, earliness, disease resistance, and grain quality. The Kanred and Cheyenne varieties are two products of that research. These breeding and veterinary experiments also contributed to the economic well-being of commercial agriculturists.

In addition to technological and scientific change, the development of the agricultural extension system became one of the most important agricultural improvements during the early twentieth century. Although farmers still attended institutes before World War I, many could not, and most refused to accept information they considered "book farming." The USDA partially overcame this problem by establishing demonstration farms where farmers could learn the newest and best techniques based on research at the agricultural experiment stations and government laboratories. Seaman H. Knapp was the most successful USDA leader in demonstration work. In 1903, on a farm near Terrell, Texas, Knapp inaugurated the demonstration technique for the dissemination of agricultural information. This effort proved a success, and the USDA established other demonstration farms across the South.

The efforts of the USDA to educate farmers about improved agricultural methods culminated with the Smith-Lever Act of 1914. Introduced by Senator Hoke Smith of Georgia and Congressman Asbury F. Lever of South Carolina, this legislation authorized the federal government to support, with matching state funds, the creation of an extension system at the land-grant colleges. Extension agents had the responsibility of demonstrating practical agricultural methods that applied to field and home and helping farm men and women learn about new techniques through the dissemination of publications. Extension activities were supplemented with the Smith-Hughes Act of 1917, which provided federal funding for teaching vocational agriculture and home economics in the high schools.

DURING THE EARLY twentieth century, railroad companies often ran special educational trains consisting of several cars in which experts, often from a nearby land-grant college, provided the latest information intended to help farmers improve their operations. In 1905, this audience on an extension train of the Burlington Railroad listened to a talk about raising better corn. Note the Holden sawdust corn-testing box, which tested the germination quality of seed corn. *Special Collections, William Robert and Ellen Sorge Parks Library, Iowa State University.*

4-H

In 1902, Albert B. Graham laid the foundation for the 4-H when he organized an agricultural boys' club in Springfield Township, Ohio. Graham used this organization to help the members pursue special projects such as testing for soil fertility and selecting the best seed corn to improve agricultural production. Graham gained support from the Ohio Agricultural Experiment Station and Ohio State University, thereby beginning a close relationship between both institutions and the USDA for educational work among young people.

In 1920, these 4-H members showed considerable group pride during fair time in Iowa. *Special Collections, William Robert and Ellen Sorge Parks Library, Iowa State University.*

At the same time, in Illinois, O. J. Kern organized local agricultural clubs, and Will B. Otwell worked with agricultural businesses to sponsor corn-growing contests for farm children. By 1908, clubs had organized in Iowa to help young farm women learn to bake and sew. Two years later, canning clubs began in South Carolina. The USDA encouraged the organization of agricultural clubs for young adults and incorporated club activities in the work of the extension service with the passage of the Smith-Lever Act.

The Smith-Lever Act of 1914 provided federal funds to the states for the organization of agricultural clubs for boys and girls as a basic extension service activity. USDA officials recognized that agricultural clubs provided an excellent means to convey knowledge from the experiment stations and land-grant colleges to the farms. Agricultural clubs also would enable farm children to "learn by doing." Special projects such as raising livestock and crops or canning and sewing would help improve agricultural techniques and rural life. While the boys and girls learned, they would also gain self-confidence, leadership experience, and a commitment to community service. Although the term *4-H Club* did not appear on official USDA publications until 1918, the cloverleaf emblem and the term were used throughout rural America by 1924 to designate agricultural clubs for young people.

By the late twentieth century, nearly half of the 4-H members lived in towns and cities, a reflection of the dramatic decline of family farms and sharp demographic changes in rural America. The organization no longer emphasized agricultural activities. The 4-H also stressed the development of citizenship and public-service skills, good physical and mental health activities, and volunteer and leadership activities. And it supported community development projects such as voter-registration campaigns and after-school child-care projects. The 4-H, however, remained reliant on the resources of the land-grant universities and the U.S. Department of Agriculture.

Membership in the 4-H is available to anyone 9 to 19; most 4-H clubs are coeducational. The federal, state, and county governments continue to contribute money to the extension service for the support of 4-H organizations. County extension agents provide leadership for the formation of local 4-H clubs, and extension agents at land-grant universities organize statewide leadership and coordinate activities. USDA officials contribute educational literature to the clubs.

The four-leaf clover with a white *H* imprinted on each leaf is the symbol of the 4-H Club. First used in 1910, the letters represented "Head, Heart, Hands, and Health."

AGRICULTURAL ORGANIZATIONS

Although the radicalism of the Farmers' Alliance and People's party died after the presidential election of 1896, the economic problems of the late nineteenth century lingered. Although most agriculturists argued that a "farm problem" plagued the nation and the rural economy, few agreed about solutions. Socialists advocated state and federal ownership of marketing and processing. Urban-based middle-class reformers known as Progressives championed federal regulation of railroads, banks, and commodity exchanges. They urged agricultural diversification and increased attention by farmers to the benefits of agricultural science and technology. Most farmers agreed that they needed alternatives to existing agricultural institutions to solve their economic problems, and many turned again to the cooperative movement.

In 1902, Isaac Newton Gresham founded the Farmers' Educational and Cooperative Union in Texas to address the farm problem. Known as the Farmers' Union, this organization advocated collective action by farmers for cooperative buying and selling. The organization also championed the improvement of agricultural education, advocated the elimination of speculation on commodity markets, and generally sought an improved standard of living for farmers, the key to which involved economic improvement. With membership limited to farmers, farm laborers, and agricultural service workers, the organization spread rapidly northward. By 1920, Kansas had become the center of activity for the Farmers' Union, especially in areas where the Farmers' Alliance and People's party had been strong and drought made economic conditions the worst.

As the Farmers' Union spread across the Great Plains and into the Midwest, the organization appealed to agriculturists who were more liberal than those who had joined the Grange during the late nineteenth century. They were also more militant in posing solutions to the economic problems of agriculture, particularly advocating aid from the federal government. The Farmers' Union also attacked the land-grant colleges because those colleges sought to help farmers increase their productivity at a time when surplus production and inadequate markets kept prices low and farmers poor. Instead, the Farmers' Union wanted the federal government to guarantee cost-of-production prices for agricultural commodities plus a reasonable profit. The organization hoped to enlist a membership large enough to permit control of agricultural marketing and to gain favorable commodity prices in a way similar to the collective-bargaining practices of organized labor.

The Farmers' Union attempted to increase agricultural prices and the standard of living, particularly by reviving the cooperative system of the Patrons of Husbandry and the Farmers' Alliance. The Union organized cooperative warehouses, gins, stores, fertilizer manufacturers, cottonseed-oil mills, and banks. The goal was to reduce production and increase prices. The Union advocated not only plowing-under a portion of the cotton crop but withholding cotton from the market to force price increases though the creation of artificial scarcities. Since approximately half the farmers in the South were tenants, however, this plan was doomed to failure. Tenants could not afford to reduce production. They needed maximum production to make up for income loss from low prices. At best, the Farmers' Union appealed to landowners in the South.

During the 1920s, the Farmers' Union divided into two camps. One group sought to improve the economic conditions of farmers through cooperative marketing and buying to raise prices and reduce costs. The other sought to improve the standard of living of farmers by controlling marketing, with the goal of bargaining for prices in a way similar to that of organized labor. Although the first group would solve their problems independently, the second faction sought assistance from the federal government to stabilize farm prices. Despite these differences, cooperatives in the form of grain elevators, retail stores, creameries, banks, insurance companies, and gasoline stations remained fundamental services the Farmers' Union provided its membership.

The most radical response to the "farm problem" came in North Dakota where in February 1915 Arthur C. Townley single-handedly organized farmers into the Non-Partisan League (NPL). Townley preached economic salvation through moderate socialist action, advocating the establishment of a state-owned terminal grain elevator to ensure the proper grading and pricing of a farmer's crop. Townley also championed the creation of state-owned banks, flour mills, and meat-packing and cold-storage plants. He won many supporters, particularly among the Norwegians in the northern portion of the state, by advocating the exemption of farm improvements from property-tax valuations, state hail insurance, and rural credit banks that would provide cheap loans to farmers. Most important, he advocated nonpartisan political support for politicians that supported NPL goals.

The NPL grew rapidly, even though membership was limited to farmers. Townley was a skilled organizer who sent recruiters out to enlist members: "Find out the damn fool's hobby and then talk about it. If he likes religion, talk Jesus Christ; if he is afraid of whisky, preach prohibition; if he wants to talk hogs, talk hogs—talk, talk, talk, until you get his God-damn John Hancock to a check for six dollars."

Townley's psychological technique of showing interest in a farmer to win his support worked. By the end of 1915, the NPL had established headquarters in Fargo and, by using its own newspaper, the *Nonpartisan Leader,* to disseminate the NPL's point of view and coordinate political action, it had become an important force in state politics. The next year the NPL effectively used the direct primary system to support candidates in either major party for state executive, legislative, and judicial offices. The NPL elected Lynn J. Frazier governor, gained control of the state house of representatives, and became a strong minority force in the state senate.

When the legislature met in 1917, it enacted the program of the NPL, including a grain-grading system, guaranteed bank deposits, and a nine-hour workday for women. The legislature also provided reduced rates of tax assessment for farm machinery and improvements, regulation of railroads to prevent rate discrimination among carriers, and woman suffrage. However, the NPL opposed American entry in World War I, and the Republican party charged it with disloyalty. At a time when the NPL planned to expand into neighboring states, politics placed it on the defensive, and it never recovered. Before the movement collapsed in 1922, however, the NPL succeeded in winning legislative approval for a state bank that provided low-interest loans to farmers, a state mill and elevator, and compulsory and more affordable hail insurance through the state government.

Although the NPL waned, it remained a liberal force in North Dakota's agrarian politics, but it never became a third party. It effectively used the direct primary to make the existing political power structure, particularly the Republican party, supportive of the will of the majority—the agricultural interests. In many respects the NPL functioned as a third party until it merged with the Democratic party in 1956 for political purposes. During its existence, the NPL united farmers as no other agricultural organization had. It successfully translated Populist ideology into a workable program, and the state bank, mill, and elevator remain an important economic testimony to its success.

The NPL appealed to the most radical farmers in the northern Great Plains, but the organization had virtually no appeal to the more financially secure and conservative farmers in the Midwest.

Those farmers gained representation of their interests with the formation of the American Farm Bureau Federation in November 1919 at Chicago, the most important attempt to organize farmers nationwide. The supporters of the Farm Bureau were not only farmers who did not have major financial problems but those who wanted a conservative farm organization that would oppose radicalism and labor unrest. The Farm Bureau actively sought the support of the railroads and the financial and business community. It strongly supported the work of the USDA through county agents and the extension system. The Farm Bureau sought increased agricultural production and opposed unnecessary government regulation of the economy, especially government ownership of the railroads and public utilities. It advocated increased federal support for the privately owned Federal Land Banks, tariff protection for farmers, and cooperative marketing.

The Farm Bureau worked closely with the Cooperative Extension Service and the land-grant universities. At first, the Farm Bureau underwrote the expense of the county agents sent from land-grant colleges to educate farmers about new scientific and technological practices, providing funds from membership subscriptions. The county agents, extension service, and land-grant colleges cultivated this support to finance their services and build a clientele. Critics of the Farm Bureau, such as the Farmers' Union, charged that the county agents, land-grant colleges, and experiment stations gave their attention to landowning, capital-intensive farmers, most of whom were members of the Farm Bureau.

The Farm Bureau countered that it was the only agricultural organization that encouraged the work of the land-grant colleges and the dissemination of new techniques to help farmers become more productive and efficient, both of which would put more money in their pockets. Moreover, while the Farmers' Union advocated production controls and withholding commodities from market, the Farm Bureau believed improved marketing at home and abroad would solve most of the farmers' economic problems.

While these agricultural organizations contested for membership and power in an attempt to improve the economic fortunes of farmers and the standard of living in the rural community, the postwar agricultural depression lingered for the remainder of the 1920s. When World War I farm prices collapsed in 1920, economist Charles A. Wiley noted that "agricultural prices fell first, fell fastest, and fell farthest." In 1921, farm prices averaged 67 percent of parity. Before the farm economy improved, it got considerably worse as the nation plunged into the Great Depression with the stock market crash of

1929. Hard times now created a desperation among some farmers in the Midwest that led to the creation of the most violent agricultural movement in American history—the Farmers' Holiday Association.

During the 1920s, foreclosures had skyrocketed in the Midwest, and many banks that had extended credit to farmers failed. Foreclosure auctions became a common sight across the region. Confronted with these problems in addition to low prices, many midwestern farmers were convinced that they needed to take "direct action." The most radical members of the Farmers' Union were prepared to go to the barricades to win by force what their desperate appeals for economic relief had failed to achieve. Milo Reno was their leader. By February 1932, many members of the Farmers' Union were demanding a farm "holiday," a time when farmers would neither buy nor sell. This withholding action, Reno believed, would quickly close food-processing plants and empty grocery-store shelves. The inevitable result would be federal intervention, guaranteeing farmers the cost of production plus a profit of about 5 percent. Once that happened, farmers would return to their fields.

On May 3, 1932, Reno, a veteran of the Populist revolt of the 1890s, persuaded some two thousand farmers assembled at the state fairgrounds in Des Moines to organize the Farmers' Holiday Association. Although Iowans predominated, delegates attended from a half-dozen nearby states. The delegates, inspired by Reno's fiery oratory and evangelical fervor, chose him president of the association and vowed to strike on July 4. Although the strike date was soon changed to August 15, it began four days earlier when dairy farmers blockaded the roads into Sioux City in an attempt to increase the price of milk. Reno had not intended the Farmers' Holiday strike to include picketing of roads, and he could not prevent or control the violence that followed. Quickly the strike spread. Barricades were thrown up on the roads that led into the major markets in Iowa and Nebraska. Farmers dumped milk, blocked roads, and interrupted railroad service. Bloody skirmishes broke out between farmers and law-enforcement officers.

But after more than two weeks of manning the barricades, the strikers began to waver, largely because the movement in general and the strike in particular had been founded upon an incredibly unsound economic assumption—that an embargo of all markets by the farm community could be maintained. Most farmers could not participate for long in such a movement. They had to sell commodities to support their families. Like everyone else, they had bills to pay. A withholding action added additional costs to their already troubled operations. The longer they boycotted the markets, the longer they had to feed their livestock, and any commodities temporarily withheld from sale would flood the market at a later date, driving prices down still further.

By the end of August, the strike had failed in Iowa, but it soon spread to Wisconsin where it turned into a "bonafide war" between dairy farmers and sheriffs and national guardsmen. A Wisconsin reporter observed after one confrontation that "blood from noses was flowing almost as freely as milk." Another observer noted that the national guardsmen lined up one group of striking farmers and "pounded the Hell of out them." When the fighting stopped, one farmer was dead, scores were injured, and three hundred arrested. Before the violence ended, several more people had died.

While the general public repudiated the violence and some farmers thought the nation was on the verge of revolution, Reno urged the members of the Farmers' Holiday Association to await the November election returns in hopes that Democrat Franklin Delano Roosevelt would be elected president and initiate a cost-of-production agricultural policy. Before that happened, however, the Holiday became associated with the "penny auction" or "Sears-Roebuck" sales in October 1932. At these foreclosure sales farmers bid a few cents for every item offered and threatened bank officers or mortgage holders with physical harm if they did not accept the returns as full payment of the debt. Quickly penny auctions became the most important activity of the Farmers' Holiday Association. Between 1932 and 1936, this activity marked the peak of the Farmers' Holiday movement, involving more farmers than any other aspect of the association's protest activity.

Reno and other leaders of the Holiday had never officially sanctioned the penny auctions, but they were swept along. Reno had labeled them a temporary measure that the association would support because they achieved specific and immediate results. The most important long-term achievement of the Farmers' Holiday Association, however, was not the withholding action or the penny auction but the creation of "councils of defense" on a county basis. These organizations brought creditors and debtors together to work out refinancing and payment of delinquent debts. The association saved many farmers from foreclosure. Despite this success, the Holiday quickly lost its credibility because of the violence whenever it picketed during a withholding action and because the newly elected Roosevelt administration dramatically changed the nature of federal agricultural policy in 1933.

In the end, the Farmers' Holiday Association could not control enough markets in the Midwest to force major price increases, and the organization held no appeal to farmers in other sections of the nation. Moreover, farmers have traditionally banded together only for brief periods to achieve specific objectives. Bacon's Rebellion in 1676, Shays's Rebellion in 1786, the Whiskey Rebellion in 1794, and the Black Patch War of 1906–1908 are examples of such limited actions. American farmers have been too independent and their interests too diverse for them to maintain a large, united organization for an indefinite period. Although the movement flickered with life after the New Deal, in 1937 it ceased to exist. By that time, however, revolutionary economic change had come to American agriculture, but it was not always equitable.

FEDERAL POLICY

During the first three decades of the twentieth century, federal agricultural policy emphasized research to help farmers increase production, supported agricultural education and extension, and expanded the regulatory activities of the USDA. When the agricultural depression began in 1920, however, farmer organizations and politicians from the major agricultural states began to advocate a new policy that would enable the federal government to intervene in the marketplace to regulate production and marketing for the purpose of controlling surpluses and increasing prices. When Congress did not move quickly to address the problems of overproduction and low prices, a group of legislators from agricultural states formed a bipartisan "farm bloc" to support legislation beneficial to farmers. This coalition organized in May 1921, with Senators Arthur S. Capper of Kansas, William S. Kenyon of Iowa, Hoke Smith of Georgia, and Congressman Lester J. Dickinson of Iowa, the most important leaders, all arguing that the nation could not be economically strong if agriculture did not prosper.

Although the farm bloc was a short-lived coalition, it achieved important regulatory legislation. In 1921, under farm bloc direction, Congress enacted the Packers and Stockyard Act, which prohibited monopoly and the restraint of trade in the slaughtering business. A year later, Congress approved the Capper-Volstead Cooperative Marketing Act, which quickly became one of the most important federal measures to aid farmers because it exempted agricultural cooperatives from antitrust laws.

The most controversial efforts to involve the federal government in a new direction for agricultural policy began in January 1924 with the introduction of the McNary-Haugen bill. This proposal called for a two-price system in which domestic agricultural prices would rise behind tariff protection while a federal export corporation would sell the surplus production on the international market at world prices. The bill called for the federal government to fund this export program by levying a tax on farmers, an "equalization fee," on each bushel or other unit sold. Ideally, the domestic price would increase to parity levels and enable farmers to earn more, even with an equalization fee.

President Calvin Coolidge opposed the use of the federal government to increase agricultural prices, and with the major agricultural organizations divided, Congress at first refused to approve the McNary-Haugen bill. George N. Peek, president of the Moline Plow Company, who was primarily responsible for the McNary-Haugen plan, now began to mobilize support for this bill. The Coolidge administration argued that the McNary-Haugen bill would increase the cost of living for the American consumer while providing foreigners with cheap food. Many farmers believed the federal government now supported the business, industrial, and labor communities at the expense of American agriculture. For many farmers, the McNary-Haugen bill became a test to determine whether "special privilege" would be given to urbanites and "special discrimination" imposed on the farmer.

In 1927, the McNary-Haugen bill again came before Congress. Southerners who had previously opposed the bill now supported it because a large cotton surplus continued to suppress prices. Although both houses passed the bill, Coolidge vetoed it because the legislation "involved the government in price fixing," would create a "cancerous bureaucracy" to administer the program, and would authorize an unconditional delegation of taxing power to Congress, aiding some farmers and harming others.

Although Congress passed the bill again the next year, Coolidge vetoed it a second time. Frank R. Kent, reporting for the *Baltimore Sun,* wrote that Coolidge's veto message "first hit the McNary-Haugen bill squarely in the nose. Then it kicked it full of holes. Finally, it swept it up in the corner and struck a match." Farmers understood that as long as Coolidge was president, the federal government would not help them improve their economic positions with surplus-control legislation.

Herbert Hoover, who became president in 1929, also opposed

direct governmental intervention in the agricultural marketplace, and when the farm economy failed to improve, he could do little more than recommend that farmers form cooperatives. In 1929, before the stock market crash that ushered in the Great Depression, Hoover supported the Agricultural Marketing Act, which authorized the federal government to loan $500 million to agricultural cooperatives so that these organizations could purchase price-depressing surpluses to keep them off the market. The cooperatives could market those commodities in an orderly manner when prices and demand warranted. The act also authorized the federal government to establish a Federal Farm Board that would supervise agricultural stabilization corporations for grain and cotton. These organizations also had the responsibility of purchasing surplus commodities to keep them off the market and distributing grain and cotton to relief agencies for processing into food and clothing. By 1932, the Federal Farm Board had failed, losing $184 million, and agricultural surpluses and low prices continued to worsen.

In the early 1930s, Hoover's great mistake was not that he did not try to aid agriculture. He did. His administration was the first to pass major legislation designed to aid farmers, and he was the first chief executive to establish an agricultural "action" agency. But he tried too little too late. The nation's farmers were in a desperate economic condition. One Indiana woman spoke for most small-scale family agriculturists when she said, "We are a sick and sorry people."

With the failure of the Hoover administration to aid agriculture adequately, farmers increasingly advocated government intervention to help restore the farm economy so that agriculturists could enjoy a comfortable living. As the crisis worsened, farmers, agricultural groups, and intellectuals began to champion a new way of looking at the farm problem—making the federal government an active supporter of the agricultural economy through a variety of direct aid and regulatory programs. Soon this new governmental approach to agriculture would chart a revolutionary course for relations between the federal government and the farmer, and it would change the nature of agriculture into the twenty-first century.

RURAL LIFE

At the turn of the twentieth century, 60 percent of the population lived in rural areas and would do so until 1920 when the Census Bureau reported that the majority of Americans lived in urban areas, in towns and cities with more than 2,500 people. In many respects, however, life for farmers and other rural inhabitants remained closely linked to the past. Horsepower and manpower predominated. Kerosene lamps, wood and coal stoves, and outhouses prevailed. Seasonal activities such as plowing, planting, and harvesting regulated daily life, and trips to town on Saturdays remained a treat for parents and children. Special social occasions that temporarily broke the work routine and isolation of rural life occurred when the Chautauqua, which provided both entertainment and instruction, came to town. By 1930, however, the Chautauqua succumbed to other forms of entertainment, particularly the automobile, radio, and motion picture.

THRESHING TIME was a social occasion as well as hard work. Often threshing crews composed of men and their sons from nearby farms aided their neighbors with the task. The owner-operator of this steam engine and threshing machine is standing on the right. *Ohio Historical Society.*

Rural life experienced great change during the first three decades of the twentieth century, but it did not come uniformly or quickly to all sections or for all farmers. The use of electricity to make farm and home tasks easier and more efficient provides an example. After the turn of the twentieth century, many scientists were exuberant about electricity's potential as an energy source for farmers. It would make farmers equal to urbanites in access to conveniences, improve the quality of rural life, and help keep young men and women on the farm. Cheaper than human, animal, or steam power, electricity would enable farmers to reduce the number of draft animals and replace forage crops with cash crops.

Electricity, however, was not available to those farmers who wanted it, and fewer still could afford it. The problem was that electricity producers were almost wholly uninterested in serving the rural market, at least until about 1912. In part, the utility companies were busy electrifying cities and large towns. Company officials contended that the volume of closely grouped urban customers reduced costs and the wide variety of city uses provided a steady and profitable demand. Few utility officials believed that farmers needed electricity, and most, with an eye on the bottom line, cringed at the thought of stringing mile upon mile of wire to provide electrical service for a sparse rural population. Consequently, the average farmer had to provide his own generating source by using a steam engine, windmill, or water wheel to power a dynamo. The arrangements were invariably expensive, and most farmers could not afford the initial investment or were unwilling to advance it against uncertain savings or returns.

As a result, farm men and women went about their labors much as their parents had—milking by lantern light, reading beside an oil or gas lamp, cleaning lamps and trimming wicks, carrying fuel in and ashes out, and making daily trips to the windmill for buckets of water. This last task was drudgery for women. Without electricity, farm women had far heavier daily burdens than their urban sisters. Stoking the kitchen stove with wood or coal was labor enough, but cooking over that stove during harvesttime in the sweltering heat of summer approximated the eternal reward the minister promised the unrepentant sinner.

Monday "washdays" meant carrying buckets of water for heating on a stove or outdoor fire and manual scrubbing, wringing, and rinsing. Tuesdays were traditionally for ironing, and the stove-heated "sad" or "flat" irons made that work hot and tiring. It was axiomatic that running water on most farms depended upon how fast some-

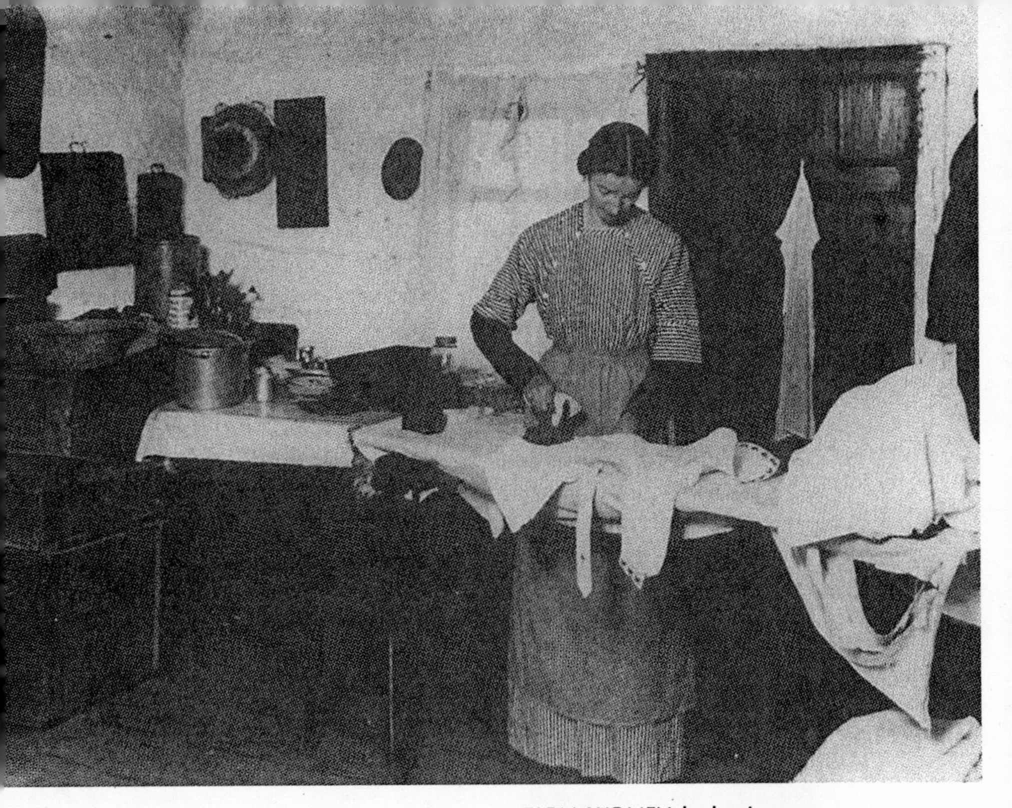

FARM WOMEN had primary responsibilities for maintaining home and family, although many country women also worked in the fields, tended livestock, and conducted other chores. In November 1919, this farm woman attended to the family's ironing near Winchester, Virginia. Note the extra flatiron heating on the stove at the left and the hot pad she used to help grasp the iron. Ironing with a flatiron was hot drudgery when the stove heated the kitchen in the summertime. *U.S. Department of Agriculture.*

one could carry it from the pump to the house. But a 4-gallon bucket filled with water weighs more than 30 pounds, and for proper balance, a carrier needs one in each hand. Little surprise, then, that most farm women were stoop-shouldered before their time. Electricity would not begin to improve rural life on a widespread basis until the 1930s with the aid of the federal government.

In addition to electricity, farmers also wanted rural free delivery of mail. In this wish, often expressed as a demand, they were more successful. Although the Post Office Department inaugurated rural free delivery on a limited trial basis in 1896, by the early twentieth century rural routes had proliferated with such popularity that Congress made it a permanent government service in 1902. Within three years more than 24,000 new rural routes had been established. Rural mail routes meant votes, and congressmen pressured the Post Office to establish them in their districts as quickly as possible.

Although local businessmen often complained that rural delivery hurt business because farmers no longer had to come to town for their mail, farmers loved it because this service brought the world beyond to their doors. Catalogs with items for the barnyard and farmhouse whetted appetites for a host of conveniences and items to improve the standard of living. Newspapers arrived in a more timely fashion to keep the household abreast of national, state, and local events, such as political reform, Prohibition, and consumer protection laws as well as world war, agricultural prices, and the activities of the farm bloc. Rural delivery also brought anticipation, which helped break the drudgery of farm and housework. Letters, newspapers, catalogs, and advertisements were equally welcome. Many rural inhabitants would have agreed with the woman who wrote, "Thank God and Government for rural free delivery."

Rural delivery not only brought the outside world to the farmer's gate but stimulated the good-roads movement in the countryside. Farmers wanted mail delivery and willingly contributed their time to the township road supervisor to help keep the roads in good condition. In 1916, Congress provided funds under the Federal Highway Act to help establish "post roads." With the establishment of rural free delivery and the beginnings of road improvements, farmers also demanded parcel-post service. Although private express companies objected, the farmers prevailed, and Congress approved a general parcel post in 1912. The mail-order catalog companies profited from these developments at the expense of local merchants, as those businessmen had feared. But rural delivery, parcel post, and good roads did more than hurt local businesses. Each contributed to the lure of city life and the drain of young men and women from the farms. Rural delivery, parcel post, and good roads were not the only contributing factors to the twentieth-century exodus from the countryside, but each played an important role.

In addition, the radio profoundly influenced life in rural America. While newspapers and rural free delivery brought the news of the world to the farm, the radio brought the sounds of the world. During the 1920s, the number of radios on the farms increased rapidly. Radios helped mitigate rural isolation, but the practical benefits of the radio were the most important. The radio brought weather reports. Farmers could go to bed knowing what the weather would likely be the next day. This helped them plan their plowing, planting, harvesting, and other work schedules. The radio provided timely market reports, which helped the farmer decide when to sell or hold

BEFORE the widespread adoption of the truck during the 1920s, farmers used sturdy wagons like these built by John Deere to transport grain, cotton, and other products to market. In addition to the pleasure of driving a tandem of new wagons, this farmer had the advantage of a paved road. *Deere & Company Archives.*

his commodities to gain the highest price. Farmers no longer had to take the word of unscrupulous middlemen for information on agricultural prices or wait for the arrival of the daily newspaper. When one radio manager asked his listeners for advice about programming, farmers essentially said, "Cut out the music and give us educational features and market quotations. We must know what our products are worth." For agriculturists, the radio transcended distance and provided an informational equality between buyers and sellers.

Daily programming across the nation featured practical agricultural advice, such as that offered on the NBC-USDA-sponsored "National Farm and Home Hour," or on local broadcasts where farmers received tips on soil conservation, successful farming, and home economics. By 1926, approximately five hundred stations reached more than 1 million farm families nationwide. Businesses such as the Ralston Purina Company and the seed and livestock feed companies also used the radio to reach farmers and offer them purchasing alternatives. Moreover, radios helped pass the time, whether fixing machinery, cleaning the barn, or ironing clothes. Farm men and women quickly came to depend on the radio for company. In 1923, one Missouri farmer wrote to a nearby radio station: "We hill-billies out in the sticks look upon the radio as a blessing direct from God. We farmers are going broke anyway, but we would like to have our radios to sorta ease the pain." Most farmers agreed.

Radios became a necessity in rural America during the 1920s. Farmers bought the set and ran a copper-wire antenna up to the barn roof or windmill from the house. Dry-cell or storage batteries provided the power, but these batteries had to be replaced or recharged frequently. Even so, once farmers bought radios, they would not give them up, even if they suffered bankruptcy during the hard times of the 1920s. Will Rogers astutely reflected on radios and country people: "The only thing that can make us give up our radio is poverty. The old radio is the last thing moved out of the house when the sheriff comes in." The radio changed the lives of those in rural America for all time.

Despite important technological changes in American agriculture, rural life remained unchanged in many respects, especially for women. During the early twentieth century, women proved crucial to the success of farm operations. Childrearing, domestic duties, and food processing, especially canning and pickling, occupied most of their time. Farm women, then as in the past, primarily served as the unpaid hired hand. Their responsibilities for churning and

FARM WOMEN often converted surplus cream into butter with a plunger churn. When they sold their butter to the local country store, they earned important income for the farm family. When this picture was taken during the early twentieth century, most farm women made butter at home only to meet family needs. Factory-made butter at creameries provided the quality controls and standardized product that urban consumers desired. *Library of Congress.*

tending chickens brought important income from butter and eggs. Long hours and hard days throughout every week of the year were an accepted "condition of life." Hard work and isolation remained a way of life for their children.

Women sometimes worked in the fields, particularly African-American sharecroppers or migrant laborers. One black sharecropper reflected that she spent her summers "choppin corn or choppin cotton" or "hoein peanuts." The days were long and hot and offered little more than unending toil. She remembered, "To be in the field hoeing, it was awful to look from one end of the row to the other after the sun gets hot. . . . I'd hate that cotton start to open; pickin cotton was the tiredest thing." With new jobs created in the northern cities during World War I, many black sharecroppers left the rural South for a better life. By 1915, the pull of wartime industry and the push of the boll weevil and poverty had begun the great black migration from the South.

In 1908, President Theodore Roosevelt appointed the Country Life Commission to suggest ways to improve rural life. Chaired by

Liberty Hyde Bailey, a renowned agricultural expert from Cornell University, the commission reported that country people suffered from poor roads, inadequate communication and social isolation. The commission found that young people often left the farm as soon as the opportunity occurred and recommended improving education to make farming more efficient, productive, and profitable, thereby keeping young men and women on the farm and out of the increasingly congested cities. The Progressives, who supported the Country Life movement, believed education would solve the social ills of rural America. Although educational opportunities for rural boys and girls improved during the early twentieth century, economics, not social planning, determined the fate of American agriculture. Profit, or the lack of it, shaped and changed the face of the countryside.

By the early twentieth century, American agriculture had become more varied than ever before. It was characterized by mechanization on an industrial scale in the Great Plains, extensive vegetable fields and orchards worked by migrant labor in the Far West, and sprawling wheat fields in the Pacific Northwest. Dairying and truck farming predominated in the Northeast; corn and livestock feeding in the Midwest. In the South, sharecropping remained a racial and class system for most cotton and tobacco growers. Many farmers abused the soil, and white planters exploited human labor. Farmers often wrung great profits from the land, but with high prices in dislocation, misery, and despair for those who had neither land nor capital to survive hard times.

Despite serious problems, however, American agriculture became more efficient and for a time more profitable than ever before. The first thirty years of the twentieth century were filled with rapid change. Still, distance remained a major hurdle for transportation and communication. Many country people probably would have agreed with a contemporary who reflected that "the end of the neighborhood was almost the end of the world."

By 1930, however, the truck, automobile, radio, and rural free delivery had foreshortened much of the distance and eliminated most of the isolation in rural America. New tractor-powered implements brought industrialization, efficiency, and productivity to the farm. But technology and science also helped bring overproduction, low prices, and increasing costs. These problems drove white and black farmers from the land. In 1923, Secretary of Agriculture Henry C. Wallace wrote that the "drift from the farms to the cities is due in part to inability to make a decent living on the farm." Little did he know that life on the farms would soon get worse. Indeed, American agriculture teetered on the brink of economic collapse. When the crash came, a new government agricultural policy emerged that changed farm life for the remainder of the twentieth century.

SUGGESTED READINGS

Benedict, Murray R. *Farm Policies of the United States, 1790–1950.* New York: Twentieth Century Fund, 1953.

Black, John Donald. *The Rural Economy of New England.* Cambridge, Mass.: Harvard University Press, 1950.

Bogue, Allan G. "Changes in Mechanical and Plant Technology: The Corn Belt, 1910–1940." *Journal of Economic History* 43 (March 1983): 1–25.

Bowers, William L. *The Country Life Movement in America, 1900–1920.* Port Washington, N.Y.: Kennikat Press, 1974.

Brown, William L. *Corn and Its Early Fathers.* Ames: Iowa State University Press, 1988.

Chan, Sucheng. *This Bitter Sweet Soil: The Chinese in California Agriculture, 1860–1910.* Berkeley: University of California Press, 1986.

Danbom, David B. *The Resisted Revolution: Urban America and the Industrialization of Agriculture, 1900–1930.* Ames: Iowa State University Press, 1979.

Daniel, Cletus E. *Bitter Harvest: A History of California Farmworkers, 1870–1941.* Ithaca, N.Y.: Cornell University Press, 1981.

Daniel, Pete. *Breaking the Land: The Transformation of Cotton, Tobacco, and Rice Cultures Since 1880.* Urbana: University of Illinois Press, 1985.

Day, Clarence. *Farming in Maine, 1860–1940.* Orono: University of Maine Press, 1963.

Field, Gregory. "Agricultural Science and the Rise and Decline of Tobacco Agriculture in the Connecticut River Valley." *Historical Journal of Massachusetts* 19 (Summer 1991): 155–74.

Fink, Deborah. *Agrarian Women: Wives and Mothers in Rural Nebraska, 1880–1940.* Chapel Hill: University of North Carolina Press, 1992.

Fite, Gilbert C. *American Farmers: The New Minority.* Bloomington: Indiana University Press, 1981.

———. *Cotton Fields No More: Southern Agriculture, 1865–1980.* Lexington: University of Kentucky Press, 1984.

Fuller, Wayne E. *RFD: The Changing Face of Rural America.* Bloomington: Indiana University Press, 1964.

Gates, Paul W. *History of Public Land Law Development.* Washington, D.C.: Government Printing Office, 1968.

Hamilton, David. *From New Day to New Deal: American Farm Policy from Hoover to Roosevelt, 1928–1933.* Chapel Hill: University of North Carolina Press, 1991.

Hargreaves, Mary W. M. *Dry Farming in the Northern Great Plains, 1900–1925.* Cambridge, Mass.: Harvard University Press, 1957.

Helms, Douglas. "Revision and Reversion: Changing Cultural Control Practices for the Cotton Boll Weevil." *Agricultural History* 54 (January 1980): 108–25.

Hurt, R. Douglas. *Agricultural Technology in the Twentieth Century.* Manhattan, Kans.: Sunflower University Press, 1991.

______. *The Dust Bowl: An Agricultural and Social History.* Chicago: Nelson-Hall, 1981.

Isern, Thomas D. *Bull Threshers and Bindlestiffs: Harvesting and Threshing on the North American Plains.* Lawrence: University Press of Kansas, 1990.

Jelnick, Lawrence J. *Harvest Empire: A History of California Agriculture.* San Francisco: Boyd and Fraser, 1982.

Kirby, Jack Temple. *Rural Worlds Lost: The American South, 1920–1960.* Baton Rouge: Louisiana State University Press, 1987.

Liebman, Ellen. *California Farmland: A History of Large Agricultural Holdings.* Totowa, N.J.: Rowman and Allenheld, 1983.

McMurray, Linda O. *George Washington Carver: Scientists and Symbol.* New York: Oxford University Press, 1981.

Malone, Michael P., and Richard W. Etulian. *The American West: A Twentieth Century History.* Lincoln: University of Nebraska Press, 1989.

Miller, John G. *The Black Patch War.* Chapel Hill: University of North Carolina Press, 1936.

Morlan, Robert L. *Political Prairie Fire: The Nonpartisan League, 1915–1922.* Minneapolis: University of Minnesota Press, 1955.

Nelson, Paula M. *After the West Was Won: Homesteaders and Town-Builders in Western South Dakota, 1900–1917.* Iowa City: University of Iowa Press, 1986.

Pisani, Donald J. *From the Family Farm to Agribusiness: The Irrigation Crusade in the West, 1850–1931.* Berkeley: University of California Press, 1984.

Rasmussen, Wayne D. *Taking the University to the People: Seventy-five Years of Cooperative Extension.* Ames: Iowa State University Press, 1989.

Reisler, Mark. *By the Sweat of Their Brow: Mexican Immigrant Labor in the United States, 1900–1940.* Westport, Conn.: Greenwood Press, 1976.

Rikoon, J. Sanford. *Threshing in the Midwest, 1820–1940: A Study of Traditional Culture and Technological Change.* Bloomington: Indiana University Press, 1988.

Russell, Howard W. *A Long Deep Furrow: Three Centuries of Farming in New England.* Hanover, N.H.: University Press of New England, 1982.

Saloutos, Theodore. *Farmer Movements in the South, 1865–1933.* Berkeley: University of California Press, 1960.

Saloutos, Theodore, and John D. Hicks. *Twentieth-Century Populism Agricultural Discontent in the Middle West, 1900–1939*. Madison: University of Wisconsin Press, 1951.

Scott, Roy V. *The Reluctant Farmer: The Rise of Agricultural Extension to 1914*. Urbana: University of Illinois Press, 1970.

Shideler, James H. *Farm Crisis, 1919–1923*. Berkeley: University of California Press, 1957.

Shover, John L. *Cornbelt Rebellion: The Farmers' Holiday Association*. Urbana: University of Illinois Press, 1965.

Street, James H. *The New Revolution in the Cotton Economy*. Chapel Hill: University of North Carolina Press, 1957.

Waldrep, Christopher. "Planters and the Planters' Protective Association in Kentucky and Tennessee." *Journal of Southern History* 52 (November 1986): 565–88.

______. "The Reorganization of the Tobacco Industry and Its Impact on Tobacco Growers in Kentucky and Tennessee, 1900–1911." *Mid-America* 73 (January 1991): 71–81.

Wessel, Thomas, and Marilyn Wessel. *4-H: An American Idea, 1900–1980*. Chevy Chase, Md.: National 4-H Council, 1982.

Wik, Reynold M. "The Radio in Rural America During the 1920s." *Agricultural History* 55 (October 1981): 339–50.

______. *Steam Power on the American Farm*. Philadelphia: University of Pennsylvania Press, 1953.

______. "The USDA and the Development of Radio in Rural America." *Agricultural History* 62 (Spring 1988): 177–88.

Williams, Robert C. *Fordson, Farmall, and Poppin' Johnny: A History of the Farm Tractor and Its Impact on America*. Urbana: University of Illinois Press, 1987.

Wilson, Harold Fisher. *The Hill Country of Northern New England: Its Social and Economic History, 1790–1930*. New York: Columbia University Press, 1936.

VETERINARY MEDICINE

Until the late nineteenth century, many farmers charged that the practitioners of veterinary medicine were "wholly unqualified." Since illness and injury to livestock were among the most ruinous problems confronting farmers, they often needed professional help, and they did not get it. Injuries, particularly to draft animals, as well as various respiratory, digestive, reproductive, and parasitical diseases were common. Periodic virulent outbreaks sometimes wiped out a locality's entire population of horses, cattle, hogs, sheep, or poultry. In the absence of qualified veterinarians, farmers were forced to doctor livestock themselves or to hire others who claimed the requisite skill.

When a farmer treated ill or injured livestock himself, he relied upon a combination of traditional practices and his own ability—an amalgam of superstition, personal experience, and common sense. Articles in agricultural periodicals and books published throughout the nineteenth century were of some help. Farm magazines and agricultural how-to books were as popular then as today, and most publications contained sections that described animal diseases and injuries and instructed farmers on the best methods for curing or healing them. Still, do-it-yourself animal doctoring left much to be desired. Farriers and blacksmiths often treated various equine diseases.

The limited ability of most farmers to diagnose illness, improve health, or even ease the suffering of livestock frequently induced them to call in a "hoss doctor," "cow leech," or "veterinary surgeon." Like the farmer, these practioners based their ability to cure livestock on traditional practices, natural ability, and experience. Although few of them had received formal education, farmers often sought their aid. Livestock investments usually were substantial, and for the small farmer, the loss of one or two animals was as disastrous as that of a large operator whose herds or flocks were decimated by an epidemic. No matter how many head of livestock a farmer owned, losses due to disease or injury—whether from accident, negligence, or cruelty—were serious.

Unfortunately, livestock under the care of untrained, self-styled animal doctors often suffered when inflicted with common nineteenth-century treatments. Hoss doctors, who believed that bleeding, burning, and blistering would cure most illnesses and diseases, were quick to lance a vein on a horse's neck and take a quart or two of blood to treat blind staggers, colic, or dysentery. The technique sometimes so weakened the animal that its death became a certainty. Although medical doctors seldom practiced this once-popular treatment on humans by the

EQUINE ORAL SURGERY, 1897. *Special Collections, William Robert and Ellen Sorge Parks Library, Iowa State University.*

end of the Civil War, most horses bore scars from bleeding until late in the nineteenth century. Hoss doctors also commonly used red-hot irons to burn or blister an animal. This practice, "firing," was often used to treat mouth and bone ailments. Less torturous treatments hoss doctors applied to the dumb if not silent patients included applying a wad of wet hay to a bruised back, pouring a mixture of brandy and salt into the ears of a horse with the staggers, and putting a frog (minus its left leg) with a mixture of salt and ale into a horse's mouth to eliminate blood in the urine.

Cow leeches, practioners who specialized in the treatment of cattle, employed many of the same treatments as hoss doctors, particularly bleeding. Cow leeches commonly believed illness stemmed from one of two major causes: "hollow-horn" or "wolf-tail." Hollow-horn supposedly caused problems in the front of the body; wolf-tail afflicted the hind portion. When a cow became ill, the leech examined its horns by drilling through the base. If the horn was hollow (leeches believed that healthy animals had solid horns), the diagnosis was complete. Although the drilling was painful, the treatment was usually worse—turpentine poured into the hole. If this did not effect a cure, the solvent was then rubbed along the animal's back from horns to tail to produce a solid blister, which cow leeches believed had a healing effect.

An unseen worm, a "wolf," supposedly caused a condition known as wolf-tail or tail-ail. Cow leeches contended that the worm ate some of the bones in the tail, thereby making it more flexible and the animal ill. The remedy was to split the tail near the end with a sharp knife and remove the portion that allegedly had been eaten away. After several inches of tail had been amputated, the leech rubbed red pepper and salt into the incision to prevent the return of the worm. If the animal lived, the leech professed that the parasite had been removed before it reached the spine. If the animal died, the leech contended that he had not been called soon enough to save the animal. Farmers tolerated, even requested, these cruel treatments for their livestock because they had no alternative.

Agriculturists considered themselves particularly fortunate to have the services of a "veterinary surgeon." Whereas hoss doctors and cow leeches made no pretense of having professional education in veterinary science, animal doctors who appended the abbreviation V.S. behind their names claimed to have had formal training in veterinary medicine from a professional school. Before the 1850s, the only professionally trained veterinarians had been educated in Europe. Before and even after the founding of several private veterinary colleges in the United States during the mid-nineteenth century, many animal doctors who claimed to hold degrees in veterinary medicine had little or no training.

Some veterinary colleges were diploma mills that admitted students with no more than an eighth-grade education. These schools frequently did not require students to attend lectures and graduated two classes annually. Veterinary colleges like these existed simply to make money from tuition and diploma fees; any contribution to the effective, humane treatment of animals was secondary to a transaction in which students exchanged money for diplomas. Veterinary surgeons learned little about scientific, humane treatment of animals. They considered, for example, that surgery was humane because animals had little feeling. Veterinary surgeons commonly treated swollen sinuses by drilling into those cavities and injecting a gallon of tepid water containing zinc chloride. These animal doctors considered the sinuses a storehouse of disease that sometimes needed to be opened and cleansed.

Armed with a diploma and a few books, the veterinary surgeon set out to turn a profit and perhaps to aid farmers. Some used their purported professional training to win business away from local hoss doctors and cow leeches. Others embarked upon an itinerant career not unlike that of the patent medicine men who catered to the human population throughout rural America. Itinerant veterinarians usually advertised their services in agricultural periodicals whose editors hesitated to lay charges of quackery and lose needed income. These traveling veterinary surgeons quickly descended on any area afflicted with epidemic disease, such as the outbreaks of Texas fever, cholera, and pleuropneumonia that occurred in the 1870s and 1880s. They were apt to be long gone by the time a farmer realized that the remedy for which he had paid did not provide the promised cure.

Professional veterinary medicine based upon science lagged in the United States, in part because veterinary education developed slowly in Great Britain, a country that continued to influence Americans. Most important, however, professional veterinary education lagged because of rural suspicion and government neglect. Rural conservatism encouraged farmers to continue traditional agricultural practices, and the economic and political philosophies of the day did not place the problem in the sphere of government concern. Substantive improvement in the treatment of livestock did not come until 1852 with the founding of the Veterinary College of Philadelphia. Other private colleges soon followed, but most private veterinary schools did not survive beyond 1920.

Two major reasons for the demise of private veterinary schools were (1) slowly advancing scientifically based instruction at public colleges and universities and (2) increasing regulation by state and federal governments. An additional factor was the early-twentieth-century replacement of urban horse transport by the internal-combustion engine. The graduates of private schools had found their most lucrative business among city horse owners. Although the best of these schools played a role in improving animal health, they failed to meet the broadest expectations of stock raisers and could not compete with publicly supported institutions.

The great breakthrough in veterinary medicine came with passage of the Morrill Act in 1862. This legislation provided for the creation of a land-grant university system that emphasized agricultural education, including veterinary medicine. By the end of the nineteenth century, twenty-three agricultural schools employed professors who taught veterinary medicine. More important, the universities established the research programs necessary to advance the field on a scientific basis.

The salient feature of the new education in veterinary medicine departed radically from the system of the entrepreneurial institutes and private colleges where students learned their profession by observing their instructors in private practice. Veterinary students in the universities gained knowledge from classroom study and laboratory experience. Anatomy became part of the curriculum. The knowledge gained through dissection, lectures, and reading was supplemented with clinical experience. They also kept abreast of medical research. By the turn of the twentieth century, veterinary medicine was far more advanced than it had been a century before, but progress was not universal. Only four institutions—Iowa State Agricultural College, Ohio State University, Cornell University, and the State College of Washington—provided complete schools of veterinary medicine.

At the same time that veterinary education was improving in both public and private colleges during the last quarter of the nineteenth century, the federal government began to show a belated interest in the improvement of veterinary medicine. Sudden and unaccountable outbreaks of disease, including hog cholera, pleuropneumonia, tuberculosis, brucellosis, Texas fever, anthrax, and blackleg, brought demands that the U.S. Department of Agriculture discover the causes and provide immediate remedies and long-range protection. The department, however, gave little attention (beyond statistical study) to the problems of animal diseases until the early 1880s. By that time the livestock population was increasing rapidly. Farmers had made greater investments in purebred or improved livestock than ever before, with a resultant concern about the animals' health. More and better livestock brought a heightened interest in improved veterinary procedures.

In 1883 the USDA responded to the needs of farmers by creating a Veterinary Division, which became the Bureau of Animal Industry the following year. Thereafter, USDA veterinarians gave increased attention to controlling and eradicating contagious animal diseases. These governmental veterinarians, veterinarians at state experiment stations, and those at the universities contributed to the increase of knowledge about veterinary medicine and improved care and treatment of farm and household animals.

By the late 1890s the patterns of twentieth-century veterinary medicine had been established. Agricultural colleges and universities had significantly increased the admission and graduation standards for veterinary students; the days of the diploma mills and other private schools were numbered. By the turn of the century, state and national veterinary associations also helped improve the standards for education and practice. These organizations worked long and hard to impose higher standards for licensing and to prevent the poorly trained or outright quack from practicing veterinary medicine.

By 1900 veterinarians had taken major steps by placing their profession on the firm foundation of scientific inquiry and formal education. The day of hoss doctors, cow leeches, and veterinary surgeons had nearly passed. By 1920 farmers were no longer dependent on the questionable skills of uneducated, self-proclaimed veterinarians. Both livestock and owners profited. Livestock on the farms and animals in the cities and towns were mute beneficiaries of the victory of science over superstition.

7 • TROUBLED TIMES

AMERICAN FARMERS ENDURED THE WORST ECONOMIC DEPRESSION and enjoyed the greatest prosperity many had ever known during the quarter century following 1930. They also experienced extraordinary change in science, technology, and government policy. Farm prices collapsed by late 1932, cotton and hogs reaching $0.03 per pound, wheat $0.12 per bushel, and cattle $4.14 per hundredweight on some markets. With this decline, a new activist administration interceded for the first time on a massive scale in the agricultural economy and revolutionized the relation between the farm and the federal government. This program did not, however, solve the farmer's economic difficulties.

World War II ended the problems of surplus production, low prices, and overpopulation in agriculture. When the war concluded, however, the problems of overproduction and inadequate prices reemerged to trouble farmers and the federal government. Even so, government agricultural policy had stretched a safety net beneath the most efficient, capital-intensive, and large-scale farmers. Federal agricultural policy did not protect all family farmers from failure during these traumatic years, but it established new programs from which neither the federal government nor farmers were willing or able to deviate for the remainder of the twentieth century.

FEDERAL POLICY

In 1932, the election of Franklin Delano Roosevelt to the presidency and overwhelming Democratic majorities to Congress brought a revolutionary change in agricultural policy. The Roosevelt administration did not believe that Herbert Hoover's policy of modest support for cooperatives and voluntary production controls was sufficient to meet the farm crisis. The administration moved quickly to gain congressional approval for a host of regulatory programs that would involve the federal government in agricultural planning, production, marketing, and financial support. The domestic allotment program became the most important feature of this new direction in federal policy. Passed by Congress on May 12, 1933, the Agricultural Adjustment Act was designed as a "farm relief" measure to restore parity purchasing power.

The Agricultural Adjustment Act created the Agricultural Adjustment Administration (AAA), which had the responsibility of paying farmers to limit the production of seven basic commodities—wheat, cotton, corn, hogs, rice, tobacco, and dairy products. Farmers produced all these commodities in great surplus, and the price of each affected other prices, such as those of beef, bread, and clothing. The administration operated on the premise that the regulation of these commodities would be easier because each needed some form of processing before it could be used by the consumer. The AAA imposed a tax on the processors of these commodities to support an acreage-reduction program. By paying farmers to reduce acreage and production in these basic commodities, the federal government hoped to reduce the surpluses and increase prices to the levels based on prices for the period 1909–1914, except for tobacco, which had parity prices indexed to 1919–1929. In addition, direct federal payments to farmers for reducing their production would put needed money into their households and help prevent bankruptcy, increase purchasing power, and keep farmers on the land. The county extension agents and local farmer committees, often organized with the help of the Farm Bureau, administered the program.

Most hog producers chose to participate in the reduction program. But because their sows had had their spring litters by the time Congress approved the AAA, the pork surplus would not decrease without drastic measures. Similarly, the cotton crop had been planted by May 1933, and the possibility of another large harvest in the autumn called for emergency action. The AAA decided to solve the pork-surplus problem by paying farmers to slaughter 6 million hogs. The agency also contracted with 1 million farmers to plow under 10.4 million acres of cotton, for which they received $112 million. Neither the pig kill nor the cotton plow-up occurred in subsequent AAA programs; each was an emergency policy. Drought made the necessary acreage reductions for wheat farmers in the Great Plains.

HENRY A. WALLACE

Special Collections, William Robert and Ellen Sorge Parks Library, Iowa State University.

Henry A. Wallace (October 7, 1888–November 18, 1965) made an indelible mark on American history as secretary of agriculture during the New Deal years of Franklin D. Roosevelt's administration. His influence as vice president from 1941 to 1945 and his foray into presidential politics as the nominee of the Progressive party in 1948 contributed to his public and political image, about which few people could be neutral. Wallace's contributions to agriculture beyond the political arena, however, were substantial. From 1921 to 1933, he served as the editor of the popular and influential *Wallaces' Farmer.* In 1926, he founded the Hi-Bred Corn Company. Nine years later it became the Pioneer Hi-Bred Corn Company and eventually the largest hybrid seed company in the United States. Wallace's career in agriculture and politics came naturally, considering his family lineage, but he also had exceptional talent, a keen mind, remarkable ability at observation, and a personal touch that opened many doors.

Born on a farm near Orient, Iowa, Wallace was the son of Henry C. Wallace, who served as secretary of agriculture from 1921 until 1924, mostly in the administration of Warren G. Harding. He was the grandson of the Henry Wallace who first edited the prestigious *Wallaces' Farmer.* After his family moved to Ames for his father to attend Iowa State College, young Henry developed an interest in plant breeding, and the family's friendship with George Washington Carver enabled him to accompany Carver occasionally on his botany expeditions. As a high school student, Wallace experimented with corn breeding to show that the appearance of seed could not indicate productivity.

During the farm depression of the 1920s, as the senior editor of *Wallaces' Farmer,* Wallace urged agriculturists to mechanize to improve efficiency and increase profits. He also favored government-initiated price and production controls to reduce surpluses and enable farmers to earn a satisfactory living. His advocacy of production controls through acreage reductions won the attention of Franklin Roosevelt and the position of secretary of agriculture. Wallace became secretary of agriculture in March 1933 and quickly marshaled the New Deal agricultural programs. He supported the Agricultural Adjustment Administration, which he called a new "social machine" for the benefit of farmers and for which he was both praised and condemned.

Wallace was a strong supporter of rural reform that would ensure the general welfare of the family farm, but he has received great criticism for ignoring the needs of sharecroppers and agricultural workers. Despite this shortcoming, Wallace understood the problems of farming better than any secretary of agriculture before him. Wallace was ambitious, idealistic, and outspoken, and he believed the federal government had an obligation to improve the lives of farmers while restraining the interests of agribusiness that were often opposed to the welfare of the family farmer. Without question, Wallace was the most visible, controversial, and important secretary of agriculture in American history. He served in that position until 1940. The next year he became Roosevelt's vice president.

The farmers who participated in the AAA program could not afford to do otherwise. Although they preferred to plant as much as possible, commodity prices had reached such low levels that AAA checks were essential for their livelihood. Indeed, AAA payments became the chief source of income for many. Although most farmers who participated in the AAA program welcomed their checks, this new policy was not without critics, who called it a "policy of scarcity." Some complained that federal policy should not permit killing little pigs merely to reduce the pork surplus and increase prices when many people were going hungry. America faced the perplexing "paradox of want in the midst of plenty."

Henry A. Wallace, secretary of agriculture, replied that it was no more immoral to kill little pigs than full-grown hogs and that farmers could not run an "old folks home for hogs and keep them around indefinitely as barnyard pets." Yet Wallace recognized the essential senselessness of destroying food in time of need. To mollify the opposition to the AAA program, he pledged that the government would purchase agricultural commodities "from those who had too much in order to give to those who had too little." Only by doing so could such a program be justified to force agricultural adjustment.

Other opponents of the AAA charged that the program gave preference to large-scale farmers because they had more land to remove from production and that the sharecroppers had been forgotten. The processors not only complained about being taxed to support the program but challenged the constitutionality of the AAA. In January 1936, the U.S. Supreme Court ruled that the processing tax was an unconstitutional extension of congressional taxing power because it permitted "the expropriation of money from one group for the benefit of another."

Quickly the Roosevelt administration and Congress revised the program with the Soil Conservation and Domestic Allotment Act. This legislation authorized the AAA to pay farmers for soil conservation rather than acreage reduction. The AAA now paid farmers to plant "soil-building" or conserving rather than "soil-exhausting" crops on acreage that had been devoted to crops in surplus, such as wheat, cotton, and tobacco. Payments for this form of crop limitation came from general revenue rather than a special tax on processors. Congress then institutionalized this program with the Agricultural Adjustment Act of 1938, which not only permitted government payments for acreage reduction but also provided crop insurance, marketing controls, and price-supporting loans. The legislation of 1938 remains the foundation for American agricultural policy on price-support and acreage-reduction programs.

The Roosevelt administration also provided nonrecourse loans on corn, cotton, and other commodities through the Commodity Credit Corporation (CCC), created in October 1933. Essentially, the CCC was the subtreasury that the Farmers' Alliance and People's party had advocated during the late nineteenth century. The CCC enabled farmers to receive price-supporting loans on harvested crops at 4 percent interest. In 1933, the CCC loaned cotton farmers money based on the price of $0.10 per pound for their crop. If prices advanced above the loan rate, the farmer could sell his crop, pay off the loan, and keep the profit. If prices fell below the loan rate, farmers forfeited their crop to the federal government, and the debt was canceled. All farmers who applied for a CCC loan had to sign production-control agreements. With the loan rate on commodities slightly higher than market prices, CCC loans served as a price-support program. Thereafter, CCC loans became the primary means of supporting farm prices, and the program expanded rapidly after World War II.

In 1933, Congress strengthened the federally financed agricultural credit system with the Farm Credit Act. This legislation authorized farmer-owned Production Credit Associations to make short- and intermediate-term loans. It also created a banking system to support the marketing, supply, and service operations of cooperatives. Farm Credit Banks, together with the Farmers' Home Administration and the Commodity Credit Corporation, quickly became the primary federal lenders for the agricultural community.

In 1935, the federal government moved the Soil Erosion Service from the Department of the Interior to the USDA and rechristened it the Soil Conservation Service (SCS). The SCS had the responsibility of helping farmers conserve their land and loaned money to help them begin soil-conservation practices such as terracing, reseeding native grasses, and contour plowing. The SCS worked closely with the states and counties to establish soil-conservation districts to support the best methods to conserve and restore soil, particularly in eroded areas such as the Great Plains. On the eve of World War II, the SCS had identified more than 200 million acres subject to serious erosion. When the war began, the problems of soil conservation remained far from solved.

The Roosevelt administration not only sharply departed from the past by making direct payments to farmers to curtail production, supporting prices, and promoting soil-conservation practices but also engaged in unprecedented government planning for the family farmer. In 1935, the Resettlement Administration (RA), followed by the Farm Security Administration (FSA) two years later, provided "rehabilitation loans" to help farmers meet daily expenses and remain on the land.

To qualify for an RA or FSA loan, a farmer had to submit his operation and family to an agency investigation. The agency inquired about the nature of his lease if he was not an owner and determined whether the farm was economically viable, with adequate tools, implements, and livestock. The government also wanted to know about the farmer's soil-conservation work, whether he faced ruin from one-crop agriculture, and whether he and his wife limited expenses by raising their own food and made the best use of the land. Upon approval of the loan, the FSA officer worked out a plan for farm operations and expenditures. This form of state planning imposed an unprecedented dependence upon the farmer. Even so, the Resettlement Administration and Farm Security Administration became the last hope of farmers who could not obtain loans or credit from banks or other financial institutions.

After the presidential election of 1932, public power advocates for rural America gained a sympathetic ear. In 1933, their pleas for cheap, extensive electric service brought action when Congress created the Tennessee Valley Authority (TVA) and President Roosevelt authorized the Electric Home and Farm Authority (EHFA). The TVA and the EHFA provided rural citizens of Mississippi, Alabama, Georgia, and Tennessee with power and the financial ability to use it effectively. The TVA developed a government-controlled hydroelectric system designed to produce electricity at rates substantially below those of private industry and to encourage cooperatives; the EHFA enabled farmers and rural residents to purchase electric ranges, refrigerators, water heaters, and other appliances. The EHFA paid the appliance dealers for the purchases, then sent low monthly billings to the customer. With an interest rate of 5 percent, half that of commercial lending institutions, these loans were within the means of many farmers, and appliance sales jumped 300 percent in the Tennessee River basin. More important, the success of the TVA and the EHFA demonstrated the effectiveness of the federal government as a provider of cheap power and better quality of rural life.

With the success of the TVA and EHFA, New Dealers began to work for the expansion of public power to all rural areas. On May 11, 1935, President Roosevelt created the Rural Electrification Administration (REA) to begin the work of providing electricity to rural America. The private utility companies refused to participate in this project, claiming that few farmers needed electricity and that rural electrification was a "social" rather than an "economic" problem. Ultimately the REA resorted to loaning money to rural cooperatives to finance stringing power lines, wiring homes, and purchasing or generating electricity. The long tradition of cooperative efforts among farmers enabled the success of this program.

To acquire funds, the local cooperative had to prove its economic viability and guarantee "area coverage"; that is, all farmers had to be provided with service, thereby equalizing costs within a compact area. Area coverage also allowed the application of labor-efficient techniques for line construction. By 1938, REA lines had become a common feature. When the year ended, more than 350 REA projects in forty-five states provided electric service to nearly 1.5

million people and approximately 40 percent of the nation's farms. The REA motto became "If you put a light on every farm, you put a light in every heart." It was corny but appropriate, especially for farm wives and appliance dealers, who were the first to benefit from rural electrification. The REA was one of the most successful New Deal agencies by the time President Roosevelt transferred it to the Department of Agriculture in 1939 to streamline administration.

New Deal agricultural policy, then, substantially changed the relationship between the farmer and the federal government, probably for all time. It placed the federal government on a new course of intervention and regulation. Thereafter, farmers increasingly demanded government aid, chiefly in the form of commodity loans and price supports to enhance their chances of (if not guarantee) economic survival. Still, New Deal farm policy was not without criticism, the most important contending that it catered to large-scale at the expense of the small-scale farmers. Certainly the federal government was most concerned with helping those farmers who had the best chance to survive, and many poor farmers necessarily gave up and left the land.

Without question, inefficiency, inadequacy, and insensitivity often characterized New Deal agricultural policy, which did not end rural poverty. World War II revived the farm economy by increasing the demand for food and fiber, boosting prices, and providing industrial jobs for farmers no longer needed on the land. Even so, New Deal agricultural policy proved a great social and economic experiment that provided more benefits to farmers than would otherwise have been possible. Moreover, New Deal agricultural policy established the foundation for all farm programs that came thereafter. Both farmers and politicians have been unwilling to depart from the security and stability it provided.

THE SOUTH

The Great Depression made life immeasurably worse for southern farmers. One southerner observed that they lived in "a world in which survival depended on raw courage, a courage born out of desperation and sustained by a lack of alternatives." In 1930, cotton

SHARECROPPING kept many southern farmers, black and white, in dire poverty. During the 1930s, New Deal agricultural programs caused a revolution in the use of southern agricultural land. Planters who owned extensive lands decreased cotton and tobacco acreage. With less acreage under cultivation, they released many sharecroppers, who often did not want to leave the land, even though they lived in abject poverty. In 1941, these sharecroppers maintained a tenuous hold on the land in Green County, Georgia, where they harvested their potato crop in the autumn for home use. World War II would take most of the remaining sharecroppers from the land. *Southern Historical Collection, The University of North Carolina at Chapel Hill.*

prices averaged about $0.10 per pound and an index of 58 based on parity prices at 100 for the period 1909–1914. In 1931, a near-record-breaking crop dropped prices by half. Tobacco, rice, and other crops also suffered severe price declines, but operating costs and taxes did not fall as quickly, and farmers complained about the "cost-price squeeze." Drought made matters worse, and many southerners who had moved to the cities for employment and found none compounded the agricultural problem by returning to their rural communities. In 1931, James Edward Rice, writing for the *New Republic*, reported that tenant farmers did not have enough to eat. "And," he wrote, "with hunger growing at their vitals they plow in earnest, because they are in a desperate situation and they exist in terrible anxiety. So they plow hard." With approximately 5.5 million whites and 3 million blacks working as tenants and sharecroppers by 1935, poverty reigned supreme.

The demands for aid overwhelmed local relief agencies, and southerners uncharacteristically looked to the federal government for assistance. It came with the Agricultural Adjustment Act and more than two dozen other agencies designed to aid the farmer. The New Deal agricultural program would in many respects do all of that, but it also revolutionized the credit system of the past, destroyed the institution of sharecropping, and substituted the federal government for the landlord and furnishing merchant in the lives of farmers across the South.

In the South, the AAA program applied primarily to cotton, rice, and tobacco. Depending on the past productively of cotton acreage, for example, the federal government paid farmers between $7 and $20 per acre to reduce their cotton acreage. Cotton farmers were desperate for money, and their cooperation with the AAA ranged from a low of 50 to a high of 79 percent of cotton farmers in North Carolina and Louisiana, respectively. Although the AAA intended for landlords to share the check for acreage reduction with their tenants and sharecroppers as a "moral obligation," usually on the same basis that they shared the harvested crop, few received any cash from the plow-up in 1933. To remedy this problem, the AAA required participating landowners to distribute acreage reductions equally among their tenants based on their cropping arrangements, let them remain on the land if they were no longer needed, and let them raise food and livestock and cut firewood free of charge. In return, the Federal Emergency Relief Administration would assume the role of the furnishing merchant by providing relief to jobless sharecroppers and tenants the landlords released when they no longer needed them.

Sharecroppers and tenants received virtually nothing from the AAA except further hardship, often in the form of eviction.

The unintended consequences of the AAA drove the sharecroppers from the land, often to nearby towns and cities and onto the relief rolls. In February 1935, Harry R. Fraser, a reporter for the Scripps-Howard newspapers, wrote that the thousands of sharecroppers along the highways in the South were "lonely figures, without money, without homes, and without hope." Their condition did not soon improve. The migration from the countryside that had begun with World War I continued. Whites and blacks moved north to the industrial cities. Many migrants to California from Oklahoma, Arkansas, and Texas encountered an industrialized agriculture and poverty similar to what they had left behind.

For the landowners who stayed in the South, the economic benefits of federal agricultural policy permitted a host of improvements. These landlords systematically took the AAA checks and released their sharecroppers. They consolidated tenant holdings into unified farms, ripped out fencerows, and bought tractors and other mechanized equipment. With fewer acres under cultivation and tractors and other implements replacing human labor, most of the tenants and sharecroppers no longer had a place in southern agriculture. Owners tended to keep the best tenants, if they needed any, or employed some of those they had released as day laborers to chop and pick cotton. By 1940, 30 percent of the sharecroppers and 12 percent of the tenants had left farming in the thirteen cotton states during the 1930s.

Increasingly, financial institutions like banks and insurance companies gained possession of great portions of southern farmland, owning 30 percent of cotton land in 1934. Indeed, financial institutions, land companies, and large-scale plantations profited most from the AAA program. Between 1933 and 1935, the Delta and Pine Land Company received $318,000 in AAA funds. In 1934, the Connecticut General Life Insurance Company earned $35,000 from 179 cotton farms under AAA contracts for acreage reduction, and the next year the Louisiana Irrigation and Mill Company received $73,659. The federal government did not impose limits on benefit earnings to gain the support of the large-scale farmers whose participation it needed to make acreage reductions work. Overall, most cotton farmers received less than $200 per year in AAA benefits, but these funds often meant the difference between survival and flight from the land.

African-American farmers fared even worse. Locally administered relief agencies systematically discriminated against blacks. Most planters also opposed government aid for blacks, especially work relief, because that assistance might drive up labor costs. In 1933, Governor Eugene Talmadge of Georgia complained that federal relief was "utterly ruining" black farmers. Often southern relief offices, administered by planters, closed during cotton-picking time to force unemployed and displaced African Americans back into the fields as cheap day laborers.

The Second World War accentuated the changes in southern agriculture that had begun with the New Deal. Military and industrial demands substantially increased commodity prices for cotton, peanuts, soybeans, and tobacco and encouraged diversification. In 1941, with the United States aiding Great Britain in a new war against Germany and with the Japanese threatening in the Pacific, Claude A. Wickard, secretary of agriculture, told southern farmers: "There's no sense in continuing to pile up cotton. The country needs milk and eggs and meat a lot worse." Moreover, when the war took southern men and women to the cities or abroad, they enjoyed greater incomes and a higher standard of living than ever before. Few chose to return to southern tenant farms when the war ended, and those who came back demanded modern conveniences such as tractors and electricity. They accepted the old sharecropping system only as a last resort.

Between 1940 and 1945, the migration from the South accelerated, and the farm population decreased 20.4 percent. White farmers, especially, left the land during World War II, which contributed in part to the region's urban population increase of 36 percent. Approximately 30 percent of the 12 million men and women who served in the armed forces were southerners. As southern farmers moved from the countryside, landowners increasingly turned to wage labor and mechanization. The number of tractors on southern farms doubled during the 1930s and nearly doubled again by 1945 when 407,400 implements, or 16.8 percent of the tractors in the nation, plied southern lands. But tractors did not so much cause the flight from the land as follow it. Those who left departed because government programs made many farmers superfluous.

World War II, however, did not benefit all southern farmers equally. The large-scale planters took the greatest advantage of increased demand and higher prices. In 1944, one Mississippi Delta planter reported, "I have twenty tractors now and will continue to mechanize. We are not buying any more mules." Yet that same year some Mississippi planters kept African-American sharecroppers locked in a state of peonage by using local law-enforcement officers to prevent them from moving away in search of higher-paying jobs and breaking their labor contracts. They also influenced draft boards to defer their best male sharecroppers. Overall, however, farm ownership, consolidation, and mechanization increased during the war years, and the landowners who remained on their farms usually prospered. Still, compared to farmers in other regions, southerners remained poor, the tenants and sharecroppers who remained clearly the "farmers left behind."

THE NORTH

In New England, the agricultural population continued to decline as farmers abandoned the land. Between 1930 and 1950, the number of New England farms decreased from 103,255 to 21,670. Only the continued migration from Oklahoma, Texas, Arkansas, and Louisiana exceeded the flight from rural New England. Many of the small-scale farms that remained were highly capitalized and devoted to high-value specialty production such as dairying and poultry raising. To support dairying, farmers devoted most of the cropland to hay. Poultry farmers raised little feed because they could buy it cheaply from agriculturists in the Midwest.

In New England, part-time farming increased, but many farmers preferred to maintain some agricultural production even though urban jobs paid more money. Rather than leave the farm entirely, farmers often held part-time jobs in town, usually in service occupations, to support their families or finance their agricultural operations. By 1950, part-time farmers ranged from 31 percent in Vermont to 52 percent in New Hampshire, a regional average of 42 percent. The federal government defined part-time farming at this time as an operation where a farmer worked off the land for one hundred or more days per year or nonfarm income exceeded the value of the products sold from the farm.

In the Midwest, farmers continued to emphasize corn, which reached the market primarily as pork and beef. With escalating operating costs, however, young farmers had increasing difficulty following the occupations of their parents. When land prices reached $200 per acre or more; young men and women required considerable capital or heavy indebtedness to begin farming; $80,000 might be needed to purchase land, equipment, livestock, and seed. Few could acquire that sum or had security to gain sufficient financing. Indeed, by the mid-twentieth century, most young people did not have much opportunity to become farmers unless they inherited land or received unusual financial support. With the few exceptions, the days when a hired hand could save enough money to buy land and become a landowning farmer were lost in the distant past.

Midwestern farmers also began using heavy applications of chemical fertilizers to mitigate the problem of soil depletion caused in part by wartime production. They increasingly applied newly developed insecticides, such as DDT, which appeared to be a "cure-all" for insect control. Although some people questioned the long-term safety of the new fertilizers, herbicides, and pesticides on public health, most farmers believed the postwar chemical industry had created a "golden age." As farmers increased their productivity with new forms of science and technology, the farm population decreased, and the number of farms continued to decline. As more people left the farms to seek a better life elsewhere, part-time farming increased. Both trends would continue for the remainder of the century.

THE GREAT PLAINS

In 1931, a bumper wheat crop accompanied by drought brought economic disaster to the Great Plains. As the price of wheat plummeted, good farmers could no longer afford to practice soil-conservation techniques such as listing, terracing, or strip-cropping. Even minimal soil exposure contributed to wind erosion, and some of the worst dust storms came in the southern Great Plains in areas where less than half of the acreage had been planted in wheat. In May 1934, a dust storm removed an estimated 300 million tons of soil from the Great Plains. By 1935, drought and wind erosion especially plagued the southern Great Plains. Portions of the five-state area including western Kansas, northeastern New Mexico, southeastern Colorado, and the panhandles of Oklahoma and Texas became known as the Dust Bowl.

Although approximately 32 million acres composed the Dust Bowl, the people in the region suffered from the storms not because southern Plains farmers grew too much wheat but because the drought prevented them from raising much wheat from 1932 to 1940. During years of normal precipitation, wheat offered excellent soil protection against wind erosion. The drought, however, prevented the wheat from growing, and the strong prevailing winds quickly

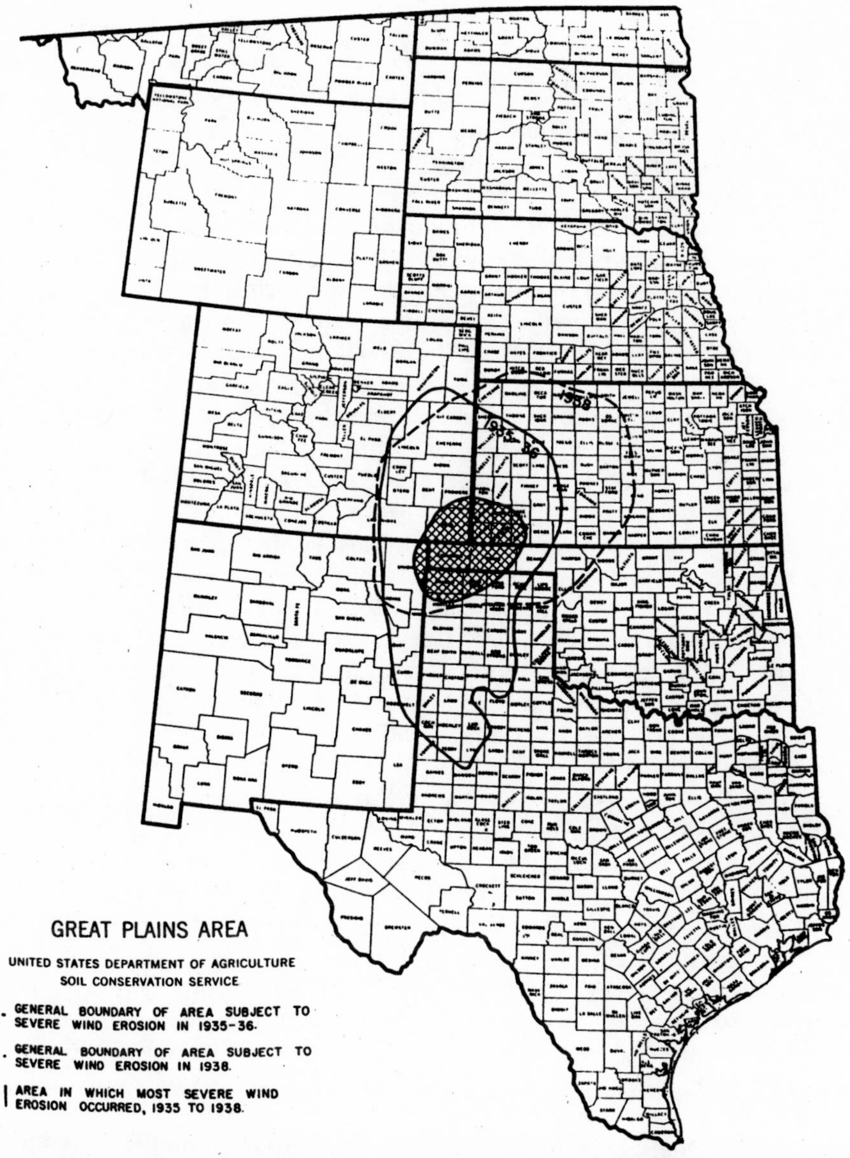

SEVERE DROUGHT and poor crop-production methods created a serious wind-erosion problem in the southern Great Plains during the 1930s. Known as the Dust Bowl, this erosion area included portions of five states. The dust storms that turned day into night, drifted soil like snow, ruined crops, killed livestock, and took human lives eventually ended with the return of near normal precipitation by 1939. The work of the Soil Conservation Service helped farmers restore their wind-eroded lands. This map shows the general boundaries of the Dust Bowl from 1935 to 1938. The checkerboard area indicates the section where the most severe wind erosion occurred. *Soil Conservation Service.*

lifted the dry bare soil over a wide area. By the spring of 1934, the vast majority of the farmers and ranchers were willing to adopt the necessary measures to prevent wind erosion.

The Soil Conservation Service was instrumental in helping Great Plains farmers control erosion by using private land to demonstrate proper soil-conserving techniques. The farmers who cooperated with the SCS signed five-year contracts in which they permitted the SCS to work conservation programs on their land and agreed to follow SCS recommendations. The Civilian Conservation Corps also established camps in the Great Plains that provided labor to aid the SCS and farmers in their conservation efforts. As Great Plains agriculturists teetered on the verge of failure and bankruptcy, federal aid designed to alleviate hard times proved invaluable in helping them to stay on the land until the rains returned.

Nearly all Great Plains farmers participated in federal agricultural relief programs. In the northern and central Great Plains, 40 to 75 percent of their income came from AAA and other economic programs such as cattle purchases, crop and seed loans, and relief grants. Although farmers preferred to plant as much wheat as possible and hope for rain, they had little choice but to accept federal aid. They needed the money. Both the SCS and the AAA provided financial relief to help farmers initiate a wind-erosion program in addition to payments received for diverting acreage from production under the AAA allotment program.

The Resettlement Administration and the Farm Security Administration also provided "rehabilitation" loans to help farmers purchase necessities—food, clothing, feed, seed, and fertilizer—to make them self-sufficient once the drought ended. Despite aid from the AAA, SCS, RA, FSA, and other agencies, Great Plains farmers and town dwellers alike had great difficulty meeting their financial obligations, and mortgage indebtedness and tax delinquencies increased. As wheat prices, property values, and tax collections fell, tenancy and nonresident land ownership increased.

New Deal agricultural programs were also designed to rehabilitate Great Plains agriculture by retiring marginal and submarginal cropland, restoring pastures, and increasing farm size to enable a more efficient and profitable economy of scale. In 1936, President Roosevelt appointed a Great Plains Drought Area Committee to recommend "practical measures" to remedy the problems of drought, dust, and depression in the region. The committee soon issued a report, *The Future of the Great Plains,* urging the development of an agricultural policy that would adapt farming practices to the semiarid climate of the region. This policy would take "certain sub-marginal lands permanently out of commercial production" and support regrassing, contour plowing, terracing, strip-cropping, and tree planting—all financed and managed by the federal government.

Although New Deal agricultural programs were fragmented on the Great Plains because of overlapping aims, inadequate funding, and disruption by World War II, they contributed collectively to improving agricultural conditions during a time of extreme economic difficulty and climatic aberration. Soon the return of normal precipitation and wartime demands lifted Great Plains farmers from economic and ecological crisis. Although many farmers left the Great Plains, others moved from their "empire of dust" to nearby towns where they could be closer to employment and relief offices. Still, perhaps 500,000 people, mostly farmers, pushed by the worst years of the drought, left the Great Plains during the 1930s. Usually they were the ones who had little money invested in their land. Most farmers who remained in the Great Plains had too much invested to leave, and the government helped them stay through a multiplicity of relief programs.

Although wheat served as the major crop, relatively few farmers relied entirely on it. Most raised some livestock and planted feed grains. Even with minimal diversification, wheat, livestock, feed grains, and government aid enabled them to retain their holdings and remain on the land until the drought ended. Even so, life was difficult at best for Great Plains farmers. No farmer got rich from government aid, but without it the agricultural community would have suffered far more. Still, only rain and increased demand, not government aid, could restore the agriculture of the Great Plains to prosperity. In the meantime, drought, dust, and depression left an "invisible scar" on those who lived through those difficult years. Long after the rains, green grass, and prosperity had returned, many could look back on the Dust Bowl years only with "heartache."

Throughout the 1940s, Great Plains farmers prospered from bumper crops and high prices for wheat, beef, and cotton brought on by high wartime demand. Precipitation returned to normal, and lush grass and bountiful crops made the dust storms of the past decade seem like ancient history. Land values skyrocketed from $3 per acre to $60 per acre in some areas. Farmers raced to plow native grasslands and plant more wheat. One plainsman warned, however: "When it turns dry again, the dust will blow again." He could not have been more prophetic. By the late 1940s, sand storms were common on newly broken cotton lands in West Texas and across the wheat fields of the central Plains. By the spring of 1950, drought had returned to the Great Plains, and wind erosion once again became serious. As a result of steadily worsening agricultural conditions, one Oklahoman said, "If this keeps up and the dust comes again, generally, I'll be a C.I.O. You known, a California Improved Okie."

Although twenty years had passed since the Dust Bowl, Great Plains farmers remembered the past and quickly began an emergency tillage program funded by the USDA. Stockmen also turned to the federal government for aid. Great Plains cattlemen, while demanding complete freedom from government regulation during good times, have never hesitated to seek federal aid when their pocketbooks become pinched. Now they sought long-term loans at low interest rates, cheap feeds subsidized by the federal government, and minimal freight rates for "bona fide" farmers and stockmen.

These demands did not go unheard. In late July 1953, the Commodity Credit Corporation began selling corn, wheat, oats, and cottonseed at below-market prices to help livestock producers maintain their basic herds. The USDA also agreed to purchase 200 million pounds of beef for the military and school-lunch program to remove cattle from parched grasslands and protect stockmen from serious economic loss. In addition, the USDA provided funds to help pay railroad charges for transporting hay into designated drought counties and gained the cooperation of several railroads in reducing the freight rate 50 percent on incoming hay. The Farmers Home Administration also provided livestock and crop loans to keep farmers' operations viable.

Although drought and wind erosion lasted until the mid-1950s and affected a larger area of the Great Plains than during the Dust Bowl years, the distressing and tragic conditions of the past did not

DURING THE 1930s, a severe drought and a government agricultural policy that paid landowners to reduce their planted acreage drove many sharecroppers and tenant farmers from the cotton area of the southern Great Plains. Collectively known as Okies, many loaded their possessions in the family car, truck, or trailer and headed to California, where they hoped to find work in the vegetable fields. Instead they encountered prejudice and abuse. The Farm Security Administration eventually built temporary camps for many agricultural refugees, but only the development of West Coast war industries during the early 1940s gave them economic opportunity, not the fertile fields and affluent growers in California. *Labor Archives and Research Center, San Francisco State University.*

return. Most Great Plains farmers now had a better understanding of the relationship between soil conservation and successful farming. Moreover, the financial problems of a serious economic depression did not plague them, and they were better able to survive the drought with government aid and properly farm their land, while they waited for the return of normal precipitation.

While they waited, many farmers in the southern and central Great Plains expanded irrigation with earnings from high wartime prices and postwar demands. They dug wells that tapped the Ogallala aquifer and pumped seemingly inexhaustible water with the aid of electric and gasoline-powered pumps. Irrigation systems became more convenient with the introduction of gated pipe and improved sprinklers. These farmers used irrigation to supplement natural precipitation but also used it to raise new crops such as sugar beets and corn where they would have been impossible before. Although diversification like this and the insurance that underground water provided during times of drought gave Plains farmers more security than ever before, water mining soon proved problematic, with the recharge rate insufficient to maintain underground levels. For the moment, few irrigation farmers cared. By the end of the Korean War, they pumped water cheaply and easily and made the Plains from West Texas to Nebraska the land of the "underground rain."

THE FAR WEST

In 1930, the agriculture of California continued to symbolize farming in the Far West in image, if not in fact, for most Americans. Agriculture in the Bear Flag State remained characterized by large estates, irrigation, migrant labor, and specialty production and marketing organizations. Although prices declined with national economic collapse, federal price supports eased financial hardship, especially for farmers who specialized in cotton, fruits, and vegetables. Yet these crops required extensive labor that could substantially increase operating costs and reduce profits. By 1945, specialty crops and large-scale holdings dominated California's agricultural economy, seventy-eight growers owning an estimated 6 million acres. In contrast to the past, however, when specialty crops had been intensively produced on small-scale operations and extensive agricultural practices had been relegated to large-scale farms, California growers now specialized with intensively grown crops on extensive holdings. Cotton became an important field crop, especially in the San

Joaquin Valley, until the 1950s. High yields, federal price supports, subsidized water, acreage allotments, and a tradition of mechanization that favored the large-scale operator made cotton a "money" crop.

Vegetable producers received the best prices for their crops, and growers expanded their acreage in lettuce, tomatoes, and asparagus to take advantage of economic opportunities, while grain and forage production declined. Increased demand for specialty crops resulted from changing consumer tastes facilitated by improvements in the canning industry, particularly the fast-freezing of vegetables during the late 1930s, followed by freezing fruits, particularly orange juice, by 1945. With food preservation significantly improved, distant markets expanded, and demand improved.

Similarly, the rapid growth of California's population stimulated by migration from other states during the 1930s and the war industry a decade later also increased the home market. Large-scale growers responded by purchasing or renting more land to capitalize on technical and demographic changes. Although small-scale holdings still predominated, the best farmland became increasingly concentrated in large-scale specialty crop producers, particularly on the Sacramento–San Joaquin Delta and in the Imperial Valley.

To maintain and control supply, food processors began to integrate their operations from the fields to the grocery stores and expand contract farming. By the end of the Great Depression, food processors and shippers largely controlled specialty-crop production. By 1935, three companies packed 40 percent of the prunes and raisins in California; three wineries handled 26 percent of the wine grapes; and the California Fruit Growers Exchange marketed 47 percent of the fresh oranges produced in the state. Three years later, four companies ginned 66 percent of California's cotton crop. As a result, corporations wielded important economic and political power and influenced marketing, labor, and water issues in California and the Far West.

Without cheap and abundant water for irrigation, however, the consolidation of agricultural lands and the emphasis on high-value speciality crops would have been impossible, and California's agriculture would have developed far differently during the twentieth century. Although irrigation had been important for California farmers since the late nineteenth century, much of that water had been supplied by private groups with limited capital. By 1939, irrigated acreage had increased considerably since the end of World War I, and the state and federal governments were largely responsible for that expansion, particularly in the Central and Imperial valleys, with the development of reservoirs by the Bureau of Reclamation. This expansion, however, occasionally deprived urban areas of sufficient water for their growing populations and laid the foundations for a turbulent relationship between growers and urbanites for the remainder of the twentieth century.

The growers believed they had the legal right to unlimited quantities of cheap water. The cities demanded their fair share and a cost structure that required farmers to pay the real cost of their water rather than cheap rates subsidized by urban charges. Increased irrigation also created problems of salinity and long-term, if not permanent, ruin of the soil. It also created continuous complaints from the growers that the 160-acre limitations and residency restrictions of the Bureau of Reclamation were punitive. The growers willingly subverted those regulations, given lax enforcement by the bureau. They also increased their reliance on state water projects for irrigation. Still, until 1960, all large-scale water projects were developed by the federal government.

California's growers, then, improved their economic gains by increasing their economy of scale through the extensive control of irrigated land. They also used government programs that favored large-scale operators. Equally important, they adamantly controlled their labor costs. In 1942, the growers used the labor shortages caused by World War II to gain access to a consistent and cheap supply of Mexican workers under the guise of "national emergency." In that year, an act of Congress and an agreement with Mexico created the "bracero program." It gave the USDA the authority to recruit, contract, transport, house, and feed temporary farm workers, known as *braceros* (strong-armed workers), for stoop labor in the vegetable and cotton fields of California. The bracero program enabled the growers to tap an unlimited labor supply in Mexico through contracting centers in California. Operating under the rationale that American workers would not do the stoop labor, the growers required workers from Mexico.

Although the bracero program enabled the growers to schedule workers in an efficient manner, to control expenditures for wages, and engage in long-term planning, in reality it permitted the growers to replace workers they considered problems and keep wages low and housing inadequate. The bracero program also displaced many African-American and Mexican-American workers from the cotton and vegetable fields. With their jobs taken by temporarily imported Mexican workers, African-American and Mexican-American workers had little choice but to flee to the ghettos and barrios of California's cities. With labor-contracting centers able to deliver hundreds of braceros within forty-eight hours' notice to fields several hundred miles from Mexico, the bracero program became a strike-breaking program sanctioned by the federal government at the behest of the

MIGRANT WORKERS have been essential to agriculture in California. By the 1930s, men and women workers of Mexican heritage became the most numerous in the vegetable fields. Growers preferred Mexican workers because they accepted the stoop labor and low wages to gain a better life. When they attempted to organize and bargain collectively, the growers crushed those efforts with force. These migrant workers are picking peas in the Imperial Valley. *California Section, California State Library.*

growers. The program continued after the war, although President Harry Truman transferred it from the jurisdiction of the grower-influenced USDA to the Department of Labor in 1948. In 1951, Congress continued the bracero program as an "emergency" measure in response to the Korean War.

MIGRANT WORKERS picked cotton in California and Arizona before the mechanization of the crop after World War II. Until scientists changed the cotton plant so that the bolls ripened uniformly, these workers picked the field at least three times. By 1949, mechanical harvesting had proved substantially cheaper than handpicking, and growers began to mechanize. *California Section, California State Library.*

LAND POLICY

The severe drought that plagued much of the Great Plains and Far West during the 1930s seriously damaged the rangelands of the public domain. To enable greater federal control of those lands for conservation, Congressman Edward T. Taylor of Colorado introduced a bill that became the Taylor Grazing Act in 1934. This legislation transferred most public land suitable for grazing and forage crops to the Department of the Interior and authorized the secretary of the interior to organize grazing districts for livestock raisers. The department would issue permits that stipulated the number of livestock that could be grazed on a certain acreage of rangeland during a specific period. The secretary would monitor the grazing districts and authorize reductions in the livestock on those lands if drought required limitations to prevent overgrazing and soil erosion. The act required stockmen to pay annual grazing fees, 25 percent of which would be used to support conservation programs on those lands.

All unreserved grazing lands under the Taylor Act were withdrawn from public entry unless those lands were capable of producing cash crops, thereby ending homesteading in the American West. By early February 1935, the public land in twenty-four western states had been withdrawn from homesteading except in locations where homesteaders had existing rights. After 1955, no further entries were made under the Stock Raising Homestead Act of 1916, although homesteading continued in Alaska under other legislation. Ultimately, the Taylor Act enabled the creation of grazing districts where the Department of the Interior and the Civilian Conservation Corps built fences to enable better rangeland management and control of grazing herds. These agencies also built farm "ponds" or water-storage "tanks," dug wells, and pursued rodent-control programs and the elimination of poisonous plants, all for the benefit of stockmen and rangeland conservation. By the end of the New Deal, more than 11 million livestock grazed 142 million federally control acres. Thereafter, strict management and government permission to use those lands became standard government policy.

At the same time that the Taylor Grazing Act closed public entry in the transmississippi West, the federal government embarked on a new land policy that involved the purchase of severely eroded lands from private owners for restoration and conservation. Although the Federal Emergency Relief Administration received funds for a "land-use" or purchase program in late December 1933, the Resettlement Administration, created in 1935, gained the responsibility for the program until it was transferred in 1937 to the Farm Security Administration, which in turn gave it to the Soil Conservation Service in 1938.

This land-use program involved federal purchase of the most severely wind- and water-eroded lands for restoration. The government planned to create demonstration projects where farmers and cattlemen could observe the best soil-conservation techniques. In the Great Plains and Far West, the federal government intended to return the worst privately owned wind-eroded lands to grazing under government management. No one knew how much land might be purchased, but it could not be taken against the owner's will. If a farmer wished to sell, the government sent appraisers, and both parties tried to arrive at a fair price based on soil conditions and improvements. Originally, the federal government intended to relocate families in "subsistence homestead communities," but inadequate funding and planning plagued this aspect of the program.

The land-use program became a grand experiment for the federal government. A soil-conservation project on such a large scale was unprecedented. The land-use projects, however, were not the panacea capable of solving all of the regional, economic, social, and erosion problems that many New Deal social scientists had hoped. As part of a broad soil-conservation program, however, the land-purchase program contributed to the efforts of the SCS and other government agencies in halting soil erosion and restoring much land to a sound agricultural base. On January 20, 1960, the Department of Agriculture created nineteen national grasslands from twenty-two land-utilization projects in eleven western states. The national grasslands remain as landmarks to a great experiment in state planning for land-use policy.

AGRICULTURAL ORGANIZATIONS

The excruciating poverty of the rural South, the bondage of the sharecropping system, and the unwillingness of the federal government to provide economic aid to tenant farmers through the Agricultural Adjustment Administration provided the foundation for a new agricultural organization, the Southern Tenant Farmers' Union (STFU). This organization had its roots in northeastern Arkansas, near Tyronza, where the sprawling Fairview Farms plantation, under new management, evicted forty families in 1934 because they had exceeded their credit limits at the corporation warehouse. Although overruns had long been standard operating procedure, a useful method to keep sharecroppers tied to the plantation, the new owner chose to release those families in violation of Section 7 of the Agricultural Adjustment Act. Hiram Norcross, the absentee landlord in St. Louis, preferred to receive as much of the AAA payment as possible and used the credit violation as an excuse to release these tenants.

The sharecroppers in northeastern Arkansas turned to H. L. Mitchell, a local dry cleaner, and to Clay East, who operated a service station, for help because both men had treated them fairly in the past and they respected their opinions. Mitchell and East were Socialists who had been involved with organizational work in the area. Both men met with Norman Thomas, the perennial Socialist presidential candidate, who advised that the Roosevelt administration would not aid them and urged Mitchell and East to organize a sharecroppers' union. With strength of numbers and collective bargaining, he believed, the sharecroppers could gain concessions from the large-scale landowners.

In July 1934, Mitchell and East met with eighteen white and black sharecroppers about their problems with the Fairview Farms management. Although the sharecroppers favored violence to gain reform, Mitchell stressed the benefits of a sharecroppers' union. One black sharecropper urged unity between the races rather than the formation of black and white unions: "There ain't but one way for us . . . and that's to get together and stay together." Poignantly, he said, "The same chain that holds my people holds your people too. If we're chained together on the outside we ought to be chained together in the union." They agreed, and with this inauspicious meeting the Southern Tenant Farmers' Union was born. With membership fees set at $0.10 per month, free for those too poor to pay, and membership open to anyone over 18 who lived from agricultural "rents, interest, or profits," including hired farm workers, the union grew. Within a year it had spread across Arkansas into Oklahoma, Texas, and Mississippi, the STFU claiming 25,000 members. Where the STFU organized, its primary intent was to gain bargaining power and a commitment from the AAA to stop evictions by landlords.

Mitchell, elected executive secretary, gave life to the STFU. He provided leadership and inspired the courage required to become a member of the union. For his activities, a local boycott ruined his business, and the opposition labeled him a "Red" Communist and Socialist agitator. He suffered threats against his life. Many STFU men and women found their meetings at a local church or schoolhouse denied or interrupted by thugs who beat them. Despite governmental reports that the planters had inflicted a "reign of terror" on the STFU with their Klan-like "night riders," and that "criminal anarchy" prevailed in northeastern Arkansas, the federal government did nothing.

In March 1935, Mitchell complained to the AAA that "when blood flows it will drip down over your Department, from the Secretary at the top to the cotton Section at the bottom." A. B. Brooking, the union's chaplain, reported that night riders "shot up" his house, but he testified, "I am not afraid to go on being a union man." Henry A. Wallace promised an investigation of the violence, but he did not want to upset the successful operations of the AAA, particularly in the South where the votes of landowners would be of vital importance during the presidential election in 1936. Moreover, the AAA needed the support of the landlords if the cotton acreage-reduction program was to be successful. The landowners were important; the sharecroppers, insignificant.

In mid-April 1935, after two and a half months, the systematic terrorization of these sharecroppers ended as the perpetrators of the violence grew tired of it. By that time, Mitchell had fled across the Mississippi River to Memphis, at least one organizer had been killed, many sharecroppers had been beaten, free speech had been denied these farmers, and the state legislature had passed a law making it "seditious" and a felony to possess five or more copies of STFU literature. The federal officials, particularly in the USDA, responded to this blatant denial of constitutional liberties only by saying that the violence was a local matter and that they did not have jurisdiction to prevent such activities. White-supremacy lynch law remained supreme, and these sharecroppers were economically, socially, and politically powerless to prevent it.

Although the federal government would not aid these sharecroppers, national press coverage of the violence against the STFU created serious public embarrassment for the Roosevelt administration, but not enough to help the sharecroppers. President Roosevelt preferred disciplined nonintervention. "I know the South," he said, "and there is arising a new generation of leaders in the South and we've got to be patient." When liberals in the Cotton Section of the AAA attempted to make the agency more supportive of the sharecroppers, Chester Davis, administrator of the agency, fired them with Wallace's approval. This "purge" ensured that AAA policy in the South would favor the landlords. Stoically embracing nonviolence and reinforcing personal courage with spiritual-like songs such as "We Shall Not Be Moved" at camp meetings, the STFU endured.

Finally, in 1936, new legislation provided that tenants would receive 25 percent of the benefit payment to landowners who reduced their cotton acreage. But this support proved harmful because landlords now had a new incentive to release still more sharecroppers to claim all the government check.

In 1937, the STFU, claiming 30,000 members in seven states, merged with the United Cannery, Agricultural, Packing and Allied Workers of America, an industrial union under the control of the Congress of Industrial Organizations. This union had little interest in the plight of the sharecroppers other than collecting their membership fees, and the STFU quickly declined in strength. In the end, the STFU's greatest victory was that it survived these hard and violent times and publicized the plight of the sharecroppers for the sake of history if not themselves.

While migrant workers remained unable to organize, the growers achieved considerable success in 1933 with the creation of the Associated Farmers of California. This militant organization of large-scale growers was sponsored by the Agricultural Section of the California Chamber of Commerce and the California Farm Bureau. Funding came primarily from the major investment, packing, and transportation corporations and utility companies. The Associated Farmers worked closely with local law-enforcement officers to drive "dangerous radicals" from the state and recruited a "citizen's army" to break up strike activities with ax handles. Growers said that they could not afford to pay higher wages and that the Communist party was responsible for efforts to unionize agricultural workers during the early 1930s. The Associated Farmers effectively countered the efforts of migrant workers to unionize.

In California, labor costs were the major expense for the growers, and the harvest season remained the bottleneck in production because the growers were vulnerable to disruption by a work force that might demand improved wages and working and living conditions. Migrant workers, however, often lacking citizenship and organization, had neither the right nor the power to force reforms. The growers controlled local politics and exerted great influence on county labor bureaus, placement offices, and law enforcement. Consequently, most labor insurgencies had failed before the Great Depression, and the economic crisis expanded the growers' access to cheap, nonmilitant labor. For the growers, Mexican Americans or Mexicans were model farm workers because they remained culturally isolated and accepted low wages and harsh working conditions.

Because these workers were engaged in seasonal labor that required high mobility, they were difficult to organize. The growers customarily overrecruited, especially at harvesttime, to keep wages low and protect themselves from labor disruptions. Because the growers hired workers locked in poverty and without consistent leadership, the migrants could not support prolonged strikes. When a migrant strike occurred, however, the growers quickly crushed it with a speed and violence that can only be described as fury.

In California, agricultural workers turned to militant unionism to improve their working and living conditions. Although only a few workers were ideologically connected to the Communist-controlled United Cannery, Agricultural, Packing and Allied Workers of America, founded in 1931, the union succeeded in improving wages and focusing public attention on the plight of migrant workers. Still, a strong agricultural workers' union, the National Farm and Labor Union (NFLU), was not established until after World War II. This organization proved most effective from its founding in 1946 until 1952.

The NFLU, with support from the American Federation of Labor, recognized that areawide strikes were impossible to organize, difficult to control, and easy to break. Instead, the NFLU attempted to build a stable membership and to strike specific estates to win limited victories that would have a collective influence on wages and working conditions over time. Strikes against the grape producers in the San Joaquin Valley in 1947, however, proved disastrous because the workers were too poor to strike, the growers used braceros to replace laborers who associated with the union, and the growers hired thugs to break up the picket lines and ransack union headquarters. The growers contended that "outside agitators" were responsible for the violence. When the NFLU gave up the strike for a boycott of the tomato growers with the support of the Teamsters Union, the Truman administration invoked the Taft-Hartley Act, which prohibited secondary boycotts, sympathy strikes, and mass picketing.

After the failure of the strike and boycott activities, the NFLU began to use a new tactic. During the late 1940s, the union attempted to enlist the support of the Mexican migrants the growers had used as strikebreakers. The organizational efforts of the NFLU now shifted away from local permanent workers to the migrants in the fields. The NFLU strike demands also narrowed to seeking higher wages and better working conditions rather than recognition of the union. By the early 1950s, however, the NFLU had not overcome the effective use of braceros by the growers to break strikes, nor had it built a political coalition that could represent the migrant workers in the halls of the state legislature or Congress. Moreover, in an age of political conservatism, Communist witch-hunting, and loyalty oaths, the NFLU had little hope of marshaling support. In 1952, Congress exempted the growers from prosecution for hiring illegal workers. Faced with this insurmountable opposition, the NFLU ceased to exist. The politically powerless migrant workers could not make a sustained challenge to the agricultural system in California.

AGRICULTURAL IMPROVEMENTS

By the early 1930s, falling agricultural prices, overproduction, and low labor costs had created economic hardships and prevented many farmers from purchasing improved equipment such as tractors. Despite these problems, by 1940, 23 percent of American farmers, mostly corn and grain farmers in the Midwest and Great Plains, owned tractors. Those regions were well suited for complete mechanization because tractors could easily pull plows, drills, combines, or corn pickers over the relatively level terrain and large fields. Southern farmers, however, were slower to adopt tractors because other technological problems prevented the complete mechanization of staple crops such as cotton and tobacco. Nevertheless, where farmers adopted the tractor, this implement enabled swift completion of tillage, planting, and grain-harvesting operations. Plowing requires the greatest power of any farm operation. At the turn of the twentieth century, a farmer could plow 1 acre in 1.8 hours with a horse; by 1938, a tractor enabled him to till that area in 30 minutes.

During World War II, high prices and labor shortages encouraged farmers to purchase more tractors. Farmers recognized that expanded production would enable them to reap large profits from wartime prices and that tractors provided the means to cultivate more acreage. With the aid of tractors, farmers increased corn and wheat production by more than 9 and 2 million acres, respectively, during the war years. By 1945, 30.5 percent of the farmers in the United States used tractors. Nearly 690,000 new tractors had displaced 2 million horses and mules between 1941 and 1945. Tractors were now adaptable to nearly every American farm.

As farmers increasingly replaced horses and mules with tractors, they gained additional land for cash crops. By 1950, more than 3 million tractors had freed an estimated 70 million acres for the production of food and fiber. As more farmers adopted the tractor, the number of horses and mules on American farms continued to decline; in 1955, tractors exceeded the number of farm horses for the first time. Not long thereafter, the Statistical Reporting Service of the Department of Agriculture stopped counting the number of horses and mules used for farming. For some farmers, however, technological change posed insurmountable problems that often resulted more from lack of education than income. Nate Shaw, a southerner, reflected: "I knowed as much about mule farmin as ary man in this country. But when they brought in tractors, that lost me." Shaw would continue to use mules into the 1950s.

Although World War II provided the great stimulus for the mechanization of southern agriculture in the form of tractors, peanut pickers, hay bailers, and dairy equipment, an affordable and reliable cotton picker was not yet available. As a result, cotton farmers abandoned marginal lands to ease the labor problem, and cotton acreage declined by 5.5 million harvested acres between 1942 and 1945, even though the price of cotton increased from $0.17 to $0.22 per pound.

While cotton farmers contended with their labor problems, agricultural engineers at the International Harvester Company (IHC) worked to develop an efficient picker, and in 1942 IHC emerged as the leader in the field of cotton-harvesting technology. By mid-November 1942, IHC had developed a cotton picker that worked well enough to merit commercial sale. The machine consisted of two rotating drums with spindles approximately 4 inches long and 1/4 inch in diameter, the spindles about 4 inches apart. As the drums rotated past a cotton plant, the wetted spindles extracted the lint from the bolls. Rubber doffers removed the fibers from the spindles, and an air conveyor blew the cotton into an overhead container. A modified Farmall tractor, operating in reverse, propelled the picker and powered the mechanism. Continued labor shortages after World War II and additional technical improvements made the machine a commercial success. Production steadily increased, and by 1949 about 1,500 spindle pickers were operating in the South and California.

Still, the cost of owning a cotton picker remained high, considering that more than half of the cotton farmers in the South planted fewer than 30 acres and produced less than 4 bales annually. Moreover, agricultural labor costs were cheaper in the South than in Arizona and California where cotton also was an important crop and environmental conditions such as dry weather during the harvest season reduced the expense of mechanical harvesting and encouraged the use of mechanical pickers. In addition, inadequate capital, insufficient financing, and the absence of preharvest mechanization slowed the adoption of the cotton picker in the South. Consequently, mechanization first became established in the West, then moved to the South.

In 1949, cotton farms in California averaged 103 acres; the average farm in South Carolina raised about 13 acres. At that time, farmers needed at least 100 acres of cotton to make the purchase of a cotton picker feasible. Consequently, southern cotton farmers did not rush to purchase mechanical pickers. In 1950, 304,469 farms raised cotton as the main crop, but farmers harvested only 5 percent of the crop mechanically, mostly in California and Arizona. With the perfection of the mechanical harvester, however, farmers could completely mechanize their cotton crop if they could afford it. They recognized that mechanization saved time and money and ended the problems caused by an uncertain labor supply. At the same time, mechanical cotton harvesters contributed to the consolidation of small-scale farms into larger holdings and to the decline of the farm population.

The self-propelled combine heralded the future for grain harvesters in the Great Plains and Far West. By eliminating the tractor to pull the combine and the auxiliary engine or power-takeoff system to operate the machine, farmers streamlined their harvesting operations even more. By removing the tractor from the harvest, farmers gained an implement for other work such as immediate plowing of the stubble field after the grain had been cut. The elimination of the tractor and auxiliary engines from the harvesting process also reduced fuel expenses because the combine's engine needed to be only large enough to power the machine. Since tractors used for pulling a combine usually had far more power than necessary, tractors used more fuel than combine engines. Moreover, self-propelled combines decreased the manpower requirements for operating the implement. Now one person could harvest the crop. A driver, who hauled the grain from the field to the bin or nearby elevator, was the only additional worker required. Because the family could usually furnish the driver, no expense was incurred if that worker did not receive a wage.

Self-propelled combines also eliminated grain losses when a farmer "opened" his field—when the combine made its first round. Because tractor-pulled combines had cutting tables on the side of the implement, the tractor knocked down much of the grain in its path when the farmer made his first round. The self-propelled combine also had greater maneuverability, enabling a farmer to swing around areas of unripened grain with relative ease in order to come back later. In addition, self-propelled combines had less difficulty passing down narrow country roads because the cutter bar was centered in front of the implement, not off to the side. Overall, then, self-propelled combines operated with greater efficiency than the larger tractor-drawn models.

During the 1940s, ownership of a combine for those farmers who could afford one provided the independence they so enjoyed. With their own combine, they did not have to endure the agony of waiting for the custom cutter to arrive while worrying whether their already dead-ripe wheat would be lost to a hailstorm or shatter from the heads in the wind. Still, change comes slowly. Relatively high costs and previous investment in tractor-drawn models limited the adoption of the self-propelled combine. Although most wheat farmers used combines to harvest their crop by the end of World War II, tractor-drawn models still predominated until 1960. By 1950, however, the combine had mechanized the wheat harvest from Texas to the Canadian border.

RECOVERY

By the late 1930s, the rains had returned to the Great Plains. With the drought broken and the dust storms diminishing in number and intensity, some farmers expected the "old-time prosperity" to return. Near-normal precipitation, combined with the onset of another European war, quickly enabled an increase in agricultural production, the depletion of surplus commodities, and high agricultural prices. Although 5 million people left the farms during World War II for military service or better-paying jobs in cities, technology (particularly tractors, combines, and corn pickers) replaced them.

As prices rose, farmers reached the elusive goal of parity. In May 1941, Congress guaranteed farmers 85 percent of parity prices on five basic commodities—corn, wheat, rice, cotton, and tobacco—and urged them to raise more soybeans and peanuts for oil, vegetables, and livestock to meet military demands at home and abroad. As agricultural prices rose above the loan rate and with high wartime demands, consumers complained about high food prices, and Congress responded by passing the Emergency Price Control Act in January 1942. This legislation provided a ceiling on farm prices at 110 percent of parity. Congress amended this act in October 1942 to protect farmers from a postwar price collapse similar to that in 1920. Known as the Steagall Amendment, this legislation required Congress to provide 90 percent parity price supports for two years after the conclusion of the war to enable farmers to adjust to peacetime conditions.

Congressional wartime agricultural policy proved highly favorable to farmers, and the major agricultural organizations such as the American Farm Bureau Federation, Farmers' Union, and Grange lobbied effectively for it. Indeed, War War II and favorable agricultural policy gave many farmers the first prosperity they had ever known. By 1946, agricultural prices averaged 123 percent of parity. Farmers responded to this new prosperity by expanding production and using their profits to purchase more land and equipment, to pay off debts, and to save. For many farmers, the war years were a time of "milk and honey." Between 1940 and 1945 net farm income rose from $2.3 to $9.2 billion. Even so, per capita farm income averaged only 57 percent of nonfarmer per capita income.

The war years brought industrialization to the farm and changed agricultural life for all time. During the war, agricultural acreage expanded about 5 percent, but productivity increased by 11 percent, largely because of the increased use of hybrid seeds, pesticides, insecticides, fertilizer, and mechanization. This new emphasis on agricultural science and technology continued the second technological revolution on the American farm. Although farmers had always been commercially minded and profit oriented, new scientific and technological developments together with high wartime demands and government price supports ended any lingering illusions that farming was a way of life. Thereafter, few could disagree that agriculture was a business and frequently nothing more. Although critics of the new agriculture still contended that farm life was morally superior to urban living, their arguments lost energy and influence. The farm increasingly became a place to live and work, not for a superior quality of life but economic gain.

The end of World War II did not bring a decrease in national and international demands for agricultural commodities. Europe faced famine, and meat remained in limited supply at home. Although the influence of the farm lobby remained strong in Congress because farmers and the agriculturally dependant rural communities marshaled an important vote, high wartime production and prices posed significant problems for farmers, politicians, and consumers. When the war ended, farmers wanted to retain the high wartime price supports rather than return to the prewar policy of flexible price supports based on production.

The problems of production controls and price supports, however, were complex, and no organization or lobbying group represented all farmers nationwide. Indeed, specialization by farmers and regions often made agricultural interests contradictory, and no single policy could solve all farm problems. The American Farm Bureau opposed the continuation of high price supports because this policy would hurt the market for corn farmers by increasing feed prices, and these farmers sold most of their crop for fattening cattle and hogs. Although the Farm Bureau supported a flexible price-support plan, its membership divided on this policy issue, and the organization began to lose its predominant voice as the representative of farmers in Congress.

By the late 1940s, consumers advocated lower agricultural price supports to control food costs. Consumers, particularly those in urban areas, had become a new and increasingly vocal and powerful voice that gained the attention of Congress; soon agricultural policy would be based partly on the needs of this constituency. Some people advocated solving the surplus-production problem through massive international relief programs like the Marshall Plan. Others favored using surplus agricultural commodities to subsidize school-lunch programs. Critics of these proposals, however, argued that such policies would only encourage farmers to keep producing surplus commodities rather than help them shift to production that would not need price supports.

In 1948, Congress attempted to balance the positions of those who advocated high and reduced price supports with new legislation. It retained high fixed price supports for one year on basic commodities such as wheat, corn, cotton, and hogs. Then a flexible price support program would pay farmers as much as 90 percent of parity, depending on the commodity, through the 1950 crop year. Congress again gave the secretary of agriculture authorization to impose marketing quotas on farmers, provided that two-thirds of those who produced that commodity and participated in the price-support program that required acreage reduction voted to approve that policy.

Congress also changed the index for determining parity prices from the 1909–1914 period for most commodities to calculations based on the relationship of agricultural and nonagricultural prices during the past ten years. Although agricultural prices declined slightly as a result of this legislation, the federal government still accumulated large stocks of agricultural commodities such as wheat, cotton, and dairy products under this price-support policy.

Despite the efforts of Congress, the farm lobby, and the USDA, no one knew how to solve the problem of surplus production without high price supports based on production controls. Some agriculturists argued that without price supports many farmers would go bankrupt and leave the land. The result would be extreme social and economic dislocation. Others contended that government agricultural policy was too expensive and that too many farmers produced too much. Before these opinions could be resolved, however, the Korean War temporarily solved the farm problem. Instead of imposing a flexible price-support program as planned, Congress passed the Defense Production Act of 1950. This legislation required 100 percent parity prices if wartime price controls were implemented to ensure that farmers reaped the benefits of increased demand. If controls were not imposed, agricultural price supports would remain at 90 percent of parity.

During the Korean War, the demand for agricultural commodities increased rapidly. In 1951, farm prices rose above parity for cattle, cotton, hogs, and rice. By the end of the year, net agricultural income rose $1 billion from 1949 earnings. When the war ended, however, Congress guaranteed that agricultural prices would be supported at 90 percent of parity prices through the 1954 crop year. Production remained high, a clear indication to some critics of agricultural policy that parity prices encouraged surplus production. Congress once again grappled with the problems of surplus production and the merits of high fixed or flexible price supports. Congressmen from agricultural states continued to favor high price supports, while the USDA and Farm Bureau advocated flexible price supports linked to production. Many agricultural economists urged Congress to maintain high fixed price supports coupled with production controls as the only means to ensure a satisfactory farm income.

In 1954, the Eisenhower administration favored a reduced and flexible price-support program based on production to help curb government spending and ease the tax burden. The administration, however, immediately encountered the difficulty of reconciling good economic policy with agricultural politics because farmers still wielded a significant vote. Although President Eisenhower asked Congress to expand farm exports to help reduce price-depressing surpluses and to approve a flexible price-support program based on production, Congress could not agree on any of these suggestions.

DURING THE 1950s extension agents began using the relatively new technology of television to reach more rural and urban residents than by the traditional methods of home visits, publications, and demonstrations before special groups. This extension home economist in Iowa is demonstrating new food preparation methods. *Special Collections, William Robert and Ellen Sorge Parks Library, Iowa State University.*

Finally on July 10, 1954, Congress approved the Agricultural Trade Development and Assistance Act, known as Public Law 480.

This important legislation authorized the federal government to sell surplus agricultural commodities abroad for foreign currency and to use surplus commodities for emergency food aid to friendly and underdeveloped nations. This legislation became highly political in time, with the emphasis on foreign policy rather than improvement of agriculture. Foreign sales quickly focused on regions and nations that the United States wanted to influence favorably. Eastern Europe, for example, became an early recipient of agricultural sales under PL 480, and India, Pakistan, and Southeast Asia became later targets to help the United States win the cold war through agricultural trade. Later in 1954, Congress also approved a flexible price-support program for major crops, but it substantially limited the government's power to restrict production.

By the mid-1950s, then, the problem of overproduction, high price supports, and increased operating costs continued to plague the American farmer. Between 1951 and 1956, for example, agricultural prices declined 23 percent, and operating costs increased or remained the same. Farmers responded by trying to produce as much as possible on their restricted acres with heavy applications of fertilizers, herbicides, and pesticides. This solution not only made the problem worse but created new difficulties for agriculture and the environment.

RURAL LIFE

Rural America experienced both continuity and change between the Great Depression and the end of the Korean War. Rural schools suffered from a declining tax base as land values plummeted during the 1930s and many farmers could not pay their tax assessments. Rural women who had completed minimal education requirements to qualify for teaching positions often vied for jobs in the country schools. If they married, however, they usually lost their positions because school boards took the patriarchal position that a woman's husband would provide for her and she no longer needed the work. Often the daughters of farmers became country schoolteachers, lived at home, and contributed their earnings to the maintenance of the farm's economic viability until more profitable times returned. Rural children, particularly African Americans, suffered the worst educational conditions in the nation, which the commissioner of education in Georgia called "terrible beyond description." Poverty epitomized life for southern children, and their condition did not soon change.

Despite the benefits of new implements, technological change did not affect farmers uniformly. Isolation prevailed in the South because of inadequate roads and an absence of automobiles. Even so, rural life across the nation began to experience fundamental change as the result of technology and government policy. Some farmers believed the REA was socialistic and were hesitant to give it their full-fledged support, even though they desperately needed and wanted electrical service. But once their farms had been wired and the power turned on, any misgivings about participation in a socialistic venture quickly faded. With REA service, the long cold trip to the windmill in subzero temperatures became a hardship of the past.

If the REA represented socialism, no farmer cared. They agreed with Senator George Norris, an early congressional supporter of the REA, that electricity meant "a hired man for the farmer and a hired girl for his wife." With electricity, much of the drudgery and drabness of rural life had been eliminated. It lengthened the day and enabled farmers to use their leisure time in ways other than going to bed at nightfall. It promoted reading and, through radio, contact with the outside world.

Rural life also became more comfortable. Electric pumps brought water indoors, and electric heaters warmed it. The traditional Saturday-night bath disappeared; with ample hot water, farmers could take as many baths as they wanted at any time. By making indoor plumbing practical, electricity eliminated the bedtime trip to the outhouse. It also helped improve dietary habits with the aid of refrigeration, ended the winter task of ice harvesting, and improved the quality of rural schools, particularly for reading instruction. Electricity revolutionized the poultry industry with lights, and heated barns saved lives when the weather turned raw at lambing time.

The day the lights came on was a turning point in the lives of farm families everywhere. Indeed, it was a history-making day, the kind of day farmers remembered by marking the date on the barn wall next to the etchings that signified other important events in their lives such as births, deaths, or the purchase of a new mowing machine or prize cow. Thereafter, rural life was never the same. No one wanted to return to the darkness of the past.

Not long after the REA began to change life in rural America, telephones improved communication. Although the traditional way for farmers in the Great Plains to send a message had been to "put a kid on a horse," by the early 1950s, telephones had ended that practice. In 1949, Congress passed the Hill-Poage Act, which enabled the REA to support the extension of telephones to rural areas. At that time, two out of three farm families still did not have a telephone, largely because the Great Depression had prevented them from owning one and the private companies had been as reluctant to provide telephone service as they had been to furnish electricity. REA support for rural telephone service brought party lines to the countryside. Party-line service meant that anyone who shared a line could listen to the phone calls of others, but farmers preferred to have this communal service rather than none. Moreover, in times of emergency, "everybody's ring" quickly signaled a need for assistance.

Farm women continued their dual roles of helpmate and

worker, and they labored to keep families and homes together during hard times. Their earnings from eggs and cream brought needed cash into the farm home and often meant the difference between survival and failure. Technological change, however, least influenced the lives of women in the South. They lived much like their mothers and grandmothers.

One tenant woman nursed her baby in the fields while she strung tobacco, then returned home to cook meals. Usually she ended her days "tired enough to die." During the late 1930s, a woman in North Carolina accepted her fate with resignation: "We ain't never had nothing and we won't never have nothing." About that time, Ann Marie Low, a North Dakota woman who grew up on a farm and taught school, reflected that she was accustomed to "everlasting poverty," but like most country people, she was optimistic.

American farmers remained the greatest "next-year" people in the world, and Low exhibited that hopeful spirit when she wrote that "we just wait year after year for good times to come again." Despite the recovery of the agricultural economy fostered by World War II, women would look back on most of the quarter century after 1930 and remember only the "terrible, heartbreaking years."

Between the collapse of agriculture during the Great Depression and the rise of more financially secure federally protected farming by the mid-1950s, great changes had occurred. Mechanization had increased, particularly the use of tractors, trucks, combines, and cotton pickers. Science produced chemical fertilizers, herbicides, and pesticides that most considered safe and beneficial, although time would prove that optimism unwarranted by the late twentieth century. Improved hybrid seeds, livestock, and poultry breeds boosted production and often profits. Yet the application of science and technology substantially increased operating costs and kept farmers in a cost-price squeeze. In turn, farmers looked to the federal government for price-support and marketing aid to ensure parity income with other Americans. The federal government responded by helping the farmers who it judged had the best chance to succeed.

As a result, the Agricultural Adjustment Administration revolutionized the landlord-tenant relationship in the South. When the federal government became the agricultural creditor of that region, the furnishing-merchant and sharecropping systems also quickly collapsed. Many farmers could not compete, however, and left the land, and in fact they were not needed, because production remained high. Farm sizes increased with consolidations. As the surplus farm population left the countryside, the political power of farmers also began to decline. By the end of the Korean War, many farmers experienced an uncertainty about their lives that would stay with them and those who came after them through the remainder of the twentieth century.

SUGGESTED READINGS

Benedict, Murray R. *Farm Policies of the United States, 1790–1950: A Study of Their Origins and Development.* New York: Twentieth Century Fund, 1953.

Brown, D. Clayton. *Electricity for Rural America: Fight for the REA.* Westport, Conn.: Greenwood Press, 1980.

Calef, Wesley. *Private Grazing and Public Lands: Studies of the Local Management of the Taylor Grazing Act.* Chicago: University of Chicago Press, 1960.

Cochrane, Willard W. *American Farm Policy, 1948–1973.* Minneapolis: University of Minnesota Press, 1976.

Conrad, David Eugene. *The Forgotten Farmers: The Story of Sharecroppers in the New Deal.* Urbana: University of Illinois Press, 1965.

Daniel, Cletus E. *Bitter Harvest: A History of California Farmworkers, 1870–1941.* Ithaca, N.Y.: Cornell University Press, 1981.

Daniel, Pete. *Breaking the Land: The Transformation of Cotton, Tobacco, and Rice Cultures Since 1880.* Urbana: University of Illinois Press, 1985.

———. "Going Among Strangers: Southern Reactions to World War II." *Journal of American History* 77 (December 1990): 886–911.

———. "The Transformation of the Rural South: 1930 to the Present." *Agricultural History* 55 (July 1981): 231–48.

Fink, Deborah. "Sidelines and Moral Capital: Women on Nebraska Farms in the 1930s." In *Women and Farming: Changing Roles, Changing Structures,* ed. Wava G. Haney and Jane B. Knowles, 55–70. Boulder, Colo.: Westview Press, 1988.

Fite, Gilbert C. *American Farmers: The New Minority.* Bloomington: Indiana University Press, 1981.

———. *Cotton Fields No More: Southern Agriculture, 1865–1980.* Lexington: University of Kentucky Press, 1984.

———. "Farmer Opinion and the Agricultural Adjustment Act, 1933." *Mississippi Valley Historical Review* 48 (March 1962): 656–73.

Grubbs, Donald H. *Cry from the Cotton: The Southern Tenant Farmers' Union and the New Deal.* Chapel Hill: University of North Carolina Press, 1971.

Hadwiger, Don F., and Clay Cochran. "Rural Telephones in the United States." *Agricultural History* 58 (July 1984): 221–38.

Hargreaves, Mary W. M. *Dry Farming in the Northern Great Plains: Years of Readjustment, 1920–1990.* Lawrence: University Press of Kansas, 1992.
______. "Land Use Planning in Response to Drought: The Great Plains Experience of the Thirties." *Agricultural History* 50 (October 1976): 561–82.
Haystead, Ladd, and Gilbert C. Fite. *The Agricultural Regions of the United States.* Norman: University of Oklahoma Press, 1955.
Helms, Douglas. "Conserving the Plains: The Soil Conservation Service in the Great Plains." *Agricultural History* 64 (Spring 1990): 58–73.
Hundley, Norris, Jr. *The Great Thirst: Californians and Water, 1790s–1990s.* Berkeley: University of California Press, 1992.
Hurt, R. Douglas. *Agricultural Technology in the Twentieth Century.* Manhattan, Kans.: Sunflower University Press, 1991.
______. *The Dust Bowl: An Agricultural and Social History.* Chicago: Nelson-Hall, 1981.
______. "Federal Land Reclamation in the Dust Bowl." *Great Plains Quarterly* 6 (Spring 1986): 94–106.
______. "REA: A New Deal for Farmers." *Timeline* 2 (December 1985/January 1986): 32–47.
______. "The National Grasslands: Origin and Development in the Dust Bowl." *Agricultural History* 59 (April 1985): 246–59.
Isern, Thomas D. *Custom Combining on the Great Plains: A History.* Norman: University of Oklahoma Press, 1981.
Jones, Jacquline. "Tore Up and a-Movin': Perspectives on the Work of Black and Poor White Women in the Rural South, 1865–1940." In *Women and Farming: Changing Roles, Changing Structures,* ed. Wava G. Haney and Jane B. Knowles, 15–34. Boulder, Colo.: Westview Press, 1988.
Liebman, Ellen. *California Farmland: A History of Large Agricultural Holdings.* Totowa, N.J.: Rowman and Allanheld, 1983.
Lowitt, Richard. *The New Deal and the West.* Bloomington: Indiana University Press, 1984.
McDean, Harry C. "Federal Farm Policy and the Dust Bowl: The Half-Right Solution." *North Dakota History* 47 (Summer 1980): 21–31.
McWilliams, Carey. *Factories in the Field: The Story of Migratory Farm Labor in California.* Boston: Little, Brown, 1939.
Malone, Michael P., and Richard W. Etulian. *The American West: A Twentieth-Century History.* Lincoln: University of Nebraska Press, 1989.
Meister, Dick, and Anne Loftis. *Long Time Coming: The Struggle to Unionize America's Farm Workers.* New York: Macmillan, 1977.
Mitchell, Harry L. *Mean Things Happening in This Land: The Life and Times of H. L. Mitchell, Co-Founder of the Southern Tenant Farmers Union.* Montclair, N.J.: Allanheld, Osmun, 1979.
Pratt, William C. "Rethinking the Farm Revolt of the 1930s." *Great Plains Quarterly* 8 (Summer 1988): 131–44.
______. "Rural Radicalism on the Northern Plains, 1912–1950." *Montana, The Magazine of Western History* 42 (Winter 1992): 42–55.
Riney-Kehrberg, Pamela. "From the Horse's Mouth: Dust Bowl Farmers and

Their Solutions to the Problem of Aridity." *Agricultural History* 66 (Spring 1992): 137–50.

Russell, Howard W. *A Long Deep Furrow: Three Centuries of American Farming.* Hanover, N.H.: University Press of New England, 1982.

Saloutos, Theodore. *The American Farmer and the New Deal.* Ames: Iowa State University Press, 1982.

______. "The New Deal and Farm Policy in the Great Plains." *Agricultural History* 43 (July 1969): 345–55.

Schapsmeier, Edward L., and Frederick H. Schapsmeier. *Henry A. Wallace of Iowa: The Agrarian Years, 1910–1940.* Ames: Iowa State University Press, 1968.

Schlebecker, John T. *Whereby We Thrive: A History of American Farming, 1607–1972.* Ames: Iowa State University Press, 1975.

Sherow, James Earl. *Watering the Valley: Development along the High Plains Arkansas River, 1870–1950.* Lawrence: University Press of Kansas, 1990.

Taylor, Paul S. "Central Valley Project: Water and Land." *Western Political Quarterly* 2 (June 1949): 228–53.

Tindall, George B. *The Emergence of the New South, 1913–1945.* Baton Rouge: Louisiana State University Press, 1967.

Walton, John. *Western Times and Water Wars: State, Culture, and Rebellion in California.* Berkeley: University of California Press, 1992.

Wilcox, Walter W. *The Farmer in the Second World War.* Ames: Iowa State College Press, 1947.

Williams, Robert C. *Fordson, Farmall, and Poppin' Johnny: A History of the Farm Tractor and Its Impact on America.* Urbana: University of Illinois Press, 1987.

Worster, Donald. *Dust Bowl: The Southern Plains in the 1930s.* New York: Oxford University Press, 1979.

8 • DAYS OF UNCERTAINTY

THE LAST HALF OF THE TWENTIETH CENTURY BECAME AN AGE OF RISK and uncertainty for farmers. Although the 1970s were a boom period for farmers, prices did not keep up with operating costs, and many agriculturists no longer earned sufficient profits to remain in business, particularly if they had overcapitalized with the purchase of expensive machinery or acquired extensive lands on credit. When they could not meet their expenses, bankruptcy and flight from the land ensued.

Government policy remained based on the Agricultural Adjustment Act of 1938 with several minor changes, but the federal government tried to reduce the cost of the farm program by paring price supports while giving farmers more flexibility in production decisions by easing restrictions. In nearly every section of the country, farmers left the land, and rural populations decreased, often with a devastating affect on local businesses and communities.

By the late twentieth century, few young people had much opportunity to acquire land by purchase or rent to begin farming, and few acquired sufficient security to qualify for the credit or loans needed to start. As a result, the farm population grew older, and young people left the farms and rural communities for the cities. Moreover, with approximately 58 percent of the farms earning less than $20,000 in gross sales, agriculturists needed more income to maintain a satisfactory standard of living. Farmers increasingly took jobs off the farm, and part-time farming had become a way of life by the early 1990s. While the agricultural population dropped to less than 2 percent of the total population, however, productivity remained strong. Each American farmer provided agricultural products for ninety-three people by the last decade of the twentieth century.

THE SOUTH

An urban rather than a rural South emerged from the Great Depression and World War II. Manufacturing, insurance, banking, transportation, and other commercial developments provided some 4 million new nonfarm jobs in the cities of the eleven old Confederate states in the decade after 1945. Many people left the land for urban areas. Others sought industrial jobs in the North. As consumer incomes increased, the market improved for fruits, vegetables, and meat, and southern farmers began to specialize in high-value cash crops and livestock. They also decreased their reliance on cotton. As a result, agriculture in the South became more like farming elsewhere in the nation. Southern farming began to lose much of its distinctiveness, the emphasis now on capitalization, mechanization, and labor efficiency.

By the 1970s, landowners rather than tenants operated most southern farms. Black farmers had virtually vanished from the land. In 1987, only 22,954 African-American farmers remained, many working the land part-time. Only 2,014 produced more than $25,000 in agricultural commodities, and 10,662 sold less than $2,500 in farm products annually. Of 22,954 black farmers, 14,954 were full owners. As farmers converted cotton acreage to grasslands for livestock, they consolidated their holdings and purchased new land. Farm sizes increased, and the agricultural population declined. Southern agriculture now became white agriculture. The black farmer on whom southern agriculture had relied largely became a relic of the past.

Although southern farmers continued to expect aid from the federal government as a matter of right, government efforts to aid the poor small-scale farmers became entangled in the civil rights movement because black farmers could qualify for aid from various new government programs. In 1967, one African-American farmer in Alabama noted that "white folks got a lot more interested in machinery after the civil rights bill was passed." Although there may be some truth to that observation, most black farmers were forced off the land by technology in the form of cotton pickers and tractors, science in the form of herbicides, and government programs that favored landowners. They simply were not needed in the fields anymore. One observer called their exodus from the land a "quiet revolution in the old Cotton South." Without education and skills, many blacks moved to the towns or to homes in rural areas where they lived in poverty.

During the late twentieth century, the USDA, agricultural colleges, and state experiment stations remained devoted to helping the most capital-intensive and economically viable farmers, and those agriculturists were invariably white. The USDA ignored black farmers because they had neither the land nor capital to maintain productive, efficient, and profitable agricultural operations, nor did the agency provide educational and development programs to help those unneeded and often displaced farmers to build a new life.

By the 1970s, however, most southerners, black and white, had no attachment to the land. Technology, government policy, and economic change made their lives urban and relegated farming to a little-known endeavor practiced by an amorphous and unseen minority in the countryside. Moreover, many whites maintained a tenuous hold on the land and often needed off-the-farm jobs to make ends meet, remain on the farm, and maintain an adequate standard of living. For the small-scale farmer, black or white, who stayed on the land, poverty remained a fact of life.

During the 1980s, many farmers faced serious financial difficulties caused by overexpansion and investment in land and technology the federal government had encouraged. At that time, the federal government removed most acreage controls and urged farmers to plant more to take advantage of increased foreign sales, particularly to the Soviet Union. When American foreign policy temporarily terminated that market for grain and soybeans in December 1979, prices fell, and farmers could not pay their debts. Most farmers inaccurately blamed President Carter's embargo of grain sales to the Soviet Union in retaliation for its invasion of Afghanistan. Many turned to the Farmers Home Administration for credit, an institution of last resort when other lenders would not extend loans to farmers. More southern agriculturists went bankrupt and left the land. One Georgia farmer poignantly reflected the depressed agricultural conditions: "I'd rather have two tickets on the *Titanic* than be in farming today."

Most southern farmers who remained by the late twentieth century believed the federal government should give greater support to family farm operations rather than nonfamily corporate farms. They contended that nonfamily corporate farms should have stricter financial reporting requirements and fewer tax breaks than family farms. They particularly disliked corporate farms because those operations represented absentee ownership and an organization that had sufficient capital to buy out family farms.

At the same time, southern farmers were usually unwilling to remain on the land and maintain an unprofitable business merely to enable their children to live in a rural environment, nor did they want their children to become farmers. With little land for sale at any price and substantial capital required to begin farming, few young southerners had much realistic hope of becoming farmers. By the late twentieth century, only relatively wealthy people who had access to enough money to own or operate farms had much opportunity to begin or remain in farming successfully. Moreover, southern farmers tended to be older.

These characteristics of southern agriculture were far different from those only a few decades before when young, poor whites and blacks with little capital, land, and technology typified farmers in that region. California and Arizona became the major cotton-producing states. Cotton no longer reigned as king in the South. Where the white fiber had once colored the autumn landscape, soybeans now became the major cash crop. Although southern farmers continued to raise the traditional crops of cotton, tobacco, rice, sugar cane, and peanuts, the production of soybeans, hay, beef cattle, and poultry were now major agricultural activities. Southern agriculture had become diversified, not for subsistence but commercial production. The old agricultural South had gone forever.

THE NORTH

By the late twentieth century, dairying had become the major agricultural industry in New England, and Boston served as the primary market for fluid milk. With half the region's population living in Massachusetts, Springfield, Worcester, Merrimack Valley, Fall River, and New Bedford provided important secondary milk markets. With improved highways and refrigeration facilities, the milkshed extended farther into the interior than ever before, and Connecticut and Vermont farmers took advantage of urban demands, particularly in the Bay State.

With few exceptions, other than dairying, poultry raising, and vegetable and fruit production, however, New England farmers still suffered from the "comparative advantage" of midwestern farmers, especially for the production of livestock. Relatively cheap feed grains raised in the Midwest enabled farmers in that region to produce beef more efficiently; higher feed, labor, and property costs in New England prevented farmers in that area from competing profitably. Consequently, about 90 percent of the red meat consumed in New England came from outside the region. As a result, specialty production, such as dairying, that did not confront unfavorable competition prevailed.

Still, not all specialty-crop farmers profited by the late twentieth century. The tobacco farmers in the Connecticut Valley of western Massachusetts were defeated by technology. This region had been important for the production of tobacco for cigar wrappers and binders since the nineteenth century. In 1955, however, the first cigars containing manufactured "binders" or inside wrappers appeared on the market. These manufactured binders required less natural leaf, and the market plummeted. Tobacco farmers in the Connecticut Valley no longer enjoyed a "preferred position" for the production of cigar leaf tobacco. Moreover, the Commodity Credit Corporation warehouses had great surpluses of this tobacco. These problems, in addition to low prices and increased production costs, caused many tobacco farmers to adjust their operations to more profitable endeavors such as vegetables or to rely more on off-the-farm employment by taking industrial jobs within driving distance from their farms.

By the late twentieth century, some New England farmers experimented with the cooperative sales of vegetables directly to supermarket chains. These farmers, such as members of the Pioneer Valley Growers Association in western Massachusetts, received higher prices from direct sales, and the grocery chains preferred to purchase from the cooperative because it provided better supply, product uniformity, grading, and packaging than the wholesale dealers in Boston, Hartford, and Springfield. Moreover, the brokers for the wholesalers charged a 15 percent commission for selling the commodities of farmers, while the cooperative assessed each member only 5 percent of returns from sales. By 1986, one of three Massachusetts farmers sold directly to consumers through nurseries, greenhouses, roadside stands, and farmers markets. The dense population of New England made diversification and direct selling possible.

Despite the success of New England farmers, new and more serious problems than ever before threatened to overwhelm them and destroy the agricultural economy of the region. Urban sprawl increasingly took their land for housing developments. As farmland rose in value near urban and suburban areas, taxes increased, and farmers often could not earn sufficient profits to merit cultivation. Developers had greater capital resources and outbid farmers for land. In addition, New England agriculture was hindered by a reverse flow of people to the countryside. In contrast to the past, urban and suburban residents who had relatively high incomes now saw rural New England as an affordable haven from the expense and hectic lifestyle of the cities. As they moved into the countryside, rural New England became less agricultural and more residential. Many of these newcomers had families, which created new rural problems for schools, solid-waste disposal, and water. Farmers and other longtime residents of rural New England now began to fight the "overdevelopment" of the countryside, usually with little success. Many agriculturists who chose not to sell to developers often let their lands return to forest if they could afford to do so. By the late twentieth century, the urban threat remained stronger than ever before, and thousands of acres of agricultural lands passed to developers.

In the lower portions of the North and across the Midwest, farmers confronted other problems. In Pennsylvania, for example, farm income was comparable with nonfarm earnings because of a combination of profitable operations that were not particularly threatened by foreign competition and were supported by off-the-farm employment by the farm operator, his wife, or both. Even so, during the early 1990s, nearly 70 percent of Pennsylvania's farmers operated on a scale too small to support their families by agricultural income alone. Moreover, 15 percent of the largest commercial farmers in Pennsylvania produced 72 percent of the agricultural commodities. This consolidation of land and production not only kept small-scale farmers at a competitive disadvantage but harbored even greater problems for the future. With production consolidated on the farms of relatively few farmers, there was a danger that the agricultural-service industry, such as tire, fertilizer, and equipment dealers, would collapse because it did not have a sufficient farm base for profitable operations.

Across the Midwest, while production remained high, prices were insufficient to enable many farmers to maintain full-time operations. Because many farmers were "long on labor and short on land," they had little choice but to work away from their farms to support their families. By the mid-1950s, farm women began to take more outside jobs. A host of home conveniences saved time, but more often their families needed the extra income to live, pay debts, and make farm improvements. With real estate prices increasing annually, not all farmers could purchase more land and increase their economy of scale. Those who could afford to do so used more technology, hoping to expand production and thereby offset low prices and high capital investments.

Poultry farmers made other adjustments by integrating; that is, signing more contracts with processors. Although tomato and sugarbeet growers were accustomed to this business method, it was relatively new to the poultry industry. Poultry producers who signed contracts had a guaranteed market and price. Most who signed contracts were satisfied with the result, particularly because contracts reduced risk. Furthermore, contracts were collateral, which enabled farmers to acquire necessary capital, expand production, and increase efficiency. Contract farming, however, had the potential of stimulating overproduction, and it caused farmers to lose some independence because the buyer established the policy. At the same time, corn and soybean production remained great because farmers increased the application of fertilizer and diverted their poorest lands from production. Some farmers who did not participate in government price-support and production-control programs expanded production by planting more acres in corn.

Most midwestern farmers remained divided over the nature of the farm problem and its solution. Essentially, they agreed that supply had to be brought into balance with demand, and most believed that government policy had to be fashioned to achieve that balance while keeping prices at an acceptable level to preserve the family farm. As a whole, however, many midwestern farmers, particularly Farm Bureau members, did not believe that polices establishing marketing quotas or land-diversion programs were in their best interests. They favored a free market and argued that net farm income would increase if the federal government would terminate all programs. One Iowa farmer expressed this belief best: "American farmers want price supports and unrestricted production." They did not want a "political solution." Despite the incongruity of his remarks and government studies that showed that a free market would cause agricultural prices to decline precipitously, he believed that expanded exports and new industrial products, such as mandatory use of corn to make ethanol for automobiles, would solve the farm problem of low prices and overproduction. But these solutions required political action. This farmer, like many others, tried to have things both ways: high prices and freedom of action, along with a benevolent federal government.

With increasing costs, relatively low prices, and heavy indebtedness, the late twentieth century proved a time of incredible uncertainty for midwestern farmers, as it did for others across the nation. Few had the luxury of long-term planning. Most lived from year to year, hoping that interest rates would decline, good weather prevail, demand increase, and prices improve. While they endured this uncertainty, many would have agreed with the Iowa farmer who blamed the federal government for most farm problems: "We now have a farm program designed by politicians, not farmers, and administered by people who know little about farming or conservation." As in the past, many of these farmers looked for villains and panaceas, but the farm problem remained complex and unresolved near the century's end.

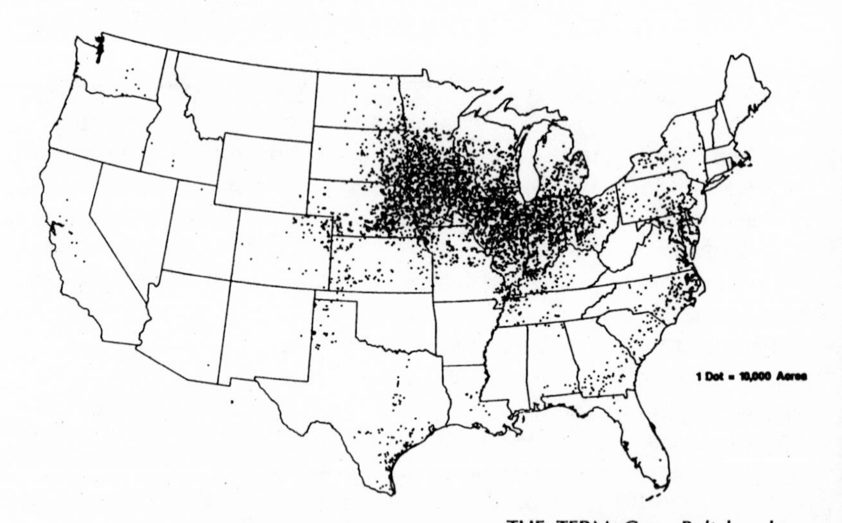

THE TERM *Corn Belt* has been used to describe the portion of the Midwest where farmers emphasized that crop and produced bountiful harvests. Most of the corn raised in the United States is fed to cattle and hogs. In 1987, the Corn Belt extended from central Ohio on the east to Nebraska on the west. *U.S. Department of Commerce, Bureau of the Census.*

THE GREAT PLAINS

By the mid-1950s, wind erosion had become a problem again for farmers in the Great Plains. Drought made the soil "powder dry." Bare soil on submarginal wheat lands that should have remained in grass blew with the nearly constant winds. The soil drifted and ruined crops in the old Dust Bowl, but the severity of the drought and the wind-erosion area extended far beyond the boundary of the 1930s. Great Plains farmers, however, were quick to begin an emergency tillage program. Some farmers had been negligent over the past decade, and "suitcase" farmers (absentee farmers who had to travel relatively long distances from home to work their fields) contributed to the problem because they were seldom available to apply the proper emergency conservation methods when their lands started blowing. Resident farmers, however, soon recognized the seriousness of the new wind-erosion menace and began to implement the soil-conservation procedures that they had learned during the 1930s.

Contour plowing, strip-cropping, and grazing management became standard farming procedures. Moreover, the economic situation was vastly different during the 1950s. Great Plains farmers had enjoyed profitable harvests and accumulated financial reserves by the time the drought and dust storms returned, nor were they still reliant on a single crop for their livelihood. Many had diversified by the mid-1950s with the help of irrigation and no longer faced financial disaster if a wheat crop failed. As a result, these farmers were able to stay on their land and practice the appropriate soil conservation techniques. By 1957, precipitation had returned to near normal, the drought had broken, and the dust storms had largely ended.

During the late twentieth century, many Great Plains farmers rapidly expanded irrigation agriculture by using center-pivot sprinklers as well as ditch and furrow methods. They believed irrigation solved the problem of drought by tapping the underground water supply. In the northern Plains, they used water from newly completed reservoirs, but in the southern Plains they tapped the Ogallala aquifer with wells. They were able to convert from raising dryland wheat and grain sorghum to irrigation agriculture where they emphasized corn, sugar beets, and alfalfa; in the southern plains, cotton.

These profitable cash crops helped farmers expand the raising of stock cattle and entrepreneurs establish commercial feedlots and packing industries on the southern Great Plains. These farmers now used water for more than supplemental irrigation—insurance during times of drought. While increasing production of high-value crops and earning substantial profits, however, they also began to deplete the underground water supply. By the late twentieth century, some wells were nearly 1,000 feet deep, and the cost of irrigation from "groundwater" became prohibitive for some farmers. In areas where pumping could not be sustained, farmers reverted to dryland practices, but this process reduced income and affected the local economy, including the ability of the land to support a larger population. With consistently high yields necessary to offset irrigation expenses, the cost of failure became high, and water mining remained a serious problem, particularly as towns and urban areas made increasing demands for water.

In the late twentieth century, northern Great Plains farmers still complained about discriminatory railroad rates. Farmers located far from terminal elevators now paid higher transportation charges for wheat than those close to markets. Often transportation rates made the wheat of northern Great Plains farmers less competitive with grain raised in areas closer to processing centers. Moreover, these farmers usually received lower prices where freight rates were high. In contrast to the late nineteenth century, railroads now charged proportionally more for a long haul than a short one. Short-haul rates were usually lower because the railroads had to compete with cheap truck transport of grain to market. Over longer distances, however, the costs leveled out, and the railroads could charge more for a long haul. Intrastate shipments also cost more because the

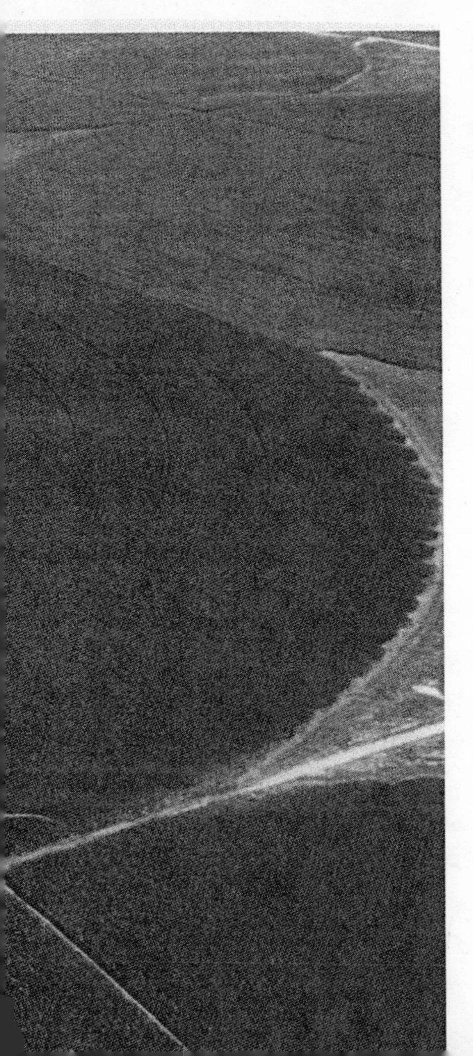

CENTER-PIVOT IRRIGATION sprinklers use water pumped to the surface from deep wells, the sprinklers traveling in a circle, and the largest can irrigate a section of land (640 acres). Center-pivot irrigation systems have enabled farmers on the Great Plains to raise corn and alfalfa where insufficient natural precipitation would make those crops impossible or difficult to cultivate profitably. These systems, however, draw more water from the ground than is restored naturally. As a result, the groundwater supply is rapidly diminishing. *Valmont Industries, Inc.*

Interstate Commerce Commission could not regulate those rates.

Although Great Plains farmers confronted unique environmental problems, they also faced other difficulties and trends. Land sales, for example, usually increased the size of farms and decreased the number of farms and the agricultural population. Health, age, and financial situations were the most compelling reasons for selling or buying land. Usually the buyers lived nearby and paid higher prices than those who lived far away and bought land merely for investment. The farmers who bought land to expand their operations usually intended to pay for it with their agricultural income.

BY THE LATE twentieth century, self-propelled combines quickly and easily harvested the seemingly endless wheat fields in eastern Washington where forty-horse hitches had pulled other combines a century before. With a self-propelled combine, one farmer and a truck driver could harvest the crop, no matter how large. Technological change like this made the grain harvest easier, but also released workers from agriculture and contributed to the consolidation of farms. *Washington State Historical Society.*

They did not purchase the land merely to rent it to someone else. Land prices were determined by a host of factors, such as the wheat allotment, acreage in cropland and pasture, availability of irrigation, the number of buildings, road quality, and distance to town. In some areas such as western Kansas and eastern Colorado, absentee owners tended to be more common.

Many farmers who could not afford to buy land rented it from local landlords. The full tenants, those operators who did not own land, tended to be young entry-level farmers, while most landlords were retired farmers. As farm size increased, Great Plains farmers relied less on hired labor and more on technology, particularly tractors. By the early 1960s, however, many farmers were giving up their combines in favor of hiring custom cutters. They now preferred to avoid the costs of buying and maintaining a combine while ensuring a timely harvest at an affordable price. In the Dakotas especially, farm size often became the "critical factor" that determined success or failure, and livestock production increased while grain production declined.

The cooperative movement remained strong in the Great Plains by the late twentieth century, particularly for grain and petroleum associations. General merchandise and grocery cooperatives had been largely displaced by supermarkets and discount stores in the major towns and cities. Population decreases continued to hurt the tax base and the maintenance of schools, churches, community services, and local merchants. Only the county seats or small cities with at least 10,000 people were able to provide adequate services to the farm community. With a rule of thumb that the loss of every seven farm families meant the loss of one business, by the 1980s many main streets had become a "disaster" across the Great Plains.

THE WEST

During the late twentieth century, California continued to epitomize western agriculture for most Americans, and farming there came to rely almost entirely on irrigation and specialty-crop production. Indeed, by 1974, 90 percent of the harvested cropland had been irrigated, primarily by the State Water Project and the Central Valley Project. At the same time, technology decreased the reliance of the growers on migrant labor, and science helped to increase production. Farm consolidations continued, and corporate agriculture expanded. Family farms decreased in number.

Although population increases provided agricultural markets, the coastal cities began to force crop and livestock production to concentrate in the Central Valley and the desert region. Many California farmers faced the same problems of those near urban areas in New England. In 1993, dairy land near San Bernadino averaged approximately $50,000 per acre, but it brought twice that price in the hands of developers. As property values rose because of metropolitan growth, farmers often could not earn a sufficient income to provide for their families and pay property taxes. Many sold their lands to the developers.

After the Korean War, the beef-cattle industry changed from

CUSTOM CUTTERS harvested much of the wheat in the Great Plains during the late twentieth century. These entrepreneurs purchased several combines and trucks and hired a crew to follow the wheat harvest. Beginning in Texas during late May or early June, they finish in North Dakota in November or December, cutting grain sorghum and sunflowers. Custom cutters freed the farmer from the expense of purchasing and maintaining harvesting implements. They charged by the acreage and also received a fixed amount per bushel. Farmers often contracted with the same custom cutters each year and welcomed their arrival when the wheat was ripe. *South Dakota State Historical Society.*

grazing to finishing beef for market at large-scale feedlots. The feedlot cattle companies preferred to integrate their operations as much as possible by raising their own feed grains and hay. Like the specialty-crop farmers, the cattle companies depended on cheap and plentiful water for their irrigated land. By 1975, approximately 42 percent of the water used for agricultural purposes in California irrigated feed crops. Still, California could not meet its own demands for beef and necessarily imported large quantities from other states. California also remained a major producer of field crops such as cotton, rice (including domestic wild rice), and sugar beets. Most of the field-crop and specialty-crop production came from large-scale owners who profited from governmental price supports and acreage limitations.

During the 1960s, large corporations expanded their investments in California's agriculture by purchasing additional land, and corporations that had not been involved in agriculture now began farming. In contrast to the incorporation of family farms for financial protection and estate planning, this form of corporate farming was designed for production and as an "investment asset." Shell Oil, Tenneco, Dow Chemical, and other corporations invested heavily in California land. The vineyards were popular attractions for major corporations. In the late 1970s, National Distillers purchased the Almaden Vineyards, and Seagram bought the holdings of the Christian Brothers and Paul Masson. As a result, the term *agribusiness* became a household word, reflecting vertical integration and the big-business nature of California agriculture.

Insurance companies such as Prudential and John Hancock and foreigners such as British Midhurst also entered California agriculture for investment purposes rather than profits based on farm production. Basically, these corporate investors operated on the old premise that land was a good investment because "they aren't making it anymore." When the price reached certain levels because of population pressures and demands, these corporate investors sold it to reap great capital gains rather than crops. At the same time, corporation-owned farms commanded greater capital and access to credit than the largest-scale individual farmers, and the corporations could marshal their resources to meet increased production costs with little difficulty or ride out hard times with less financial stress than individual farmers. If production did not meet costs, these entities could recover their losses with tax write-offs, while individual farmers usually had to get another loan or sell out and leave the land.

Individuals often pooled their capital resources to invest in land to circumvent the 160-acre limitations for the purchase of irrigation water and to draw federal payments for acreage reductions. This technique enabled them to hire a manager to operate the holding as a single farm while technically keeping each individual's parcel of land separate for various benefits such as water allotments and government programs. Large-scale and corporate farms were essentially "factories in the fields" with high capitalization, intensive specialization, and ready access to cheap federal water provided by Bureau of Reclamation projects and more expensive but still affordable water supplied by the State Water Project. Many growers also held long-term contracts for federal water in the Central Valley that kept the price far below actual water costs. Urban areas that had increasing demands for water paid considerably more for it and therefore subsidized the water used by irrigation farmers in California. As the twentieth century drew to a close, this problem became increasingly serious and often bitter.

The large-scale growers also continued to use cheap *bracero* labor to plant, cultivate, and harvest their crops. The number of braceros working in the United States peaked at 192,438 in 1957, when they formed nearly one-third of the nation's farm labor force. As late as 1960, braceros composed approximately 25 percent of the agricultural workers in California and the Southwest. More than 80 percent of the braceros worked for the large-scale growers, who raised lettuce, tomatoes, asparagus, and strawberries. Despite legal provisions regulating wages, hours, and living conditions, braceros were little more than migrant peons, locked in a life of perpetual poverty and stoop labor in the fields. The bracero program depressed wages, ruined the bargaining power of the workers, and drove away local labor, both Anglo- and Mexican-American.

By perpetuating the lowest wages and the worst working conditions, however, the growers were able to keep domestic labor from the fields and thereby plead the need to maintain the bracero program, which provided cheap, docile, and "dependable" workers. The growers would continue to use braceros until the civil rights movement of the early 1960s brought national shame to this federal policy. Although the bracero program ended with the 1964 crop year, illegal or "wetback" workers and legal temporary workers with "green cards" soon continued the tradition of cheap labor in the vegetable fields, orchards, and vineyards of California.

THE NEW COUNTRY

In 1959, Alaska and Hawaii joined the Union. Agriculture had been practiced in both areas long before the territorial period or statehood, but farming in "Seward's icebox" and the "Aloha State" now became officially a part of American agricultural history. Before World War II, Alaskan farmers conducted little more than subsistence farming, although small-scale and temporary commercial production developed during the heyday of the mining camps in the early twentieth century, particularly in the Tanana Valley near Fairbanks and the Matanuska Valley after the construction of the Alaska Railroad. World War II, however, stimulated federal support for farming in Alaska to help ensure adequate food supplies in case "stateside" shipments were disrupted by the enemy.

With the development of Alaska as an important military outpost and the coming of the Korean War, the military created a market for Alaskan farmers. The harsh climate and the market mandated that farmers specialize their production, particularly with feed grains such as barley and with dairying for fluid milk sales. By 1960, Anchorage, with 37 percent of the state's population, provided the largest agricultural market, which farmers in the Matanuska Valley worked to supply. Fairbanks provided the second most important market, which Tanana Valley farmers served. By 1967, approximately 75 percent of Alaska's agricultural commodities by value came from Matanuska Valley farms, while the Tanana Valley provided 13 percent. Milk accounted for approximately 70 percent of that production value; barley and red meat provided 21 percent, vegetables and potatoes 9 percent.

Many people believed that Alaska, with its 15 million arable acres, could develop an important agricultural economy almost overnight, evidently assuming that a direct transfer of agricultural knowledge from the lower forty-eight states and technology, crops, capital, and institutions would occur without adaptation or exorbitant cost. This assumption seemed justified during the early 1970s when the construction of the Alaska pipeline created a boom economy as the population expanded and increased the demand for agricultural commodities. By the mid-1980s, however, the boom had collapsed, and fewer consumers spent less money for farm products.

The problems were many. Federal law prevented farmers from acquiring more than 160 acres of the public domain. As a result, farmers located near rapidly expanding cities and towns could not adequately expand production to take advantage of increased de-

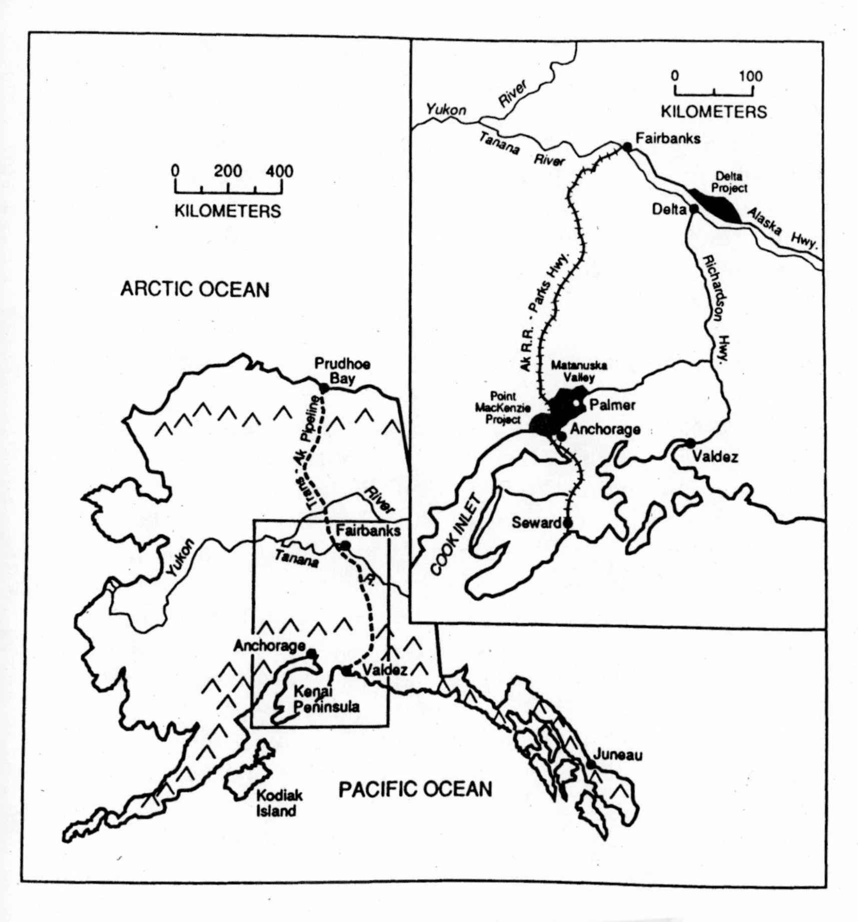

ALASKAN FARMERS have been most successful in the Tanana Valley near Fairbanks and the Matanuska Valley near Anchorage. The short growing season and the urban and military populations of these areas encouraged dairy farming and raising some vegetables. The history of agriculture in Alaska has been marked by hardship and unfulfilled expectations. It remains America's agricultural frontier. *Roger W. Pearson and Carol E. Lewis, "Alaskan Agribusiness: A Post-Statehood Review," 5 (July 1989); copyright 1989 by John Wiley & Sons, Inc.; reprinted by permission of John Wiley & Sons, Inc.*

mand. As they lost markets and profits, their costs remained high. Other farmers were unable to develop profitable operations because they were too remote from their markets and the absence of adequate transportation and marketing organizations and facilities made commercial production too expensive. The expense of clearing forests and the cool, short growing season further discouraged successful farming.

Many Alaskan farmers became part-time operators, but their absence from the farm often hindered the development of the best managerial practices. Few were able to acquire capital through savings or borrowing to expand their acreage, specialize, and overcome transportation and marketing problems. Moreover, the high costs of production and transportation to outside markets meant that Alaskan farmers had to depend on limited local markets. Few experienced and well-capitalized farmers went to Alaska, and entry-level farmers lacked both money and experience. Because most farm supplies had to be imported, prices were high, and delivery took time. Little wonder that many quickly failed.

By the early 1970s, dairy farmers had developed a marketing system that provided access to military and civilian markets. Even so, dairy farmers entering the business needed capital resources of approximately $150,000 to ensure success. Despite high capital requirements for the purchase of land and equipment, many agricultural experts saw great potential for Alaskans farmers not only for dairying but for livestock raising and the production of forage and hay crops. In 1977, the state of Alaska began a program to aid agricultural development, selling farmers rights to use its lands and requiring development and production. The government was particularly interested in developing barley farms near Fairbanks in the interior and dairy farms near Anchorage on the periphery of the state. In 1978, the state sold 103,000 acres and provided loans to help these new farmers clear the land. The farmers had to promise to produce grain and milk for market in three years. By 1984, however, the international market for barley had collapsed, and the state still had not built a promised terminal grain elevator. Dairy farmers could not compete with milk shipped from Seattle, despite the added charges for nearly two thousand miles of transport. By the early 1990s, the state bought much of the milk it had helped pay farmers to produce. One farmer called this fiasco "negative farming." Because these new farmers did not have clear title to their lands, some claimed that farmers were "basically indentured servants to the state." Ultimately, the state foreclosed on $40 million in loans to about ninety farmers, virtually all those involved in the new projects.

On October 21, 1986, the federal government terminated all homesteading in Alaska. The settlement of free public land in the United States had come to an end. By the early 1990s, few homesteaders remained. The most stable farmers in Alaska were those who capitalized on urban and military markets for fluid milk and those who raised vegetables on truck farms, particularly in the Matanuska Valley where they took advantage of the long daylight hours during the brief summer months and traded in the Anchorage market. Still, farming on the "economic fringes" of the high latitudes continued to be a risky business, and Alaska remained an agricultural frontier.

Hawaii had a longer and more economically important agricultural history. In 1778, when Captain James Cook made the first European contact with the islands, Hawaiians practiced farming, particularly raising taro with supplemental irrigation in pondfields. During the nineteenth century, sweet potatoes, introduced from South America, sugar cane (the first plantation was established in 1835), pineapples, bananas, coconuts, and "Indian corn" became important crops. Cattle introduced to the Waimea region in 1793 provided the basis for a ranching industry.

By the late twentieth century, cattle raising was hindered by limited lands and a lack of hybrid feed grains. Large-scale farmers, who raised sorghum feed grains, had great difficulties reaping satisfactory yields because of crop damages from birds, insects, and diseases. Bird-resistant varieties of sorghum grains that contained high quantities of tannins to discourage the birds also reduced the quality of the crop. Consequently, farmers could not take full advantage of the long growing season to reap the largest crops possible and earn the most profit from their labor. The warm, moist climate also compounded insect and storage problems. Although Hawaii's climate created conditions for great agricultural versatility, the environment provided limitations that farmers worked to control or learned to accept.

While agriculture, particularly sugar cane and pineapple raising, had shaped the Hawaiian economy since the early twentieth century, declining markets and increasing costs, together with greater competition for land by developers, placed it in jeopardy. In addition, Hawaii's agricultural economy became limited by the growth of the tourist industry and military expansion. By 1960, tourism had attained the same economic status as the sugar industry, and defense spending exceeded the total combined sales of sugar cane and pineapples. Both industries increased local incomes and lured workers from the fields, although many found jobs in agriculturally related industries such as food-processing plants. Urban areas expanded, and land values increased, often beyond the ability of farmers to keep their acreage.

By the mid-1970s, Hawaii's agricultural sales could not cover the costs of the state's farm imports, and by the early 1980s agriculture employed only 3 percent of the population. Moreover, agricultural income had declined to 2.2 percent of Hawaii's personal earned income. As in most agricultural endeavors, only capital-intensive farmers had much chance at survival by the late twentieth century. Although Hawaiian farmers began to diversify with other specialty production such as macadamia nuts, papayas, and floral and nursery crops after 1960, by the early 1990s sugar remained the leading export with pineapple products a distant second. For the time at least, Hawaii remained the "Sugar Kingdom."

AGRICULTURAL POLICY

Since World War II, Congress has been pressured by urban consumers demanding lower food prices, social-service agencies needing government food surpluses for distribution, and agribusinesses advocating maximum production to hold down operating costs while expanding sales. At the same time, the rural population continued to decrease, and representatives from farm districts held fewer seats in the House. With less than 2 percent of the population engaged in farming by the late twentieth century, Congress could ignore major agricultural organizations, commodity groups, and the farm vote more than ever before. USDA proposals also depended less on the wishes and support of the farm lobby because it could no longer marshal political retaliation against an administration or members of Congress. Members now looked to the specialty commodity groups to provide reliable information about constituent needs because these groups commanded the necessary technical information and provided money for election campaigns. Moreover, commodity groups supported moderate agricultural prices and favored neither a laissez-faire nor interventionist government in agriculture. Congress also listened to the specialists at the land-grant colleges and experiment stations. Congressmen from the farm states particularly supported funding for research at these institutions.

By the 1960s, rural congressmen also depended on the support of their urban colleagues for farm legislation. Compromise became more important than before, rural congressmen usually promising to support food-stamp and other social programs in return for urban congressional support for agricultural legislation. Failure to support agricultural legislation no longer meant that a congressman would face difficult reelection odds. Large-scale farmers, usually represented by the Farm Bureau, urged Congress to reduce government intervention in agriculture and favored a free-market economy. Other farmers, usually represented by the National Farmers' Union, favored high price supports. Specialized commodity organizations urged Congress to support programs specific to their needs. Members of Congress tried to address these pressure groups while responding to the needs of their constituents. Disagreements and vacillation often prevailed before compromise could be reached on agricultural policy.

Democratic-controlled congresses and Republican presidents often clashed after World War II over agricultural policy. Republican congressmen generally favored flexible price supports and the removal of the federal government from agriculture to allow the marketplace to dictate demand and prices, thereby influencing production and supply. They also favored decreasing expenditures for the purchase and storage of agricultural commodities that had accumulated in government warehouses since the Agricultural Adjustment Act of 1938 had instituted a policy that encouraged abundance rather than scarcity in agricultural production. Democratic congressmen advocated governmental intervention to achieve economic and social ends for agriculture and rural communities. Congressmen from both parties increasingly relied on the president and the secretary of agriculture to propose farm legislation, but they did not automatically approve administration proposals. Often they rejected suggestions for policy changes because of a host of political considerations.

Following the Korean War, for example, Congress attempted to solve the farm problem with the Agriculture Act of 1956, authorizing the two-part Soil Bank Program. This program enabled the federal government to rent land from farmers to remove it from production. The short-term Acreage Reserve Program permitted farmers who participated to receive federal payments for reducing their acreage of basic crops. This program differed from previous policy, however, because those farmers could not plant the withdrawn acreage with other crops. The second part provided for the long-term Conservation Reserve Program to encourage marginally successful farmers to leave agriculture. This feature authorized the government to pay farmers for reducing their cultivated acreage, provided they put that land in a long-term conservation program, thereby withdrawing it from production.

The Soil Bank Program, however, did not substantially reduce production because farmers idled their worst lands and increased production with fertilizer on their planted acreage. It did make a significant ecological contribution to conserving the environment, but the program soon became too expensive to continue. In addition, leaders, merchants, and businessmen in the rural communities complained that idled farmland ruined the local economy because farmers earned less money from idling lands in the Soil Bank Program than from crop production and spent less in town. Congress abandoned this program in 1959.

During the late 1950s, Democratic congressmen continued to advocate high price supports with production-control policies to maintain adequate incomes for family farmers. Republican congressmen, however, favored a reduction in federal expenditures for farm programs and sought export and market expansion and decreased government interference in agriculture. By 1960, moreover, science and technology had defeated all efforts to reduce surplus production and increase prices. At the same time, Congress became increasingly concerned with urban problems, and a consumer movement opposed price-support programs because they increased food costs while guaranteeing a certain level of farm income. Agricultural policy no longer remained the domain of farmers and their lobbying groups, particularly after the U.S. Supreme Court ruled in 1964 that congressional districts required apportionment based on population. This ruling gave increased congressional representation to urban areas and decreased representation in the House for agricultural and rural areas.

In 1961, President John F. Kennedy supported mandatory production controls to reduce surplus production and higher price supports to maintain farm income. Democratic congressmen, led by Hubert H. Humphrey, agreed and provided a "supply-management" bill designed to gain high agricultural prices by restricting farm production. The Farmers' Union supported the Democratic plan, and the Republican party continued to favor the Farm Bureau's preference for flexible price supports based on production. Congress responded by giving Secretary of Agriculture Orville L. Freeman the authority to initiate a payment-in-kind policy (expanded in 1983) that enabled farmers who raised grain crops to reduce their acreage and receive price supports for half their production in the form of negotiable certificates. Congress authorized the Commodity Credit Corporation (CCC) to redeem those notes with grain from its warehouses, which farmers could then sell—thus the term *payment-in-kind.* Few farmers actually received grain in payment, however, because they chose to allow the CCC to market it and send them a check. With the Agricultural Act of 1961, Congress expanded the school-lunch program, and in 1964 Congress approved a food stamp program to increase the consumption of agricultural commodities the government had purchased through it price-support programs.

By the mid-1960s, partisanship over farm policy began to fade as Congress turned its attention to voluntary rather than mandatory acreage reduction for basic commodities while still favoring high price supports. Congress also consolidated and expanded its agricultural programs with the Food and Agriculture Act of 1965. The title of this act clearly indicated that Congress had agricultural concerns beyond the farm and sought wider political support by using agricultural programs to aid nonrural constituents and communities. This legislation permitted voluntary acreage controls for feed grains, wheat, and cotton, with price supports established near world market levels and an economic development program to fight rural poverty. Tying domestic prices to world prices gave foreign production a considerably greater influence on domestic prices. With a 1 percent increase in production capable of triggering a 7 percent decline in prices, American farmers became increasingly influenced by world production. And with world production varying as much as 3 percent a year from weather conditions alone, American farmers now experienced greater uncertainty than ever. In contrast to the past when Congress approved short-term agricultural legislation each year, Congress now extended its agricultural program to four years. This policy ended the annual political fighting over an agricultural bill.

In 1970, Congress responded to the Nixon administration's request to reduce government spending by limiting each farm to $55,000 per year in price-support payments for farmers participating in acreage-diversion programs. This policy was intended to protect the family farmer while reducing the support for large-scale corporate farms. Three years later, Congress limited price supports to $20,000 annually for grain and cotton farmers. This legislation, known as the Agriculture and Consumer Protection Act, replaced the concept of price supports with "deficiency payments" whenever commodity prices dropped below a certain target price. Beginning in 1976, those target or parity prices were based on an index of production costs, that included taxes, interest rates, wages paid, and other operating costs. This program responded to world crop shortages, a falling dollar, and increased foreign demand that helped expand exports and reduce government reserves.

By the early 1980s, food prices had inflated rapidly, and Congress responded by refusing to increase most agricultural price supports, even though farmers still received low prices compared to their operating costs. In 1981, President Ronald Reagan and Secretary of Agriculture John Block asked Congress to help reduce the budget deficit by increasing agricultural exports and further decreasing price-support payments.

Congress responded with the Agriculture and Food Act, which reduced agricultural spending and limited payments to $50,000 per year to each farmer who participated in government agricultural programs. This legislation, which also expanded commodity donations abroad under Public Law 480, required the secretary to compensate farmers at the rate of 100 percent of parity for financial losses incurred during times of economic sanction, as during Jimmy Carter's embargo of agricultural sales to the Soviet Union from January 1980 to April 1981. With the exception of the Farm Bureau, all agricultural groups opposed this legislation. Clearly, Congress no longer depended on the farm lobby for advice and support.

Despite the adjustments in agricultural policy since the mid-1950s, by the 1980s, overproduction, low prices, and high government costs continued to plague the farm program. In 1985, Congress once again attempted to solve the farm problem, with the Food Security Act. This statute tried to maintain farm income while reducing costs and covered five crop years. During that time, target prices would fall closer to market demands and help reduce government spending. It also provided for acreage reductions if agricultural supplies became excessive. Like the agricultural legislation before it, this act did not solve the problem of overproduction and low prices. In 1990, Congress tried to resolve this difficulty once again with the Food, Agriculture, Conservation, and Trade Act. This act continued to provide loan rates linked to market prices and authorized price supports based on target prices, but the act also increased environmental regulation. By the early 1990s, however, this program had not solved the problems of an imbalance of supply and demand, an inadequate income for family farmers, or an escalating national debt. Agricultural policy now depended on the support of urban liberals.

Since the Korean War, then, the basic features of federal agricultural policy have been export expansion, increased food consumption at home through various agencies, expanded foreign aid, support prices near world levels for basic commodities, direct income payments to farmers (who participate in production-control programs), acreage controls, voluntary production controls (linked to payments and specific loan rates), target prices, and deficiency payments. If farm policy was not entirely successful, it remained complex and often confusing.

AGRICULTURAL TECHNOLOGY

By the late twentieth century, the tractor and the mechanical cotton picker remained the most important technological "hardware" in American agriculture on a national and regional basis, respectively. Each had contributed to profound changes in farming and rural life. The tractor, for example, became the key to the mechanization of the farm; it also helped cause the industrialization of agriculture and the technological revolution that led to the modern agribusiness industry. Moreover, the tractor helped end farming as a way of life. When a farmer purchased a tractor and equipment specially designed to meet the draft capabilities of this new implement, he often borrowed from the bank. To be received favorably by loan officers, farmers had to prove they were good managers of capital and land. Large loans mandated that farmers keep better financial records, increase production, and decrease unit costs.

The diversified farms of the past required a greater variety of implements than one-crop agriculture, but tractors and accompanying implements required a larger capital investment. Consequently, farmers who bought tractors increasingly concentrated on the production of one or two crops such as wheat, corn, or sorghum grains. These crops produced the greatest volume, thereby reduced unit costs, and provided the highest returns on their investment. While farmers increased the production of these grains, they also drastically reduced their acreages of forage crops for draft animals. Indeed, after the mid-1950s, farmers had virtually no need for horses or mules. Increased yields due to better timing for planting, cultivating, and harvesting with the aid of a tractor, however, frequently caused overproduction and low prices. Farmers then tried to offset low prices by planting more acres.

Tractors also decreased the need for hired labor and tenants for most aspects of crop production. While economists calculated a tractor's savings in labor and money, these implements had a more personal influence on people than the sterile statistics indicate. Indeed, the savings of time and money meant that many farm workers left the land to seek jobs elsewhere. Whether they improved their condition no one can accurately say. Some, no doubt, advanced their fortunes; others probably regretted that technological change had altered their lives forever.

The tractor had other effects on American agriculture during the late twentieth century. Tractors enabled farmers to plant and harvest more acres, but small-scale farmers often could not compete with farmers who could afford this implement and had sufficient land to make it profitable. And increased production, resulting from acres planted with a tractor, drove prices down. As a result, farm sizes continued to increase as small-scale farmers withdrew from agriculture. Moreover, tractors helped decrease the need for large families to help with the farm work.

Smaller families, larger farms, and fewer agricultural workers and farmers had a corresponding influence on the small towns in the countryside. These changes, together with new forms of transportation such as cars and trucks, changed marketing and purchasing patterns in rural America. Although the tractor is not entirely responsible for the demise of small towns, this technology helped eliminate many people from the land. Fewer people meant that local merchants often could not earn sufficient profits to remain in business.

Tractors also compacted the soil, which retarded drainage and root penetration. Tractors, however, enabled those who remained on the land to extend their working lives longer than before. All of these changes did not happen overnight, but these effects became increasingly apparent and important following the rapid adoption of the tractor after World War II, and during the late twentieth century they were readily apparent. By the early 1990s, approximately 4.5 million tractors gave an average of more than two tractors per farm. Many of those had sufficient power to pull multipurpose rigs, including plows, grain drills, and liquid fertilizer, pesticide and herbicide applicators.

By the late twentieth century, the tractor had not solved all farm problems; in fact, it had created new ones that troubled life in rural America. Nor had tractors been able to supply all power requirements on the farm. Trucks provided a cheaper and more convenient and efficient way to haul grain and livestock to market than tractors and wagons. Consequently, farmers used tractors primarily for pulling implements through the soil, as in the past. With the exception of improvements such as cabs, roll bars, air-conditioning, and stereos, the basic structure of the tractor had not changed substantially since the early 1920s. Even so, it remained the most important piece of technology on the American farm.

By the late 1950s, the mechanical picker had achieved beltwide appeal and affordability in the cotton South. It was no longer a machine primarily found in the cotton fields of California. By 1957, mechanical harvesting costs had dropped below hand-labor expenses in the South and remained lower than hand picking. But although the differences in cost between hand- and machine-picked cotton are important for stimulating or retarding technological change, the analysis of these costs alone places too much emphasis on the influence of the cotton picker for displacing agricultural labor and not enough on the importance of the tractor's reduction of the preharvest labor force. One can easily overlook another important technological "push" on agricultural workers in the South.

The slowness to adopt the cotton picker reflects the influence of cultural institutions such as small-scale production and the propensity of plantation owners to use annual labor contracts to guarantee an adequate number of tenants throughout the crop year. Still other problems involved inadequate capital, insufficient financing, and the absence of preharvest mechanization. Plant breeders were able to change the plant sooner in California than in the South to help farmers mechanize.

In California, they shortened the growing season by as much as thirty days to permit harvesting before the onset of wet autumn weather, and they raised the bolls by 4 inches to reduce the tendency of the mechanical harvester to collect dirt and weeds. Agricultural scientists also discovered the advantages of close planting to facilitate machine picking and developed defoliants to reduce the plant material on the stalks at harvesttime. New herbicides eliminated the need for frequent cultivating. An efficient extension service spread the knowledge about the benefits of mechanized harvesting throughout California's cotton-producing region. As a result, California led all of the states in the mechanical picking of cotton and became the first state to mechanize the cotton harvest fully during the 1950s. In contrast, the South did not completely mechanize the cotton harvest until 1975.

With the perfection of the mechanical harvester, farmers could mechanize their cotton crop entirely, if they could afford to. Mechanization saved time and money and eliminated the problems caused by an uncertain labor supply. At the same time, mechanical har-

FARMERS IN ARIZONA and California had mechanized the cotton harvest by the late 1960s. Cheaper labor, small fields, and less production per farm slowed the adoption of the mechanical cotton harvester in the South. By 1975, however, the cotton harvest had been mechanized in that region, largely because of farm consolidation, increase of field size, and science, which changed the cotton plant to make it easier to harvest. *Deere & Company Archives.*

vesters contributed to the consolidation of small-scale farms into larger holdings. By 1974, the number of cotton farms had decreased to 89,500, and the crop had been essentially mechanized. By the late 1980s, the number of cotton farms dropped still further to 42,971 on which 42,914 pickers and strippers harvested the crop. This decline, however, was not due entirely to the adoption of the mechanical cotton picker but also to the expansion of the synthetic clothing industry.

By the late twentieth century, a new development in technological hardware also brought profound change to another segment of American agriculture. World War II had created a demand by growers and processors for a mechanical tomato harvester. High wartime prices and an inadequate labor supply encouraged agricultural engineers to develop a machine that could harvest the crop quickly and eliminate the growers' dependence on migrant workers, who picked the fields three or four times at intervals of one or two weeks. Tomatoes, however, posed greater problems for the development of a mechanical harvester than most crops. The vines jammed moving parts, and the fruit did not ripen uniformly. Moreover, thin-skinned tomatoes bruised and cracked easily, which meant economic loss because the processors paid lower prices for damaged fruit. Any tomato harvester had to be a once-over machine because it would damage the plants so much that additional pickings would be impossible.

The complicated nature of tomato production led to a "systems approach" in which agricultural engineers, horticulturists, and agronomists pooled their collective talents to build a mechanical harvester and breed a tomato that could withstand machine harvesting. They developed new tomato varieties that set and ripened uniformly and resisted bruising and cracking during rough mechanical handling. They also changed planting, cultivating, and irrigation practices—all of which made mechanical harvesting possible. Because the new tomato varieties were not uniformly round, however, consumers rejected them. Consequently, machine-harvested tomatoes were suitable only for processing into catsup, juice, or paste.

Tomato processors supported the development of the mechanical harvester by adjusting production schedules to meet the large quantity of fruit harvested at one time. The processors also lowered their purchasing standards to accommodate the growers who used the new varieties and technology. This subsidization by processors encouraged growers to adopt or expand planting the new tomato varieties and to use mechanical harvesters. Between 1961, when growers first used the mechanical harvester commercially, and 1967, when they picked virtually the entire processing crop mechanically, the machine-harvested acreage increased more than 90 percent, the number of implements 80 percent. The systems approach enabled the scientific and technological problems of the tomato harvest to be solved quickly compared to other crops such as cotton.

The mechanical tomato harvester, like the mechanical cotton picker, contributed to the large-scale technological displacement of farm workers. Thousands of workers lost their jobs. The rapid mechanization of the tomato harvest in California was largely stimulated by the growers' desire to resolve a perceived labor-shortage problem after the termination of the bracero program in 1964 rather than by per-acre or tonnage savings, although those were important considerations. With the termination of the bracero program, the growers sought mechanization to replace that cheap source of labor.

In California, the mechanical harvester also encouraged growers to expand production. The tomato harvester did not increase yields per acre, but this implement reduced harvesting costs by substituting capital for labor. Only the large-scale growers could afford the $150,000 tomato harvesters, and small-scale growers could not compete with those who purchased these machines. As a result, the tomato harvester helped to decrease the number of growers nationwide while it helped to increase the acreage planted. In addition, the completion of the irrigation system in the San Joaquin Valley encouraged growers to specialize in tomato production and to convert other agricultural lands to this farming endeavor.

The systems approach to the mechanization of the tomato harvest combined basic and applied scientific research. The results of that research aided the large-scale growers and processors while eliminating most small-scale farmers and migrant workers from tomato production. The mechanical tomato harvester is an excellent example of the way technology can help some groups while harming others. Regarding the development of a mechanical tomato harvester and other forms of agricultural hardware by the late twentieth century, some critics argued that technological unemployment resulted because research was often supported by a vested "class." As a result, the new technology aided large-scale, capital-intensive rather than small-scale agriculturists, the latter most often associated with the family farm. Critics also contended that chemical fertilizers, pesticides and herbicides, expensive hybrid seeds, and machinery were the products of interest-group technological research. In contrast, other critics advocated more technological improvement to increase agricultural production and lower food prices.

In general, technological innovation by the late twentieth century served the farm owner rather than the agricultural worker. Consequently, farm workers usually lost to commercially oriented inventors, who in turn were influenced by capital and profit-oriented commitments of private industry. Certainly, some innovations, such as the tractor, improved the work of those who operated them, but in the case of the tomato harvester, an estimated 32,000 workers lost their jobs in California, and approximately 75 percent of small-scale growers sold their land to large-scale operators because they could not afford to compete. Thus technological change has sometimes solved old problems while creating new ones.

Technological change has invariably been based on economic motives that have caused lasting social consequences for small-scale farmers and agricultural workers. By the late twentieth century, technological development did not necessarily mean progress for everyone; for some, it boded technological unemployment. No easy solutions existed for this complicated problem because it involved matters of ethics and philosophy, not merely economics and hardware. In the final analysis, however, technology remained neutral, although the way farmers used it was both beneficial and harmful and those who employed it were both arbitrary and magnanimous. At the very least, technological development in late-twentieth-century agriculture was rapid and complex, and it fostered great economic and social change.

AGRICULTURAL ORGANIZATIONS

In 1955, an undetermined number of young farmers in the Midwest organized the National Farmers Organization (NFO), with headquarters at Corning, Iowa. These farmers sought relief from high capitalization and operating costs through collective action. The NFO intended to negotiate with the food processors in a manner similar to that of organized labor to win contracts that would guarantee cost-of-production prices and return a small profit. If the food processors refused to sign contracts that provided these prices, the NFO proposed staging a withholding action until the manufacturers agreed to their terms. The NFO contended that although farmers made up 9 percent of the population, they received only 4 percent of national income. It also argued that Americans enjoyed cheap food prices and that consumers could and should pay farmers more for their bountiful food supply. The NFO believed that through collective strength the organization could establish prices for agricultural commodities. To do so, the NFO estimated, the organization needed to organize 25 percent of the farmers in a given area and control at least 60 percent of a commodity to force the food processors to pay the NFO's prices.

Although the NFO organized primarily in the midwestern states, where farmers emphasized corn, hog, and soybean production, the organization also spread into the wheat-producing states of the Great Plains. Quickly the NFO gained a reputation for radicalism. One roadside sign near Ellis in western Kansas read: *The NFO Says Higher Prices or Else*. Certainly, the great withholding action in 1962 did little to dispel that association. Symbolically on Labor Day weekend, the NFO began a withholding action for cattle and hogs to force the meat-packers to bargain collectively and issue contracts to the organization. The major purpose of the withholding action was not to raise prices alone. Essentially, the NFO planned to stay home, much like their predecessors in the Farmers' Holiday Association, until the meat-packers eventually accepted the organization's demand for sales based on prearranged contracts. These contracts would provide farmers with advanced knowledge about their profits and enable them to plan their agricultural operations accordingly. Moreover, the contracts would guarantee a sufficient profit to ensure a comfortable standard of living.

Soon after the initiation of the withholding action, the meatpackers began laying off employees, and while wholesale commodity prices remained the same, grocery prices increased. In addition, NFO members did not merely stay at home and refuse to sell their livestock. Many went to the roadsides and physically prevented other farmers from taking their cattle and hogs to market. When livestock prices increased, many NFO members, after staying away from the markets, could no longer participate in this withholding action. Like their predecessors, they needed money to meet daily living expenses. Livestock had to be fed and mortgages paid. By early October, the withholding action had collapsed, and the leadership suspended it until further notice. Farmers now flooded the market with cattle and hogs, and prices fell to prewithholding levels.

Despite the failure of the withholding action in 1962, the NFO sponsored a strike by dairy farmers five years later. Amid scenes of farmers pouring milk in roadside ditches, the NFO again appeared to represent the most desperate and radical farmers. A year later, in 1968, the NFO sponsored a cattle kill in which several hundred calves and mature cattle were destroyed to emphasize that farmers could not afford to raise livestock with below-cost-of-production prices. During all these actions, violence not only divided the NFO membership but kept other farmers from joining the organization and repulsed many consumers nationwide. Nevertheless, the NFO continued to appeal to many farmers. The organization dramatized the inability of farmers to control the process that would enable them to receive fair prices for their commodities.

The essential problem of the NFO remained its inability to gain enough control of a single commodity to force the processors to meet its demands for collective bargaining. As in the past, farmers still lacked unity, organization, and discipline. Moreover, no one could determine cost-of-production prices with any accuracy. Higher grain prices eventually caused increased costs for livestock production and land, which placed the farmer back in his original position. The NFO's goal of increasing commodity prices while maintaining low food prices was impossible given the nature of the free-market system. In the end, the NFO's withholding actions failed because its membership could not fall back on the economic security of unemployment compensation or welfare payments. They had ongoing operational costs to meet. They could not simply withdraw from farming and reenter at a later date when a collective-bargaining agreement had been reached with the food processors. Unlike the Farmers' Holiday Association, the NFO did not collapse with the failure of its withholding actions during the 1960s. It remained active and succeeded in signing contracts with certain buyers on behalf of its membership. It maintained secrecy about its membership rolls, however, probably because it had far fewer members than it needed.

In California, the large-scale growers, represented by individuals and agribusiness, continued to fight the organization of migrant labor. They believed that the perishability of their fruit and vegetable crops made every harvest a crisis that could be solved only by a surplus of cheap labor. The growers also maintained that wages were the only costs that they could control and that because prices were low, they could not raise wage rates. Despite opposition, some migrant workers continued to press for the unionization of the agricultural labor force. After the demise of the National Farm Labor Union, migrant workers sought unionization with the establishment of the Agricultural Workers Organizing Committee (AWOC) in 1959.

Supported by the American Federation of Labor (AFL), the AWOC targeted Anglo workers in the fruit camps as well as other migrants, particularly the most seasonal. Although the AWOC had sponsored more than 150 strikes by the autumn of 1962, the organization suffered from a lack of unity and planning due in part to an absence of grass-roots organization by the workers. The paid professional organizers were primarily devoted to increasing the ranks of the AFL rather than improving the lives and working conditions of migrant farm workers.

The growers fought the activities of the AWOC with bracero labor, court injunctions, and the jailing of its leaders. As a result, the AFL temporarily concluded that migrant farm workers could not be organized into a union and terminated financial and organizational support. Yet the movement to unionize agricultural workers did not die. In fact, 1961 inaugurated the hallmark decade of the civil rights movement, and the organization of migrant labor became a part of it.

In December 1961, Cesar Chavez began a new effort to organize farm workers. In September 1962, those efforts culminated with the formation of the Farm Workers Association (FWA), with the Aztec black eagle the symbol of the organization. Chavez based the organizational efforts of the Farm Workers Association on the need to provide a community benefit association that would help workers contend with the everyday difficulties of poverty-stricken workers such as obtaining medical care, shelter, food, and clothing. The FWA worked closely with the Catholic church and provided immigration advice, citizenship classes, and welfare counseling. It offered social and cultural support to the migrants by sponsoring festivals and celebrations. Chavez intended to build an organization that had

MIGRANT FARM WORKERS struggled to organize and bargain collectively without much success during most of the twentieth century. During the 1960s, however, the United Farm Workers union organized and gained recognition of many growers. Contracts signed by the union achieved higher wages and better working conditions for the migrants, and the organization also provided a host of social services. Here striking United Farm Workers are urging other migrant laborers to leave the vineyard and join them. *Archives of Labor and Urban Affairs, Wayne State University.*

unity and stability. This organization would train organizers and provide experience for the members in collective action, the long-term goal being recognition of the union and collective bargaining with the growers.

By September 1965, the FWA had achieved great organizational success through its bread-and-butter social programming. On August 22, 1967, the FWA and the AWOC merged to form the United Farm Workers Organizing Committee (UFWOC). With the support of leading philanthropists, professionals, middle-class Protestants, and the Catholic church, it began to pursue the long-term goals of the the union when it supported a strike of grape workers. When the growers obtained an injunction that restricted the number of participants on UFWOC picket teams, Chavez embarked on a new tactic for agricultural organizers to win recognition of migrant workers' right to form a union. Chavez charged that the injunction violated the free-speech rights guaranteed by the First Amendment of the Constitution. When UFWOC members were arrested on the picket lines, Chavez called for outside support. He received it from students at the University of California at Berkeley who were actively involved in the national civil rights movement and from the AFL, United Auto Workers, and Longshoremen's unions.

CESAR CHAVEZ provided the leadership that led to the founding of the United Farm Workers of America. Chavez skillfully used the mass media, especially television, to gain sympathy and support for migrant workers. He also organized the boycott of wine and grapes at grocery stores across the nation to force the growers to meet union demands regarding wages, working conditions, and recognition. Chavez gained a loyal following among migrant workers and other supporters, but his leadership style also earned him bitter enemies. *Archives of Labor and Urban Affairs, Wayne State University.*

Chavez also effectively organized a nationwide grape boycott, and after a 300-mile march that crisscrossed the Central Valley with television reporters trailing along, he arrived at the steps of the state capitol to request legislative support for the farm workers. Not long thereafter, the major growers of wine grapes signed contracts with the UFWOC. The United Farm Workers Organizing Committee next targeted the growers of table grapes, but the strike failed because the migrants working in those vineyards were too poor to participate. Eventually Chavez organized secondary boycotts across the nation to encourage grocers to take grapes off their shelves.

Although success did not come quickly or without internal dissension, the UFWOC won a victory in 1970 when the growers negotiated 150 contracts that protected 10,000 workers. By the early 1970s, due largely to the new organizational basis and boycott tactics of the UFWOC, the growers could no longer ignore the migrant farm workers' union, which claimed 150,000 members. And in February 1972 the organization became a full affiliate of the AFL-CIO as the United Farm Workers of America (UFW). Even so, the UFW could not always strike effectively. The success of the UFW was due primarily to boycotts outside California, wealthy sponsors, and the Democratic party that ended the bracero program and improved economic and social services. The majority of migrant workers did not belong to the union.

During the early 1970s, internal fighting and opposition from the Teamsters' Union weakened the UFW. Efforts to organize lettuce workers and the labor force of the Gallo wine company brought strikebreakers, violence, mass arrests, and more boycotting. By 1974, moreover, the civil rights movement had waned. Most Americans were tired of the crusade that had become entangled with opposition to the Vietnam War. Although the California legislature passed a collective-bargaining law in 1975 that included agricultural workers, the UFW movement had peaked.

By the mid-1970s, many unorganized workers were more impressed with the tough-talking and highly paid Teamsters than the UFW. Many farm workers were not interested in joining the union or participating on picket lines or in boycotts. Some wanted union benefits without the obligations of membership; others resented outsiders like college students who had played a major role in the efforts to gain union recognition. As a result, the organizational pace of the UFW slowed during the 1980s. The UFW suffered from low morale, high staff turnover, political infighting, and charges that Chavez would not delegate power. Given these problems, together with lingering strikes and violence, many people began to view the UFW as merely another labor union.

Despite the problems of the UFW, the organization had effectively increased wages and improved working conditions, particularly with the abolition of the short-handled hoe that kept workers stooped over in the fields. It had gained the prohibition of arbitrary firings by the growers, and seniority rules stabilized employment. Employers also contributed to health-care, disability, and insurance benefits for UFW members. Even so, by the late twentieth century, less than one-third of California's farm workers were members of the UFW. Illegal ("wetback") labor or legal "green-card" workers diminished the power of the UFW to organize and bargain collectively with the growers. Despite these difficulties, the UFW remained the most successful farm workers' union in America.

In 1977 a new farmers' organization, the American Agriculture Movement (AAM), emerged on the Great Plains. Organized by large-scale wheat farmers and farm-related businessmen with headquarters in Springfield, Colorado, the AAM protested against a political system that did not provide economic relief after urging farmers to increase production in the 1973 farm bill. With inflation and increasing interest rates making the cost-price squeeze worse than ever, the AAM intended to organize local chapters similar to the American Farm Bureau Federation and to establish state and national offices to seek economic change.

With a remarkable inability to learn from the past, the leadership planned to hold a farm strike during which they would neither sell food and fiber nor plant or buy until the federal government guaranteed 100 percent parity prices in new agricultural legislation. Naively, they believed the family farm was the critical link in the national economy, that a strike would force the federal government to act, and that the AAM would educate the American public about the significance of the family farm. If the government did not meet their demands, they were certain, the family farm would soon succumb to large-scale corporate agriculture. This free-flowing organization did not charge dues, maintain membership rolls, formally elect officers, or provide rules. Anyone could speak at an AAM rally, provided he or she supported the goals of the organization.

By December 1977, the AAM had organized 1,100 locals in forty states, according to its leaders. Using the tactics of many college students who participated in the civil rights movement and anti–Vietnam War demonstrations, they demonstrated in their local communities with a tractor parade, dubbed a "tractorcade," and carried signs that proclaimed, *Hell No, We Won't Grow*. Quickly the AAM captured the attention of the local and national press as well as television stations and networks. One supporter said, "We may be stupid but at least we're smart enough not to buy TV time." The AAM also used peer pressure and intimidation to force other farmers to acquiesce in if not join the movement. Their rallies were designed to show anger, activism, unity, resolve, and the possibility of violence if the federal government did not meet their demands.

After Congress rejected the demands of the AAM and the majority of the wheat farmers continued to plant during the autumn of 1977, the AAM planned a major tractorcade to the nation's capitol. In January 1978, some 3,000 farmers paraded their tractors in Washington, D.C., and met with congressional leaders. Congress responded with a moratorium on foreclosures by the Farmers Home Administration and an 11 percent increase in price-support payments. The AAM leaders, however, contended that they had been betrayed and that they would reduce planting by 10 percent, but the rank and file failed to respond. The agricultural strike, such as it was, had virtually no effect on the production of food and fiber across the nation.

The failure of the AAM to hold a major farm strike and gain 100 percent parity prices caused the leadership to place a new emphasis on militant protest. In February 1979, the tractorcade was back in Washington, D.C., but only radicals remained in the organization. Although only half as many participants demonstrated compared to the previous year, they did not do so peacefully. AAM farmers used their tractors to block traffic, drove over government lawns, released goats on the capitol grounds, and threw tomatoes at Secretary of Agriculture Bob Bergland. Eventually, the police confined their activities to the mall area.

Congress and the USDA remained unreceptive to their demands for 100 percent parity prices, and the newspapers and television reports portrayed these farmers as belligerent and threatening. Instead of gaining public support, the AAM alienated it. By June, an estimated 60 to 90 percent of the AAM locals had ceased to exist, and only twenty-eight states claimed even minimal activity. Although the AAM planned to establish an office in Washington, D.C., hire a lobbyist, formally organize the movement, and call for another tractorcade in February 1980, these efforts disintegrated.

The AAM was similar to other agricultural movements in which farmers joined an organization that offered a specific plan or quick fix for an immediate problem. The AAM identified the enemies of farmers and used dramatic techniques to gain public attention and achieve its goals. The AAM glorified the family farm and criticized the federal government, the public, and other agricultural organizations like the Farm Bureau for not understanding the problems of family farmers and forsaking them for the preferences of agribusiness and industrial and urban America. In the end, the public grew tired of the AAM's radical activities and pronouncements. When the AAM no longer made the news, the movement essentially died. Its most ardent members continued to feel excluded from the political system, and they dreamed of resurrecting the movement. Without the support of the commercial farmers and the major agricultural organizations, however, the AAM was relegated to history, if not oblivion.

RURAL LIFE

American agriculture remained geographically diverse, and specialization meant that most farmers bought their food at grocery stores like everyone else. They drove automobiles over improved roads to take part in social activities in town. Television, radio, telephone, and even computer technology and other conveniences made their lives little different from those of urbanites. They sent their children to college but seldom to study agriculture not because they still opposed book farming but because agriculture required too much work for too little economic gain. Thousands of young men and women left their homes in the countryside, never to return. Their parents frequently encouraged them to go, and many of them left as well.

At the same time that many farmers were leaving the land, rural life became increasingly attractive to many metropolitan residents. With a high level of affluence, access to good roads, and automobile ownership, a growing retirement population chose to live more cheaply in the country. Senior citizens preferred the slower pace and relative safety of rural life. Younger families sought those benefits as well as an escape from urban social problems. Now, instead of fleeing farms for the cities, people fled the metropolitan areas for the countryside. Although many intended to raise gardens, few wanted to become farmers. With the out-migration of farmers and the in-migration of suburban and urbanites, in some areas rural America began to undergo fundamental change.

Rural life also changed in part because the cost-price squeeze prevented many farmers from maintaining full-time operations, and

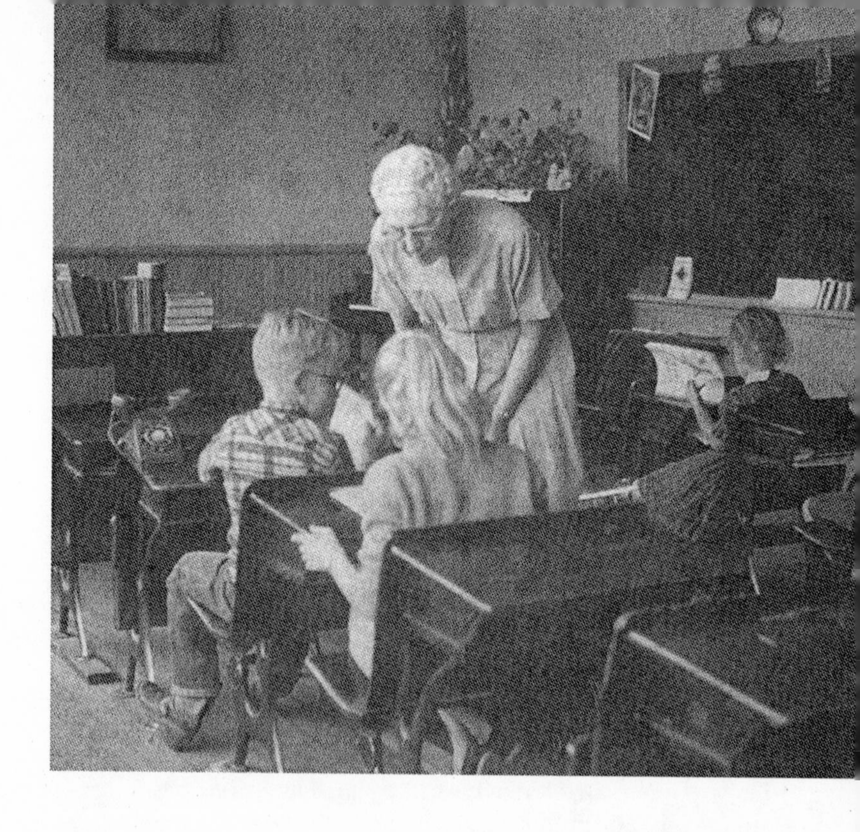

BY THE LATE twentieth century, few country schools remained. Most rural schools had consolidated with a modern building in a central location or a nearby town. Where country schools still operated, they were most often located in sparsely populated areas of the Great Plains. These students are at work on their lessons in a country school in South Dakota. *South Dakota State Historical Society.*

they necessarily worked away from the farm to support their families and qualify for health benefits and retirement plans. Automobiles and good roads enabled farmers to take advantage of better opportunities off the farm. This additional income helped them pay debts, make improvements, and remain on the land. Part-time farming continued to increase and for many became a way of life. Indeed, part-time farming no longer indicated a period when an individual worked in outside employment to accumulate the capital to begin full-time farming. Usually it meant the difference between survival and failure. Yet while off-the-farm employment provided economic stability for the family, it did not signal a transition out of farming.

Farm women increasingly took outside jobs because their families needed the extra income. In 1980, a survey indicated that approximately 36 percent of agricultural women worked off the farm and that 33 percent were also dissatisfied with farm life. Agricultural women who took off-the-farm jobs usually worked in "typical" areas for women such as clerical, secretarial, nursing, and teaching. Southern women more than women from other regions took off-the-farm jobs. In general, however, well-educated women who had high earning potential and a babysitter were most likely to work full time off the farm. Those employed full time usually did not participate in the farm operation. Farm women, however, made up only 2 percent of the total female work force.

Rural life did not change for women in some respects. They continued to share work and decision making with their husbands, but they usually identified themselves as "housewives" rather than "farm wives." Like their husbands, farm women were preoccupied with economic problems. They believed the federal government should end its regulation of agriculture and improve commodity prices. Still, their lives changed more than they remained the same. While technology altered farm activities for men, it also affected women. Indeed, technology displaced women as well as men from agricultural production. During the early twentieth century, for example, women had the responsibility for raising poultry. By the late twentieth century, vertical integration stimulated by technological change made poultry raising an industrial activity that governed 90 percent of that production. Farm women no longer raised chickens for frying, roasting, or eggs.

By the late 1970s, about 128,000 or 5.2 percent of all farmers were women. More women served as sole or primary operators of farms than ever before. More than 50 percent of these women operated in the South. Women who operated farms generally did so on a smaller scale than men, averaging 285 and 423 acres, respectively. Women farmers had inherited or received as a gift 43 percent of the land they farmed and with it the type of farming emphasized. Approximately 80 percent of these women farmers were full owners, but they usually grossed less than $20,000 annually, mostly from the sale of livestock and cash grains. The average male-operated farm grossed approximately $26,000. Women operators also averaged about 60 years in age, compared to 50.5 for the average male farmer, due largely to inheritance of the farm at the death of their husbands. In addition, 10 percent of all minority farmers were African-American women, while only 5 percent of white farmers were females.

Despite the great changes in rural life, many farmers remained on the land because they believed the "values of rural living" superior to those of an urban environment. Despite serious economic problems, farmers remained convinced that the farm was the best place to raise their children until they could leave and take care of themselves. As in the past, however, profits determined whether a farmer would be successful. Most farmers, despite sentimental attachments and nostalgic views of farming as a way of life, regarded agriculture as a business in which only the most efficient survived. The days of the small-scale poor farmer were gone, although "hobbyists" increased in number. Government loans, price supports, and acreage reductions favored large-scale farmers who could produce more for less than the smaller landowners. Many small-scale farmers went bankrupt or quit.

During the late twentieth century, farmers increased production on fewer acres with the application of new forms of science and technology, coupled with better farm-management practices. Hybridization boosted crop production; improved feeding and breeding practices produced better livestock. Fertilizers, herbicides, and pesticides, together with better tillage and harvesting methods, helped as well.

These trends also touched the lives of the men, women, and children who worked on the farms. Migrant workers, for example, felt the effect of technological change during the decade even more sharply than the landowners who used it to replace them. Usually improved technology meant that farm workers would be released from their jobs. Although some continued to live in rural areas, many sought refuge and employment in the cities.

Farm life continued to change rapidly. Many farm enterprises now had greater capital investments than businesses in town. The axiom of the age became "Get bigger, get better, or get out." Unable to control prices and unwilling to organize and bargain collectively, most farmers believed they had no choice but to increase production to lower unit costs and thereby maintain a profitable business. Increased production, however, exacerbated the problems of surpluses and low prices, and in the long run offered no solution to the farm problem. Indeed, unlike other businesses, farms were penalized in the marketplace for their efficiency. While farmers complained about the cost-price squeeze and government policy, some got bigger and better, but thousands left their homes for a more economically secure life in urban America.

Farmers felt the loss of political power that comes with minority status. Individually, and often collectively, they could do little to influence farm policy, in part because they wanted government aid if it would be advantageous but opposed it if that help threatened to decrease their independence or potential profits. Most farmers contended that economic conditions could be improved if the middlemen were eliminated from the food chain, thereby enabling farmers to market more directly while keeping food prices low. They also believed that consumers and small-scale businessmen were potentially their best allies, while environmentalists, big business, organized labor, and the urban poor represented threats to the agricultural community and rural life. Many farmers experienced "extreme" financial difficulty that made their life in rural America increasingly problematic.

Overall, the cost-price squeeze became the most important reason for the rapid decline in the number of farms. With prices equal to 100 in 1977, farmers received an index price in 1990 of 150 on all farm products, but they paid a collective price of 171 for their expenses. Given the disparity between income and expenditures, the sharp decline in farm population continued. In 1990, the American population totaled approximately 250 million, but the farm population had declined to only 4.5 million who lived on 2.1 million farms, down from a peak of 6.8 million farms in 1935. Farm men, women, and children composed only 1.9 percent of the total population.

Agricultural income often fluctuated greatly, depending on the weather, foreign production, and domestic and world demands. In 1991, net farm income declined by 12.5 percent to $44.5 billion, the first drop since 1984. This decline resulted from price decreases for crops and livestock and a 11.7 percent fall in direct government payments; the latter had been dropping since 1987 when they peaked at $16.7 billion or more than twice the 1991 level. At the same time, production expenses remained unchanged. Farmers now suffered a serious cost-price squeeze. As a result, approximately 90 percent of farm operators received some income from off-the-farm sources, although about half those farmers still claimed agriculture as their major occupation.

Even so, 44 percent of farm operators had a nonfarm job as their principal source of income. In 1990, off-the-farm income averaged

$33,265, or 85 percent of total household income. The average net worth of a farm operator was $411,681, 85 percent of which was tied up in the farm. In 1990, three-fourths of all farmers earned less than $50,000; after they paid expenses, they averaged a net of about $16,000. At the same time, only 6.2 percent of the farms sold more than $250,000 in commodities annually, but those farms produced slightly more than half of all agricultural products. In 1992, the USDA defined a farm as any place that sold or normally produced $1,000 or more in agricultural commodities each year. If the federal government had defined a farm as those operations that earned at least 50 percent of family income by agriculture, the number would have been reduced even more.

As the United States approached the twenty-first century, farmers continued to need large amounts of capital, extensive land, and better management skills to remain in business. They continued to depend on government price supports, purchases, and marketing, which despite their complaints, they considered a matter of entitlement. Entry into farming by young people remained difficult, given the capital and financial requirements to begin a profitable operation. Although farmers could begin from scratch by purchasing land, implements, and livestock, their odds of success were not high. Regional differences involving land prices and availability as well as the proximity of new farms to markets played a major role in the success or failure of new operations.

The farm problem of overproduction and low prices remained unsolved, in part because weather and inflation as well as foreign exchange rates and foreign policy made planning and programming extremely difficult. Farmers hoped the "safety net" of government income and price-support programs would keep them on the land.

SUGGESTED READINGS

Ahearn, Mary C., Janet E. Perry, and Hisham S. El-Osta. *The Economic Well-Being of Farm Operator Households, 1988–90.* U.S. Department of Agriculture, Agricultural Economic Report 666, 1993.

Browne, William P., and John Dinse. "The Emergence of the American Agriculture Movement, 1977–1979." *Great Plains Quarterly* 5 (Fall 1985): 221–35.

Clark, Ira G. *Water in New Mexico: A History of Its Management and Use.* Albuquerque: University of New Mexico Press, 1987.

Clark, Jeffery T. "Continuity and Change in Hawaiian Agriculture." *Agricultural History* 60 (Summer 1986): 1–22.

ield, Gregory. "Agricultural Science and the Rise and Decline of Tobacco Agriculture in the Connecticut River Valley." *Historical Journal of Massachusetts* 19 (Summer 1991): 155–74.

ite, Gilbert C. *American Farmers: The New Minority.* Bloomington: Indiana University Press, 1981.

———. *Cotton Fields No More: Southern Agriculture, 1865–1980.* Lexington: University of Kentucky Press, 1984.

riedberger, Mark. *Farm Families and Change in the Twentieth Century.* Lexington: University of Kentucky Press, 1988.

riedland, William H. *Manufacturing Green Gold: Capital, Labor, and Technology in the Lettuce Industry.* Cambridge, England: Cambridge University Press, 1981.

Galarza, Ernesto. *Farm Workers and Agri-Business in California.* South Bend, Ind.: University of Notre Dame Press, 1977.

Gates, Paul W. *Land and Law in California: Essays on Land Policies.* Ames: Iowa State University Press, 1991.

Green, Donald E. *Land of the Underground Rain: Irrigation on the Texas High Plains, 1910–1970.* Austin: University of Texas Press, 1973.

Grubbs, Donald H. "Prelude to Chavez: The National Farm Labor Union in California." *Labor History* 16 (Fall 1975): 453–69.

Hansen, John Mark. *Gaining Access: Congress and the Farm Lobby, 1919–1981.* Chicago: University of Chicago Press, 1991.

Harl, Neil E. *The Farm Debt Crisis of the 1980s.* Ames: Iowa State University Press, 1990.

Hightower, Jim, and Susan DeMarco. *Hard Tomatoes, Hard Times: A Report of the Agribusiness Accountability Project on the Failure of America's Land Grant College Complex.* Cambridge, Mass.: Schenkman, 1973.

Hundley, Norris. *The Great Thirst: Californians and Water, 1770s–1990s.* Berkeley: University of California Press, 1992.

Hurt, R. Douglas. *Agricultural Technology in the Twentieth Century.* Manhattan, Kans.: Sunflower University Press, 1991.

———. "Ohio Agriculture Since World War II." *Ohio History* 97 (Winter–Spring 1988): 50–71.

Ingram, Helen. "Water Rights in Western States." *Proceedings of the Academy of Political Science* 34 (1978): 134–43.

Isern, Thomas D. *Custom Combining on the Great Plains.* Norman: University of Oklahoma Press, 1981.

Jelinek, Lawrence J. *Harvest Empire: A History of California Agriculture.* San Francisco: Boyd & Fraser, 1982.

Jenkins, J. Craig. *The Politics of Insurgency: The Farm Worker Movement in the 1960s.* New York: Columbia University Press, 1985.

Kalbacher, Judith Z. *A Profile of Female Farmers in America.* USDA, Rural Development Research Report 45, 1985.

Kromm, David E., and Stephen E. White, eds. *Groundwater Exploitation in the High Plains.* Lawrence: University Press of Kansas, 1992.

Liebman, Ellen. *California Farmland: A History of Large Agricultural Holdings.* Totowa, N.J.: Rowman and Allanheld, 1983.

Meister, Dick, and Anne Loftis. *A Long Time Coming: The Struggle to Unionize America's Farm workers.* New York: Macmillan, 1977.

Pearson, Roger W., and Carl E. Lewis. "Alaskan Agribusiness: A Post-Statehood Review." *Agribusiness: An International Journal* 5 (July 1989): 367–84.

Rosenfeld, Rachel Ann. *Farm Women: Work, Farm, and Family in the United States.* Chapel Hill: University of North Carolina Press, 1985.

Sachs, Carolyn E. *The Invisible Farmers: Women in Agricultural Production.* Totowa, N.J.: Rowman and Allanheld, 1983.

Schlebecker, John T. *Cattle Raising on the Plains, 1900–1961.* Nebraska: University of Nebraska Press, 1963.

______. "The Great Holding Action: The NFO in September, 1962." *Agricultural History* 39 (October 1965): 204–13.

Shover, John L. *First Majority–Last Minority: The Transformation of Rural Life in America.* DeKalb: Northern Illinois University Press, 1976.

Socolofsky, Homer E. "The World Food Crisis and Progress in Wheat Breeding." *Agricultural History* 43 (October 1969): 423–37.

Talbot, Ross B., and Don F. Hadwiger. *The Policy Process in American Agriculture.* San Francisco: Chandler, 1968.

Walton, John. *Western Times and Water Wars: State, Culture, and Rebellion in California.* Berkeley: University of California Press, 1992.

Williams, Robert C. *Fordson, Farmall, and Poppin' Johnny: A History of the Farm Tractor and Its Impact on America.* Urbana: University of Illinois Press, 1987.

9 • TWENTIETH-CENTURY CLOSURE

THE HISTORY OF AMERICAN AGRICULTURE HAS BEEN MARKED BY many achievements since the founding of the Republic. The systematic, if not always equitable, distribution of the public lands; technological change that eased the drudgery of farming; a land-grant college, state experiment station, and extension systems that developed, taught, and shared improved methods with farm men and women; scientific change that enabled bountiful harvests; and federal policy that provided a safety net for farmers are only a few of the vast changes that have occurred in American agriculture since the eighteenth century. Agricultural slavery ended with the violence of the Civil War, while sharecropping disappeared because of federal policy during the twentieth century, and technological change pushed and pulled many farmers from the land. As the agricultural population declined, fewer people had much contact with farm life, and their lexicon and mental reference points about farming changed. By the late twentieth century, most Americans would not have understood noontime radio reports from the past that said: "Sows opened weak in Chicago," or "Wheat rose in Kansas City."

By 1990 farmers had become a minority of the population, and agriculture provided the foundation for the local economy in approximately 500 counties nationwide, down from 2,000 counties in 1950. Yet for the USDA to classify a county as "farm dependent," agriculture had to contribute only 20 percent of its income. Merely one in six counties remained dependent on farming for its economic base. Only 46 of 435 congressional districts were "farm oriented" in 1990. Off-the-farm income also exceeded that earned from the sale of agricultural commodities and constituted 95 percent of family earnings on farms grossing less than $40,000 annually, an income that represented the bare minimum needed to support a family. Yet, only 600,000 farms and ranches reached or surpassed that income. By 2000, many agricultural leaders believed that farmers had to sell at least $100,000 worth of commodities annually to be a full-time farmer. Sales below that amount indicated part-time farming and off-the-farm income for survival. As a result, less than 1 percent of the population lived on full-time farms of any kind. As farmers and ranchers left the countryside, no one took their places, and few new jobs were created. Consequently, rural businesses, schools, churches, and hospitals continued to decline. Consolidations and closings became common in rural America.

SUSTAINABLE AGRICULTURE

By the last decade of the twentieth century sustainable agriculture was becoming an important concept, if not practice, in American agriculture. Some environmentalists urged farmers to think of agriculture not as an extractive industry but rather as a sustainable process. Essentially, they urged people to return that which they took from the land and to end the wanton exploitation of the environment, caused by using chemical fertilizers, herbicides, and pesticides, which one environmentalist called "chemotherapy on the land." Sustainable agriculture for them meant harmony between humankind and nature. But not everyone agreed about the meaning of sustainable agriculture. The Board on Agriculture of the National Research Council defined sustainable agriculture as a "goal rather than a distinct set of practices," that improved productivity, produced safe, nutritious food, ensured an adequate net farm income and an acceptable standard of living for farmers, and complied with community standards and social expectations. The Rodale Institute understood this concept to mean "regenerative agriculture," by which farmers practiced the methods that benefited the land, people, and profit margins. Others understood sustainable agriculture to mean "agrarian fundamentalism," by which farmers forsook chemical technology. Some believed the concept meant simply sustaining food production to ensure a well-fed world, while others interpreted it as "environmental conservation."

In time, some environmentalists attempted to merge several definitions by urging the development of an agriculture based on perennial grains that did not require anhydrous ammonia, pesticides, and herbicides, thereby creating a new agriculture that would mimic the ecology of the particular region farmed. Other environmentalists added the matter of "social equity" to the definition, arguing that sustainable agriculture could not permit the exploitation of races, ethnicity, class, or gender. And, they contended, it must provide "Equal access to decision making for those involved in all aspects of the food and agriculture system." Not to do so, they argued, would ensure the "Poverty of Sustainability."

Although some of the definitions of sustainable agriculture were rather ambiguous, those who believed the concept could accommodate both profits and productivity while maintaining an ecologically sound agriculture received considerable criticism from environmentalists with a different conceptualization about the preservation of nature. By the late twentieth century, sustainable agriculture and environmental quality became important and highly visible concerns for farmers and nonfarmers, and the balancing of environmental considerations and agricultural economics became increasingly difficult. Soon the environmental preservationists separated into two groups—economic and aesthetic, with the economic environmentalists advocating the preservation of nature to ensure sustainable agriculture, that is, to protect the environment but still use it for economic gain. In contrast, the aesthetic environmentalists urged individuals and organizations to develop a personal, ethical relationship with the land and eschew economic considerations. The aesthetic environmentalists championed the concept that nature should be preserved for its own sake, inviolate by humankind. They also contended that the economic environmentalists believed technology could still solve all environmental, social, and economic problems. As a result, the aesthetic environmentalists maintained, this version of the old concept of conservation meant nothing more than a continuation of past "ecological conquest."

When economic environmentalists argued that agriculture was not a form of technology, the aesthetic environmentalists agreed. But when they advocated a sustainable agriculture that did not "deplete soils or people," the aesthetic environmentalists felt embarrassed by this lack of understanding and sophistication, contending that those who held such positions were locked in an "old-fashioned way of thinking" and merely supported "another measure of production." For the aesthetic environmentalists, the concept of "sustainable development" offered an all too easy, and false, solution for environmental problems. It was nothing more than a well-intentioned, but misguided, panacea that allowed the permissive use of the environment. In this context, sustainability was merely an economic concept that held nature to be nothing more than a "pool of resources" for exploitation.

For the aesthetic environmentalists, then, sustainable agriculture had to be based on a political compromise between developers and preservationists. However, they rejected all compromises that endangered nature, arguing that "We must preserve all species, subspecies, varieties, and communities, and ecosystems that we possibly can." The aesthetic environmentalists, then, essentially contended that the time had come for people to stop trying to conquer the environment and begin to change their agricultural economy and cultural relationships with nature. The preservation of nature, they argued, was a moral imperative.

In reality, the differences between the economic and aesthetic environmentalists were of degree rather than ideology. Both advocated institutional and "attitudinal" changes rather than new applications of science and technology to preserve nature and to end the traditional exploitation of the land. Both groups also agreed that an "ecological agriculture" was needed. The use of nature beyond its capability to sustain agriculture and earn profits, they claimed, would not only damage the environment but also harm the "public good." For the aesthetic environmentalists, however, wealth should not be the primary goal of farmers, and they sought a revolutionary change in the public mind to achieve the preservation of the environment.

Although both the economic and aesthetic environmentalists advocated the development of a "moral consciousness" by farmers to preserve nature, and while both advocated the superiority of the community over the individual in relation to the use of the land, the aesthetic environmentalists set themselves apart by supporting increased government control to ensure private ecological responsibility and a "commitment to community." For the aesthetic environmentalists, the preservation of nature required "deprivitizing" some lands, particularly in the Great Plains. They spoke of public ownership or management, that is, control of the land by the federal government to ensure the preservation of the environment and to assure "broader communitarian values."

Public ownership, the aesthetic environmentalists argued, could be a shared arrangement between federal, state, and local governments with management under the direction of nonprofit organizations. In this respect, they were ideologically similar to the "light and gas" socialists of the Progressive Era, who advocated government control of utilities for the public good at the expense of private ownership and enterprise. The aesthetic environmentalists did not advocate acquisition of land by the government's power of eminent domain, but rather the governmental purchase of land in the "public name." Public ownership of land, they contended, would help preserve some areas from ruthless economic exploitation. Furthermore, preservation of the environment could not be achieved until the public developed a "sense of belonging to the larger community of nature, a community that has many interests and claims besides our own."

By 2000, the concept of environmental preservation continued to cause debate, because it mixed economics, science, politics, and philosophy, and because it involved individual liberty and social responsibility and the idea that "nature has rights that humans should respect as part of an extended code of ethics." Some environmentalists even argued that

'Natural rights should extend to and encompass the rights of nature." By the turn of the twenty-first century, however, the economic considerations of environmental preservation most concerned farmers. Indeed, for a universal, sound environmental ethic to develop, it had to be a profitable concept for farmers. Put differently, if farmers adopted environmentally sound agricultural practices, they had to be convinced that those practices were reasonably profitable, or more profitable than conventional agricultural practices. Moreover, farmers tended to equate sustainability with sufficient profitability to remain on the land. Essentially, if farmers believed a particular environmental policy or concept would be more profitable than their current practices, they were willing to adopt it, but cost was always the ultimate consideration.

SCIENCE AND TECHNOLOGY

Near the turn of the twenty-first century, American agriculture began the third revolution of its long history. Whereas the first revolution occurred with the rapid adoption of horse-powered equipment during the late nineteenth century, and the second with the application of mechanical power, chemical fertilizers, pesticides, herbicides, and hybrid seeds during the early twentieth century, genetics brought the third revolution to the American farm near the century's end. A new word, *biotechnology,* reflected the merging of agricultural sciences with engineering and the promise of improved production. Critics, however, charged that biotechnology threatened to create a Pandora's box from which new ills would escape to trouble the land and society.

The advocates of biotechnology supported the scientists who worked to develop drought-resistant wheat and soybeans, herbicide-resistant tomato plants, cotton immune to caterpillars, and corn that repelled armyworms. Agricultural scientists developed bacterial sprays that lowered the freezing point of some crops to prevent frost damage; and growth hormones injected into hogs and cattle enabled better gains on less feed for reduced costs and greater returns. Scientists worked to splice or transform genes to improve the starch, oil, and protein content of corn, thereby making it more efficient for livestock feeders and more profitable for farmers. This "custom design" of corn varieties by genetic manipulation created the possibility that a farmer could soon select the precise variety to meet all physical and environmental conditions. Gene splicing had the potential to change crop and livestock raising as drastically as the development of hybrid seeds affected farming more than a half-century earlier.

Advocates further contended that it would reduce if not eliminate the risks and high costs of farming as well as the use of dangerous chemical

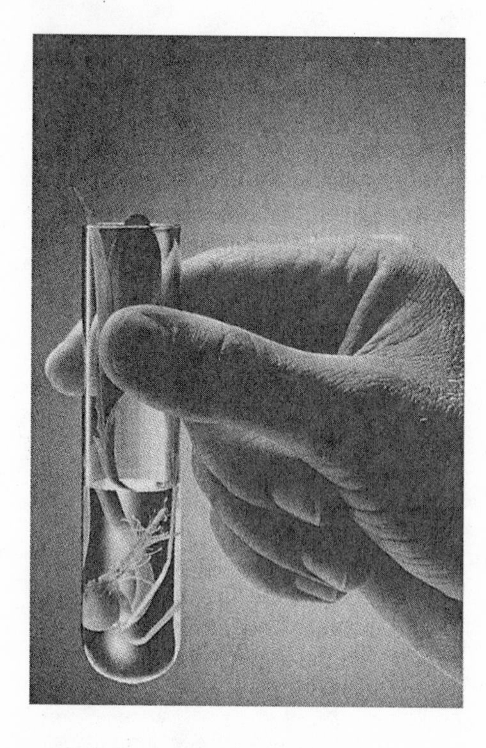

Agricultural science, particularly in relation to biotechnology, had dramatically altered crop production, marketing, and food processing by the turn of the twenty-first century. Consumers debated the safety of biotech crops, but farmers generally consider them safe and an aid to efficient, profitable operations. *Photo courtesy of Pioneer Hi-Bred, Inc., a DuPont Company*

herbicides and pesticides. But biotechnological developments did not offer better possibilities without creating new problems. Genetically engineered corn, for example, while of exceptional value, had to be stored separately from other corn for sale to poultry, hog, and livestock feeders in order to maintain its higher price, because the Food and Drug Administration had not approved its use for human food. But the separate storage, shipping, and marketing of genetically improved corn increased costs. By the end of the century, farmers, seed companies, grain dealers, and livestock feeders were increasingly concerned about the prospects and problems of biotechnological change. They particularly worried about whether there would be sufficient markets for biotech crops and beef produced from cattle that had been treated with hormones or fed genetically modified feeds. They were also concerned about potential economic problems created by genetically modified corn and soybeans, because many foreign and domestic consumers feared biotech agricultural commodities and foods and refused to purchase them, and some foreign nations even prohibited the importation of those products.

Although many scientists and farmers claimed that biotech crops and growth hormones, with the exception of DES (diethylstilbestrol) for beef cattle, had not been proven harmful to public health, consumers often considered scientifically developed agricultural products to be "unnatural" products of the laboratory rather than the products of nature. The development of hybrid corn provides a contrast. During the 1930s, when hybrid corn breeders developed strains for commercial production, no public outcry emerged to confront this strange new science. Neither farmers nor consumers could see any harmful effects of hybrid corn on neighboring crops and livestock or people. Instead, the public welcomed any efforts to ensure ample food at affordable prices, particularly after having experienced privation and want during the Great Depression. By 2000, the public considered hybrid seed corn a product of "natural methods" in the fields, as scientists used nature and the breeze to transfer pollen and develop specific corn varieties for particular soils and climates and to resist disease and increase production.

However, many consumers and members of the nonfarming public also believed that experiments to develop hybrid corn by "unnatural" methods—that is, by using gene guns and tissue cultures in the laboratory—threatened to create "monsters." They feared that these new hybrids could not be controlled and that the corn from these new plants could endanger public health. Often the objections of the nonfarming public to hybrids were driven by nonprofit consumer and environmental advocates, not by profit-seeking farmers and seed companies. Without proof that genetically modified corn and other commodities such as soybeans and potatoes were safe, many consumers refused to purchase food products manufactured from these laboratory creations. Farmers remained uncertain about the productive benefits of genetically modified organisms, because many consumers at home and abroad rejected foods processed from those crops. Moreover, biotech corn and soybeans did not bring higher prices, and the seed and chemical companies, not farmers, reaped the primary benefits from the new technology.

Few scientists and farmers believed that genetic engineering would entirely eliminate the need for chemicals in agriculture. Rather, it would enable farmers to tailor their crops and chemical use to meet specific needs without endangering the environment or public health. Many consumers and environmentalists often disagreed with that contention. For them, genetic engineering at best might produce a new weed that would spread like wildfire, while they saw biotechnology at worst as the proverbial Pandora's box. They did not consider it a useful tool but something to fear. For some, biotechnology became nothing less than the "doomsday process." As a result, while scientists proclaimed the safety of their agricultural experiments, they worked under the pressure that one mistake that damaged the environment or human health would instantly end their credibility.

The critics of biotechnology demanded strict government regulation to ensure proper testing of the various forms of genetic engineering before farmers would be permitted to raise the new crop varieties and improved livestock or use sprays of recombinant DNA products. Some critics also charged that the chemical companies could easily engineer plants that would require the use of certain chemicals and that government oversight was essential. Others warned that mergers of petrochemical and pharmaceutical companies might force scientists to develop plants that resisted damage from their company's herbicides, thereby encouraging rather than discouraging the continued use of chemicals. Such mergers would also force farmers to purchase seed and herbicides from the same company, thus decreasing competition and increasing prices for farmers. Some charged that corporate control of genetic engineering would enable those institutions to use genes for economic and political power, just as they used land, minerals, and oil. In addition, some livestock raisers worried that the biotechnology that enabled them to produce leaner meat, and more of it on fewer cattle, would reduce the need for large numbers of cattle and producers. Moreover, it might lower meat prices to the benefit of consumers but the detriment of producers. Fewer cattle would also mean less need for feed, animal suppliers, and veterinarians. Once again, progress posed more questions, if not problems, than it offered immediate profits.

While scientists worked to improve agriculture through biotechnology or genetic transformation, the previous revolution in American agriculture began to show serious negative side effects. Chemical fertilizers, pesticides, and herbicides increasingly polluted groundwater supplies, particularly the herbicide atrazine, along with the phosphates and potassium from fertilizers. Consumers worried about the adhesive qualities of pesticides on fruit and vegetables that might have a chronic health effect on their children. Migrant farm workers protested against the heavy use of highly toxic herbicides, some so dangerous that serious illness and death occurred when they entered fields long after spraying. In addition, the National Cancer Institute reported that farmers had elevated risks of contracting several illnesses, such as Hodgkin's disease, leukemia, and stomach and prostate cancer, and they suspected that exposure to pesticides was the cause.

Some farmers rejected both bio- and chemical technology and sought to develop a "sustainable" agriculture by using organic fertilizer (manure) and crop rotation to improve the nitrogen content of the soil and thus maintain satisfactory and profitable yields. They also gave greater attention to soil conservation. Until the late twentieth century, however, most farmers preferred the advantages of quick chemical fixes. One Iowa farmer remarked: "I can replace all the plant nutrients needed by buying sacks of them in town. I save all the time and labor in cultivating for weeds, and the cinch bugs and grasshoppers don't bother me."

By 1990, however, many farmers had learned that agricultural chemicals polluted farm, rural, and city water and that plowing often caused top-

Seed companies, such as Pioneer Hi-Bred, have developed corn crops that resist disease, tolerate drought, and produce abundant yields, among other genetic changes. Tightly spaced rows and plants such as these indicate a reliance on herbicides, that is, chemical technology, rather than cultivators, or hardware technology, to control weeds. Scientific change has increasingly supported intensive, specialized production, which helps make farming a business. *Photo courtesy of Pioneer Hi-Bred, Inc., a DuPont Company.*

soil to wash or blow away. As a result, some farmers used minimum tillage to protect their land as much as possible from water and wind erosion, but they were still a minority. Moreover, these farmers often did not receive much support from the USDA. In 1992, Edward Madigan, secretary of agriculture, said that the basic principles of the Republican Party were "More money, higher income, more markets. String all the environmentalists up." At the close of the twentieth century, then, most farmers preferred to use chemical fertilizers, herbicides, and pesticides because chemicals substantially improved production and profits.

Moreover, even though chemicals remained expensive, they were cheaper than other farming costs such as land, labor, and equipment. Few farmers considered nontraditional production or "alternative" farming—such as raising herbs, rabbits, and Christmas trees or the development of pick-your-own fruit and vegetable operations—economically viable. Even though "niche opportunities" could be quite profitable if geography and markets provided farmers with a "comparative advantage," government agricultural programs gave farmers a guaranteed income and protection against the failure of traditional crops. In addition, traditional crop farmers had better access to credit than the nontraditional agriculturists because of the higher risks of departing from staple-crop production.

At the same time, a strong dollar and foreign trade barriers, particularly subsidies granted by European nations for the benefit of their farmers, limited agricultural exports and the ability of American farmers to compete in the world marketplace. Corn farmers turned once again to manufacturing ethanol to raise prices by increasing demand. They argued that ethanol, when mixed with gasoline to produce "gasohol," reduced harmful emissions, and they wanted the federal government to make the use of ethanol mandatory. The petroleum industry objected that this would reduce oil profits, in part because ethanol was somewhat difficult to blend and the industry preferred to use petroleum-derived methane if it had to mix at all. Moreover, the Environmental Protection Agency contended that while ethanol might help reduce carbon-monoxide pollution in some areas, it would actually contribute to air pollution in urban locations that suffered from ozone smog.

While farmers sought new products from agricultural commodities, they also began to listen to the demands of consumers as eating habits changed. Health concerns about eating red meat made cattle raisers more interested in producing less marbled (fatty) grain-fed beef. Instead, breeders and feeders declared a "war on fat" and worked to produce a flavorful leaner beef that would cause consumers less concern about fat and cholesterol. The public-relations departments of the cattle organizations then promoted the idea that beef was necessary for a healthy, balanced diet.

AGRIBUSINESS

While many small-scale farmers struggled, the agribusiness industry—businesses that financed, marketed, transported, distributed, processed, manufactured, and sold agricultural commodities and food and equipment—usually prospered. By integrating vertically and horizontally, the food industry gained control of the production, processing, marketing, and distribution of some agricultural commodities. The industry drove many small-scale grocers out of business with its supermarkets. Vertical integration gave farmers some stability because the food companies provided financing and established procedures to help them produce certain commodities. Food companies also offered contracts that guaranteed a market and a satisfactory price for a certain amount of vegetables, fruits, poultry, or swine. Contract farmers furnished only the necessary equipment, buildings, and labor, but they lost their freedom to make decisions.

The growth of contract farming also made agriculturists think about the need for a labor organization that would represent their interests with the food processors at contract time. Increasingly, contract farmers saw themselves as wage laborers. Although some contractors, such as Farmland Industries in the Midwest, claimed that it began this type of association with farmers to ensure the delivery of high-quality hogs to its slaughterhouses, farmers grew increasingly concerned about fair treatment. Other observers of rural America saw growing environmental dangers with contract farming, particularly with the concentration of livestock in small areas and the problem of waste disposal.

The vertical-integration activities of agribusiness companies also subjected farmers to the dictates of the food companies. In 1978, for example, the Farm Labor Organizing Committee (FLOC) in Ohio struck the fields that had been contracted by the Campbell Soup Company and Libby's. In an

The development of the tomato harvester required plant breeders to produce a fruit that ripened uniformly and withstood the once-over harvesting operation of machine picking. Consumers, however, rejected machine-harvested tomatoes for the table because they were tough and pear-shaped. By 1965, mechanical tomato harvesting became economical if a grower harvested at least thirty tons per acre. Farmers who raised tomatoes for the table still use hand labor to harvest their crops. *USDA.*

effort to gain recognition of the union, the FLOC struck those fields again in 1979 and 1980. The agribusiness processors reacted by forcing their contract growers to adopt mechanical tomato pickers to ensure an adequate and timely delivery of the crop at harvest time. The Ohio growers had no choice. If they wanted to raise tomatoes, they needed a market; and to gain a market, they needed a contract. To get a contract, the growers had to promise to mechanize their harvest. The agribusiness industry, then, used its power to break the organizational back of the FLOC in the Midwest.

The agribusiness industry used vertical integration to control virtually all the broiler and vegetable and most of the citrus fruit, potato, and turkey production. Although multinational and multidimensional agribusiness corporations did not own extensive lands across the country, many farmers grew increasingly apprehensive about the economic and political power of these firms. Because agribusiness firms were established to earn profits, they tried to eliminate competition and maximize income. Neither goal was in the best interests of farmers or consumers. Many farmers, equating agribusiness with corporate farming, believed it threatened the continued existence of the family farm. In 1993, Jim Hightower, one-time commissioner of agriculture in Texas and long-time critic of agribusiness, charged, "Some of these guys are getting so rich they could air-condition hell. The way they're treating farmers, they better be putting some money aside."

By the late twentieth century, corporate mergers also had become commonplace in the agribusiness industry. Chemical and pharmaceutical companies acquired seed and implement companies, while Dutch, British, and German firms purchased American implement and seed companies. As the corporate competition dwindled, farmers had fewer choices among the seed, chemical, and implement companies. At the same time, large-scale corporate livestock producers increasingly confronted state and county governments and nonfarmers over issues of water and air pollution. State and county governments sometimes contested regulatory authority, while neighbors of hog and poultry confinement facilities and cattle feed lots frequently complained that such operations reduced their property values and made life unpleasant for nearby residents. Many violations of environmental law taken to court remain unresolved.

In some areas, large-scale agribusiness firms, such as egg producer DeCoster Farms, hired illegal immigrants and either ignored their status or sheltered them from the Immigration and Naturalization Service. Other firms, particularly meatpackers, exploited this cheap labor force. With rural wages averaging about 25 percent less than urban wages during the 1990s, U.S. workers usually rejected the hot, heavy, and hard work in the meatpacking plants, which meant the employment of foreign workers. These low-wage jobs ensured rural poverty, which contributed to social problems, such as spousal and alcohol abuse, drug addiction, school dropout, and crime of various sorts. Agricultural organizations, however, gave little attention to rural policy because those organizations feared they might lose federal dollars for farm programs that benefited their members.

As agribusinesses aggressively worked to expand and gain a larger share of the market in their respective spheres, these companies became increasingly sensitive to their public image. Early in 2002, Cargill, the nation's largest privately held company, introduced a new logo intended to bridge the gap between farmers, food processors, and consumers. It also reflected Cargill's expansion from a company that engaged in trading raw commodities, such as grain, to one involved with food processing. By jettisoning the logo that the company had used since 1966, Cargill wanted the public to know that it cared about "nourishment, growth, and making connections" as well as helping customers solve problems. With revenues of nearly $50 billion, Cargill anticipated great profits in the twenty-first century by integrating commodity trading and food processing as the foundation for its corporate existence.

AGRICULTURAL ORGANIZATIONS

During the 1980s and 1990s, the Farm Bureau maintained its traditional philosophy, which supported the development of markets to increase agricultural exports and income. The American Agriculture Movement sputtered, however, while the National Farmers Organization sought to change its image of radicalism and distance itself from a bitter and turbulent history. Willis Rowell, an early NFO member and organizer, remarked that in the past "there were those who thought it would cure everything from ingrown toenails to communism. And, there were those who thought it could cause all of those things." John Garland, vice president of the NFO, reflected: "We have changed. We are changing. But people still believe we dump milk every Wednesday night and shoot hogs every Thursday afternoon."

By the early 1990s, with $750 million in annual sales, the NFO had clearly changed. But because many people still associated the NFO with radicalism, in 1992 the organization contemplated changing its name to the United Agri-Marketing Group to emphasize that it considered marketing more important than protest. It also substituted the goal of "group marketing" for "collective bargaining," which held negative and narrow connotations for some people. Now active in 35 states with 124 collection points for the marketing of grain, meat animals, and milk, the NFO believed that it needed only to control 15 percent of a commodity to influence market prices. The NFO achieved the most success in marketing milk, with about

5 percent of the national supply under its contracts, and it controlled 25 percent of the white-corn market. At the same time, the farmers in the NFO who sought the name change adamantly defended their past, but now the organization worked for "market power" by increasing its membership. The organization hoped to achieve this by appealing to farmers who did not want to leave their current agricultural organization but who favored the marketing approach of the NFO and would join it with a "dual membership," that is, become members of both organizations.

While the NFO leaders plotted, the Farm Bureau remained the largest agricultural organization. During the 1990s, its membership totaled more than 4 million. Farm Bureau insurance companies held combined assets of $1.7 billion, which led critics to charge that the organization gave more attention to that empire than to the needs of farmers. Although the Farm Bureau continued to lobby for farmers in state legislatures and Congress, particularly for tax relief, it also provided farm-management services and marketing information. In addition, it created an agricultural education division to help inform the public about agricultural and farm needs. The Farm Bureau also contemplated establishing a "war chest," that is, a legal fund to help its members fight environmental regulations imposed by federal and state governments that interfered with the right of farmers to work the land as they thought best.

During the early 1990s, the United Farm Workers continued to struggle. At the time of Cesar Chavez's unexpected death on April 23, 1993, the UFW membership had declined to about 20,000, a substantial drop since the 1970s, when membership peaked at approximately 109,000. Upon Chavez's death, his friends, such as Jerry Brown, former governor of California and presidential hopeful, called him a "visionary" who helped improve wages and living conditions of migrant workers. In contrast, the California Farm Bureau could only voice a faint, grudging respect when Bob Vince, an organizational leader, remarked: "He was a worthy advocate for his cause. He was no saint, but he certainly changed the face of California's agriculture." Chavez had once said, "If the union falls apart when I'm gone, I will have been a miserable failure, and it will have been a terrible waste of a lot of time by a lot of people." As the twentieth century drew to a close, only time would make a final judgment about the success or failure of the UFW.

THE UNITED STATES DEPARTMENT OF AGRICULTURE

By the early 1990s, some farmers suggested that the USDA be moved to the midwestern heartland to free it from the clutches of bureaucrats and politicians in Washington and make it more responsive to farmers. With the

USDA the fourth-largest government spender and the sixth-largest government employer, with more assets than all but three U.S. corporations, it seemed too bureaucratic and too distracted with other responsibilities to meet the needs of the family farmer, especially since only 20 percent of its 111,000 employees administered programs directly related to farming. Moreover, its failure at pesticide regulation resulted in that responsibility being transferred to the Environmental Protection Agency and shoddy meat and poultry inspections caused great embarrassment for the agency. In addition, even with a multibillion-dollar rural housing portfolio, USDA critics charged, the agency had not made a commitment to rural economic development and essentially served a rural America that had ceased to exist. Critics urged the consolidation and streamlining of the agency and a reorientation to serve the needs of rural America, not just the agricultural community.

At this time, some critics advocated a name change for the USDA because about 65 percent of the agency's budget supported food, nutrition, and related programs rather than price-support payments or loans to farmers. Michael Espy, the secretary of agriculture during part of the Clinton administration, a southerner and the son of a black extension agent, said: "I don't think there is a department in more dire need of being reinvented than the USDA." Espy pledged to make the agency "tougher, leaner . . . smaller without any diminishing of services to producers and to consumers" and "farmer-friendly, consumer-friendly and just plain friendly." Secretary Espy also planned to support increased USDA expenditures for rural water and sewer projects, low-income housing repairs, child-nutrition programs, food stamps, and environmental research. But little programmatic change could occur because the USDA operated under the Food, Agriculture, Conservation, and Trade Act of 1990, which remained in effect through 1995.

For many USDA critics, the "farm problem," however it was defined, required the agency to operate far differently than it had in the past. Some critics charged that farm programs hindered rather than aided farmers because they created agricultural debt by encouraging farmers to make large expenditures for chemical fertilizers, herbicides, and pesticides. Near century's end, the USDA saw the Food and Drug Administration encroaching on its role to guarantee high-quality healthful food products, while the Environmental Protection Agency increasingly concerned itself with a host of biotechnology regulations that influenced agriculture. As a result, the political power of the USDA declined, and the agency's attempt to reach nonfarmer constituencies alienated many farmers.

Increasingly, USDA critics demanded that rural policy take precedence over farm policy and that the agency put people rather than commodities first and stop assuming that rural problems could be solved by higher prices for wheat, corn, cotton, livestock, and other commodities. Less emphasis on high government price supports, USDA critics argued, and more attention to sustainable agriculture and the quality of rural life would end the consolidation of farms and benefit moderate-scale family farmers. A policy change such as this meant abandoning the foundation of the New Deal agricultural programs that emphasized economy of scale, research, and education to increase production and improve efficiency.

The USDA also continued to benefit white over black farmers. African-American agriculturists received only $0.51 for every dollar loaned to white farmers through agricultural programs. Put differently, black farmers averaged $21,000 less in loans ($21,986 compared to $42,898) than white farmers. The USDA claimed these loans were less because black farm operations were smaller and African-American agriculturists owned only 1 percent of the nation's farms. The USDA also vehemently denied discriminatory practices. Jerry Pennick, an official with the Federation of Southern Cooperatives Land Assistance Fund, which aided black farmers, countered, "Even though they have smaller operations, they're not getting enough to expand or make the best of the operations." Whatever the truth, USDA lending policy appeared to be discriminatory, and black farmers such as Luther Marable, Jr. in southwestern Georgia spoke for the few African-American farmers who remained during the early 1990s when he claimed that the inability to get USDA loans was "just racially motivated."

He was correct. USDA studies showed that African-American farmers were underrepresented on local committees that made loan decisions for various departmental programs, particularly for operating expenses, and by the Farm Service Agency, which had driven many black farmers into bankruptcy through discriminatory practices. Certainly, black farmers had virtually no representation on the local committees that made loan decisions; only 36 of 101 counties that had the largest number of black farmers had minority representation. One USDA investigation found that local officials were "rude and insensitive to black farmers," and that the projected crop yields used to determine loan amounts were calculated differently for black and white farmers and that African-American farmers were occasionally rejected because of "computational errors."

The Civil Rights Commission, which also studied the problem, reported that black farmers believed they were "subjected to disrespect, embarrassment and humiliation" by USDA officials. In the late 1990s, although Secretary of Agriculture Dan Glickman promised reform, black farmers demanded the return of foreclosed farms, the suspension of pending foreclosure proceedings, and damages paid to farmers against whom the USDA had discriminated. After considerable discussion with black farmers, the USDA reported, "The department did not treat African Americans and other minorities fairly in the area of lending and program delivery and the servicing of loans. That's just a fact."

In 1997, the National Black Farmers Association filed a class action suit for $2 billion to force the USDA to address its grievances and force the agency to resolve a backlog of approximately 900 complaints of African-American farmers over the denial of loans and other benefits since 1983. Although the USDA argued that the statute of limitations should nullify all complaints prior to 1994, in early January 1999, the federal government announced the settlement of the lawsuit for $375 million. The deal provided tax-free payments of $50,000 each to approximately 3,000 black farmers, and the federal government agreed to forgive debts on previous loans, which ranged as high as $100,000 for the farmers involved.

President Bill Clinton considered the settlement a major achievement in "ongoing efforts to rid the Agriculture Department of discriminatory behavior and redress any harm that has been caused by discrimination against African American family farmers." Secretary Glickman contended: "It is an agreement that will close a painful chapter in USDA's history and open a more constructive front in our efforts to see this department emerge as the federal civil rights leader in the 21st century." Although the settlement ended the claims from 1983, when the Reagan administration disbanded the Office of Civil Rights, until 1997, when it was restored, many black farmers had already lost their land due to financial distress. Those who remained claimed they had been "farming with the tips of our finger nails," and they did not expect their economic situation to change with the settlement. Black farmers remained small-scale marginal operators, with more than 36 percent tenants. They averaged about 127 acres each with 43 acres in crops, compared to an average farm size for whites of 434 acres with 162 acres in crops. Little wonder, then, that African American operators accounted for less than 1 percent of all farmers by the late 1990s, after a decline of 67 percent since 1900. Like Indian farmers, by the end of the twentieth century, African-American farmers had nearly ceased to exist.

AGRICULTURAL POLICY

By the 1980s, farmers also faced increased criticism from urbanites and consumers about the multibillion-dollar annual cost of the agricultural program that enabled farmers to receive payments from the government. Some critics advocated a complete abandonment of agricultural subsidies in favor of a free market, unprotected by tariffs for agricultural commodities. Others favored reform by eliminating payments based on production and marketing and the substitution of payments based on short-term economic need.

Many who supported government subsidies for agriculture advocated limiting that aid to the family farmer rather than permitting program participation by large-scale corporate farms or agribusinesses. Those who favored this position failed to understand that fewer than 3 percent of all farms were organized as corporations and operated by someone other than the family. Moreover, many farmers had been incorporated by the family for financial protection from bankruptcy. Even among the largest farms—those producing more than $500,000 in gross sales annually—only 6 percent were nonfamily corporations. Approximately 87 percent of all American farms were single-family owned and operated. Indeed, a family farm could total 100 or 100,000 acres. Size alone did not preclude family operation or arbitrarily create a corporation or agribusiness. Nevertheless, because 70 percent of all American farms grossed less than $40,000 in annual sales, many depended on government programs for survival.

Government programs remained controversial and expensive for the public, and not equally profitable for farmers. In 1985, a USDA survey of 1.6 million farms showed that approximately 15 percent of all federal agricultural payments went to the 2 percent of farmers who produced more than $500,000 each year and that 35 percent of the farms received an average of $24,000 in direct payments and price-supporting loans. Overall, 41 percent of the payments went to 31 percent of the farmers who did not experience financial distress. Put differently, with the top 15 percent of the farmers reaping as much federal subsidy as the bottom 80 percent, approximately $17 billion in payments went to a few hundred thousand of the nation's large-scale commercial farmers.

Inequities in the farm program seemed to abound. By 1987, large-scale commercial farms with annual sales over $500,000 made up only 1 percent of all farms but produced 38 percent of all sales. Yet these farms averaged $56,700 in payments for participation in federal commodity programs. Because eligibility to participate in federal programs did not depend on financial need, program costs remained high, and taxpayers increasingly questioned the subsidization of agriculture. Still, while the federal government intended agricultural programs to increase or keep prices at an acceptable level, farmers received only 25 cents of each food dollar. Even so, consumers charged that agricultural programs did little more than encourage surplus production and increase their food bills. Critics of government agricultural policy charged that farmers produced for the government rather than the market and that agricultural subsidies were merely "welfare" payments for farmers.

By the early 1990s, farmers and consumers alike advocated a major reform in government agricultural programs, but usually for different reasons. Some critics argued that a change in policy was needed and cited as an example that the USDA paid farmers for producing a crop that killed thousands of people each year while other federal agencies were warning about the dangers of smoking tobacco and tried to limit its use. Congressmen from the tobacco states, however, skillfully protected subsidies for their farmers, and the Government Accounting Office reported that the tobacco trade was at "cross purposes" with national health policy objectives. In 1992, tobacco production reached an estimated 1.68 million pounds, the highest since 1984. With prices at $1.72 per pound, it remained a profitable crop for those farmers who had allotments to raise it. Although cigarette smoking declined in the United States, exports of unmanufactured leaf to foreign processors increased about 12 percent, exceeding 226,500 metric tons. Foreign demand for cigarettes with a blend of American tobacco promised to keep the export market strong.

Farmers were particularly unhappy with soil conservation regulations. Government farm policy still required farmers to practice soil-conserving methods, at least in theory, to receive their subsidies. But in contrast to the past, the Soil Conservation Service—which became the Natural Resources Conservation Service in 1994—now had the responsibility of monitoring and enforcing the compliance of farmers. Those farmers who did not meet their soil conservation responsibilities faced the loss of their subsidy payments. Some farmers complained that soil conservation standards were too high, while others objected to governmental efforts to preserve wetlands by preventing drainage and the use of those lands for agricultural production.

The American Farm Bureau Federation became particularly vocal in expressing the opposition of many farmers to increased regulation by proclaiming that the federal government intended to impose a "super land-use control plan" and "how-to-farm" regulations to appease the environmentalists, who wanted to make the nation "completely pure and pristine" at the expense of agriculture. Farmers did not misuse the land, the Farm Bureau insisted. Farmers were the best stewards because their livelihoods depended on good conservation practices.

Although the Food, Agriculture, Conservation, and Trade Act of 1990 essentially maintained previous agricultural programs, this legislation expired in 1995, and a Republican-controlled House and Senate moved to radically change agricultural policy by ending production control payments while letting the market economy determine farm prices. The resulting Federal Agriculture Improvement and Reform Act of 1996, also known as the Freedom to Farm Act or FAIR Act, provided the foundation for a major change in agricultural policy. This legislation authorized a descending level of payments for production and price-supporting loan programs regardless of supply and commodity prices. While farmers continued to receive federal payments for participating in various programs, they could now produce as much as they desired without acreage limitations. Farmers, then, could have maximum production, plus government payments for participating in various programs. Conservation programs already under contract that required the retirement or idling of land would remain in force. At the end of seven years, however, the commodity price-supporting payments would end, and agricultural prices would depend on supply and demand. Presumably farmers would produce less of surplus, low-priced commodities and more products for which surpluses did not exist and which, therefore, brought higher prices, which meant increased farm income.

Upon signing the Federal Agriculture Improvement and Reform Act in early April 1996, President Bill Clinton said he did so with reservation because the legislation "fails to provide an adequate safety net for family farmers." He also contended, "The fixed payments in the bill do not adjust to changes in market conditions, which would leave farmers, and the rural communities in which they live, vulnerable to reductions in crop prices or yields." President Clinton's observation proved correct. In 1998, the high prices that farmers received for their crops and which exceeded government price supports collapsed due to overproduction and economic recession in Asia and Latin America, which sharply reduced foreign sales.

As agricultural prices fell below target prices, Congress responded by making "emergency" direct cash payments to farmers along with increased deficiency payments. In 2000 alone, farmers earned $32.3 billion in government subsidies—about 40 percent of their income. Eighty-one percent of those payments went to large- and medium-scale farmers, particularly those who raised wheat, corn, soybeans, rice, and cotton. Seven percent of the farms with at least $250,000 in sales received 45 percent of the subsidies, while 17 percent of the farms with sales from $50,000 to $250,000 received 41 percent of the payments. The large-scale farmers averaged $65,000 annually in government subsidies, while medium-size farmers garnered $22,000 each year. Small-scale farmers averaged only $4,000 annually in government payments.

By 2001, the federal government had spent $71 billion in direct cash payments to farmers for emergency aid, deficiency and conservation payments and other programs, or more than three times the money they were scheduled to receive, since 1996. Although farmers received large government subsidies, however, they garnered only 21 cents of each dollar that Americans spent for food by 2000, down from 32 cents in 1990. At the same time, food prices increased. Bread averaged 79 cents per loaf in 1995 but 87 cents in 1997, even though the cost of wheat per loaf dropped from 6 cents to 5 cents. The USDA attributed the increase in food prices to escalating labor costs; wages and salaries accounted for 49 percent of the cost of processing and manufacturing food by the late 1990s. Even so, in 1997, consumers spent only 10.7 percent of their income on food, down from 11.6 percent in 1987, and 22 percent in 1947.

In early 2002, as Congress debated and approved new farm legislation to take effect the next year, payments for commodity reduction as well as various conservation practices returned as basic features of agricultural policy. Supporters believed that an important "safety net" had been reestablished to protect if not guarantee farm income. Opponents believed the new agricultural legislation would encourage the production of commodities already in surplus, benefit only large-scale producers, and burden the federal treasury with exorbitant costs. They were correct on all counts. Most of the government subsidies still went to the largest-scale farmers, who needed it the least, that is, to the 20 percent of the farmers who produced approximately 90 percent of the nation's food. With the new farm legislation, known as the Farm Security and Rural Investment Act, providing $82 billion for various programs over ten years, farmers entered the twenty-first century still reliant on the federal government for economic support.

RURAL LIFE

At the close of the twentieth century, the Census Bureau classified only 782 counties nationwide as nonmetropolitan, that is, they did not contain a city of at least 2,500 inhabitants. Only 196 of those counties received more in crop subsidies than in Social Security payments, and 40 percent received more financial aid from federal food stamps than from farm programs. Moreover, while farming contributed at least 20 percent of the earnings in 877 rural counties in 1969, thirty years later it accounted for that percentage in only 258 counties. Furthermore, most federal money paid to farmers did not stay in rural communities, because the seed and equipment purchased with it was manufactured beyond the local area. Moreover, while some agribusiness firms created low-wage jobs in packing and processing plants, they did not instrumentally stimulate the economy of rural America, where a considerable number of people were either underemployed or unemployed.

In addition, by the 1990s, rural people constituted approximately 25 percent of the population. The largest share of the rural population lived in the South, where 44 percent of the population resided in communities with fewer than 2,500. By the early 1990s, 90 percent of rural workers were employed in nonagricultural work. With the decline in the farm population, rural areas suffered economic distress and a decline of services, particularly if a town of at least 20,000 was not relatively accessible.

As people and capital left the countryside, the rural economy became an "exit" economy. Moreover, 25 percent of all rural children lived in poverty, and

High yields of corn and other crops have made American farmers the most productive and efficient in the world. While abundant crops and government policy have kept food prices low, surplus production has also depressed farm prices. Many farmers have left the land for higher-paying jobs in the town and cities. *Photo courtesy of Pioneer Hi-Bred, Inc., a DuPont Company.*

infant mortality in the 320 poorest rural counties exceeded the national rate by 45 percent. Overall, the poverty rate of rural America exceeded the poverty rate of the cities. The rural poor made up 30 percent of the population that lived below the poverty line but received only 20 percent of federal, state, and local aid designated to help the poverty-stricken. Rural poverty became worse because the national economy suffered a severe and long-lasting recession. Low-tech manufacturing, mining, and oil jobs declined, and the median rural income fell to 73 percent of urban per capita earnings. No longer could rural poverty be associated with Appalachia and the South; it did not have regional boundaries, and some called rural areas America's Third World.

The rural poor received little support from the USDA or other government agencies, other than subsidies for school meals, food stamps, commodity handouts, and limited low-rent housing. The rural poor usually held low-paying jobs that provided just enough income to curb their eligibility for Medicaid and other benefits. And an independent spirit and stubborn pride often kept many from taking advantage of the limited services that were available. Because the rural poor lived in sparsely settled areas, they usually did not have ready access to relief offices and job-training programs. Food and housing became priorities, rather than medical care. Hunger and want were not strangers in a bountiful countryside.

Despite these problems, most Americans considered farming and rural life preferable to urban jobs and city living. Many people, not necessarily farmers, still believed Thomas Jefferson was correct when he wrote that "those who labor in the earth are the chosen people of God, if ever he had a chosen people." A poll showed that 58 percent agreed that farm life was more honest and moral than life in the rest of the country; 64 percent of those interviewed believed that farmers were more hardworking than other people; 67 percent responded that farmers had closer ties to their families than most other Americans; and 39 percent agreed with all three beliefs. Those who had incomes below the poverty line—less than $12,500 annually—were most likely to believe that farmers were especially moral and honest. Those interviewed who were 65 and older and whose living depended on agriculture or resided in communities with populations ranging between 10,000 and 50,000 usually contended that farmers had closer family ties than other Americans. Overall, the poll indicated that the poor and less educated would prefer to live on a farm if they could make a living, while 53 percent of the men and 44 percent of the women polled stated that they would like to live on a farm.

In 2000, 2.1 million farms averaging 434 acres supported an agricultural population of approximately 4.4 million, or 1.6 percent of the American people. Few urbanites had any possibility of living Jefferson's ideal of a virtuous agricultural life, and not many farmers, while revering their freedom and independence, had much hope of remaining on the land. American agriculture had become so productive that even fewer farmers were needed to meet the nation's food needs. Rich land, technological developments, scientific discoveries, and a supportive government, as well as risk-taking farmers driven by the desire for economic gain, had made American agriculture respected if not envied around the world. At the same time, the era of 160-acre farms was gone forever. American agriculture had changed substantially during the twentieth century. Relatively few farmers met the food needs of the nation and contributed significantly to foreign trade. Although few farmers wanted to think about the consequences of their continued high productivity and efficiency, the most significant ramification was that even fewer farmers were needed on the land.

In retrospect, then, the history of American agriculture has been the story of nearly constant change, for better or worse. By the turn of the twenty-first century most farm men and women sought profits through commercial production in order to enjoy a standard of living comparable to urbanites, and they worried about obtaining and keeping enough land to farm profitably. Access to markets, transportation, interest rates, and technology occupied their thoughts. They wanted education and good health for their children. Women still helped, but their roles primarily involved providing advice in the decision-making process, such as commodity marketing, rather than hard physical labor, and they often held jobs in town. Moreover, after the terrorist attacks in September 2001, farmers worried about the possible terrorist introduction of mad cow and hoof and mouth diseases that would destroy herds or disease-causing pathogens that would ruin crops, all of which would endanger public health.

At the turn of the twenty-first century, no one could say with certainty where the course of American agriculture might lead. Some farm men and women looked back to the past and saw bountiful years; others saw hardship. If the past was prologue, farmers could look to the future with confidence and expectation as well as desperation and fear.

SUGGESTED READINGS

Ahearn, Mary C., Janet E. Perry, and Hersham S. El-Osta. *The Economic Well-Being of Farm Operator Households, 1998–90.* U.S. Department of Agriculture, Agricultural Economics Report, 666, 1993.

Barnett, Barry J. "The U.S. Farm Financial Crisis of the 1980s." *Agricultural History* 74 (Spring 2000): 366–80.

Broehl, Wayne G. Jr. *Cargill: Trading the World's Grain.* Hanover, N.H.: University Press of New England, 1992.

Browne, William P. *The Failure of National Rural Policy: Institutions and Interests.* Washington, D.C.: Georgetown University Press, 2001.

Carlin, Thomas A., and Sara M. Maze, eds. *The U.S. Farming Sector Entering the 1990s: Twelfth Annual Report on the Status of Family Farms.* USDA, Economic Research Service, Agricultural Information Bulletin 587, 1990.

Changing Character and Structure of American Agriculture: An Overview. Washington, D.C.: General Accounting Office, 1978.

Danbom, David B. *Born in the Country: A History of Rural America.* Baltimore: Johns Hopkins University Press, 1995.

Effland, Anne B. "When Rural Does Not Equal Agricultural." *Agricultural History* 74 (Spring 2000): 489–501.

Fink, Deborah. *Cutting into the Meatpacking Line: Workers and Change in the Rural Midwest.* Chapel Hill: University of North Carolina Press, 1998.

Fite, Gilbert C. *American Farmers: The New Minority.* Bloomington: Indiana University Press, 1981.

Friedberger, Mark. *Farm Families and Change in Twentieth-Century America.* Lexington: University Press of Kentucky, 1988.

Giebelhaus, August W. "Farming for Fuel: The Alcohol Motor Fuel Movement of the 1930s." *Agricultural History* 54 (January 1980): 173–84.

Grove, Wayne A. "Cotton on the Federal Road to Economic Development: Technology and Labor Policies Following World War II." *Agricultural History* 74 (Spring 2000): 272–92.

Hadwiger, Donald F. "The Politics of Agricultural Abundance." *Agriculture and Human Values* 4 (Fall 1986): 99–107.

Holley, Donald. *The Second Great Emancipation: The Mechanical Cotton Picker, Black Migration, and How They Shaped the Modern South.* Fayetteville: University of Arkansas Press, 2000.

Hurt, R. Douglas. *Problems of Plenty: The American Farmer in the Twentieth Century.* Chicago: Ivan Dee, 2002.

_____. *The Rural South Since World War II.* Baton Rouge: Louisiana State University Press, 1998.

_____. *The Rural West Since World War II.* Lawrence: University Press of Kansas, 1998.

Kirkendall, Richard S. "Up to Now: A History of American Agriculture from Jefferson to Revolution to Crisis." *Agriculture and Human Values* 1 (Winter 1987): 4–26.

Lear, Linda J. "Bombshell in Beltsville: The USDA and the Challenge of 'Silent Spring.'" *Agricultural History* 66 (Spring 1992): 151–71.

Marcus, Alan I. *Cancer from Beef: DES, Federal Food Regulation, and Consumer Confidence.* Baltimore: Johns Hopkins University Press, 1994.

Shover, John C. *First Majority—Last Minority: The Transformation of Rural Life in America.* DeKalb: University of Northern Illinois Press, 1976.

APPENDIX

TABLE A.1. Agricultural price index (1910–1914 = 100)

	Prices received[a]	Prices paid[a]	Parity ratio
1910	104	97	107
1915	99	105	94
1920	211	214	99
1925	156	164	95
1930	125	151	83
1935	109	124	88
1940	100	124	81
1945	207	190	109
1950	258	256	101
1955	232	276	84
1960	239	300	80
1965	245	322	76
1970	274	382	72
1975	463	613	76
1980	614	950	65
1985	585	1,117	52
1990	681	1,265	54
1999	607	1,531	40

[a]Includes prices received on all commodities and prices paid on all items, including payment for services, interest, taxes, and wages.

SOURCES: Cheryl D. Johnson, *A Historical Look at Farm Income,* U.S. Department of Agriculture, Statistical Bulletin no. 807, 1990, 24; *Agricultural Statistics, 1992,* 382–84; *Agricultural Statistics, 2001,* IX–30, IX–32.

TABLE A.2. Farm characteristics, 1900–2000

	Farm population (thousands)	Percentage of total population	Number of farms (thousands)	Acres in farms (thousands)	Average size farm (acres)
1900	29,875	41.9	5,740	841,202	147
1910	32,077	34.9	6,366	881,431	139
1915	32,440	32.4	6,458	917,335	142
1920	31,974	30.1	6,454	958,677	149
1925	31,190	27.0	6,372	924,319	145
1930	30,529	24.9	6,295	990,112	157
1935	32,161	25.3	6,812	1,054,515	155
1940	30,547	23.2	6,102	1,065,114	175
1945	24,420	17.5	5,859	1,141,615	195
1950	23,048	15.3	5,388	1,161,420	216
1955	19,078	11.6	4,654	1,201,900	258
1960	15,635	8.7	3,962	1,175,646	297
1965	12,363	6.4	3,356	1,139,597	340
1970	9,712	4.8	2,954	1,102,769	373
1975	8,864	4.1	2,521	1,059,420	420
1980[a]	6,051	2.7	2,439	1,038,885	426
1985	5,355	2.2	2,292	1,012,073	441
1990	4,591	1.9	2,140	987,420	461
2000	4,400	1.6	2,172	942,900	434

[a]Reflects a definition change for a farm.

SOURCES: *Historical Statistics of the United States: Colonial Times to 1970,* pt. 1, 457; *Agricultural Statistics, 1989,* 371, 382; *Agricultural Statistics, 1990,* 365; *Agricultural Statistics, 1992,* 353; *Agricultural Statistics, 2001,* IX–2.

TABLE A.3. Farm income, 1910–1999

	Gross (millions)	Net (millions)	Average gross per farm
1910	$ 7,707	$ 4,176	$ 1,155
1920	$ 16,632	$ 7,795	$ 2,467
1930	$ 11,203	$ 4,259	$ 1,712
1940	$ 11,340	$ 4,482	$ 1,786
1950	$ 33,103	$13,648	$ 5,858
1960	$ 38,587	$11,211	$ 9,701
1970	$ 58,818	$14,366	$19,807
1980	$149,273	$16,136	$62,000
1990	$196,000	$51,000	$39,007
1999	$235,521	$43,397	$64,347

SOURCES: Cheryl D. Johnson, *A Historical Look at Farm Income,* U.S. Department of Agriculture, Statistical Bulletin no. 807, 1990, 4; *Agricultural Statistics, 1972,* 566; *Agricultural Statistics, 1982,* 432; *Agricultural Statistics, 1991,* 353; *Agricultural Statistics, 1992,* 352; Mary C. Ahearn, Janet E. Perry, and Hisham S. El-Osta, *The Economic Well-Being of Farm Operator Households, 1988–90,* U.S. Department of Agriculture, Agricultural Economic Report no. 666, 1993, vii; *Agricultural Statistics, 2001,* IX–39, IX–41.

BIBLIOGRAPHICAL NOTE

In addition to the suggested readings for each chapter, the introductory reader who wants to pursue more detailed and specialized research in American agricultural and rural history should be aware of several important indexes and bibliographies. No matter what the topic may be, the researcher should consult the annual index of *America: History and Life,* published by ABC-CLIO. This index provides the key to locating articles and abstracts, book reviews, and dissertations on an annual basis. It is particularly good for locating journal articles published since 1964. The index is available in hard copy as well as on-line. ABC-CLIO also has some abstract information available on CD-ROM. Researchers in agricultural history will also find the *Geographical Abstracts, Human Geography* (formerly *Geographical Abstracts D: Social and Historical Geography*) quite useful. Nor should advanced researchers overlook the *Journal of Economic Literature, Sociological Abstracts,* the *U.S. Serial Set Index,* and the *Agricultural Engineering Index.* The researcher should also consult the quarterly and annual bibliographies published in various journals such as *Journal of Southern History, Western Historical Quarterly, Technology and Culture,* and *Journal of American History.*

Researchers at college, university, and other institutional libraries may also be able to search the holdings of the National Agricultural Library via hard copy, on-line, or CD-ROM. The computer data base is known as AGRICOLA. Currently, CD-ROM disks contain material catalogued since 1970. No matter which format researchers may choose to use, they can search by author, title, or subject.

The researcher also should consult the indexes for *Agricultural History,* the journal of record for the field. Those indexes, *Agricultural History: An Index, 1927–1976,* and *Agricultural History: An Updated Author, Title and Subject Index to the Journal, 1977–1989,* have been published by the Agricultural History Center at the University of California, Davis.

This center also has published a number of topical bibliographies. Although some of these bibliographies are dated, the researcher will still find them useful. The most important of these bibliographies are listed below.

Bowers, Douglas. *A List of References for the History of the Farmer and the Revolution, 1763–1790* (1971).

Bowers, Douglas E., and James B. Hoehn. *A List of References for the History of Agriculture in the Midwest, 1840–1900* (1973).

Davis, Mary, Morton Rothstein, and Jean Stratford. *The History of California Agriculture: An Updated Bibliography* (1991).

Dethloff, Henry Clay, and Worth Robert Miller. *A List of References for the History of the Farmers' Alliance and Populist Party* (1988).

Edwards, Helen H. *A List of References for the History of Agriculture in the Southern United States, 1865–1900* (1971).

Guttadauro, Guy J. *A List of References for the History of Grapes, Wines, and Raisins in America* (1976).

McDean, Harry C. *A Preliminary List of References for the History of American Agriculture During the New Deal Period, 1932–1940* (1968).

Nordin, Dennis S. *A Preliminary List of References for the History of the Granger Movement* (1967).

Olsen, Michael L. *A Preliminary List of References for the History of Agriculture in the Pacific Northwest and Alaska* (1968).

Pickens, William. *A List of References for the History of Apiculture and Sericulture in America* (1975).

Rogers, Earl M. *A List of References for the History of Agriculture in the Great Plains* (1976).

______. *A List of References for the History of Agriculture in the Mountain States* (1972).

______. *A List of References for the History of Fruits and Vegetables in the United States* (1963).

Rossiter, Margaret W. *A List of References for the History of Agricultural Science in America* (1980).

Smith, Maryanna S. *A List of References for the History of the United*

States Department of Agriculture (1974).
Whitehead, Vivian B. *A List of References for the History of Agricultural Technology* (1979).

In addition to the bibliographies published at the Agricultural History Center at the University of California, Davis, advanced students should consult R. Douglas Hurt and Mary Ellen Hurt, *The History of Agricultural Science and Technology: An Annotated International Bibliography* (New York: Garland, 1994). This bibliography includes scientific and technological literature that historians will find useful and more traditional sources. It also provides an extensive review of recent literature on veterinary medicine as well as a variety of other topics such as public policy, ethics, and the Green Revolution. See also John T. Schlebecker, *Bibliography of Books and Pamphlets on the History of Agriculture in the United States, 1607–1967* (Santa Barbara, Calif.: ABC-CLIO, 1969). For materials published between 1900 and 1929, see Everett E. Edwards, *A Bibliography of the History of Agriculture in the United States,* USDA, Miscellaneous Publication no. 4, 1930.

Researchers interested in the agricultural history of important minority groups should consult Cecil L. Harvey, *Agriculture of the American Indian: A Selected Bibliography,* USDA, Bibliographies and Literature of Agriculture no. 4, 1979. See also Everett E. Edwards and Wayne B. Rasmussen, *Bibliography of the American Indians,* USDA, Miscellaneous Publication no. 447, 1941; Francis Paul Prucha, *A Bibliographical Guide to the History of Indian-White Relations in the United States* (Chicago: University of Chicago Press, 1977); *Indian-White Relations in the United States: A Bibliography of Works Published 1975–1980* (Lincoln: University of Nebraska Press, 1982); and Imre Sutton, *Indian Land Tenure: Bibliographical Essays and a Guide to the Literature* (New York: Clearwater, 1975). For an introduction to the agricultural literature of African Americans, see Joel Schor, *A List of References for the History of Black Americans in Agriculture, 1619–1980* (Davis: University of California, Davis, Agricultural History Center, 1981), and Irwin Weintraub, *Black Agriculturists in the United States, 1865–1973, An Annotated Bibliography,* Penn State University, Bibliographical Series no. 7, 1976. For sources on the history of women in agriculture, see Vivian B. Whitehead, *Women in American Farming: A List of References* (Davis: University of California, Davis, Agricultural History Center, 1987).

Every student of American agricultural history should know about the published census materials. In 1840, the federal government took the first agricultural census as part of the sixth population census of the United States. Although it did not include a wide variety of categories and the accuracy of the tabulations can be questioned, it at least gives an approximation of the statistical status of farming in America at that time. Thereafter, until 1950, the census of agriculture was included in the decennial census. In 1925, 1935, and 1945, however, a separate middecade census of agriculture was taken and continued every five years from 1954 to 1974.

In 1976, Congress authorized the agricultural census to coincide with the economic census for manufacturing, mining, construction, transportation, retail and wholesale trade, and the service industry. That change meant the agricultural census would be conducted cyclically for the years ending in 2 and 7. The agricultural census provides excellent tabulations for crops, livestock, and characteristics of farmers as well as summaries of population, technology, and other farm-related matters. This multivolume series includes coverage of each state and a summary volume. The researcher, however, must use caution when consulting the agricultural census because the definition of a farm has changed from time to time. Since 1979, for example, the federal government has defined a farm as any operation that produced at least $1,000 worth of agricultural products during the census year. This definition can include 2-acre strawberry patches or 20,000-acre cattle ranches.

The researcher should also consult George F. Thompson, *Index to Annual Reports of the United States Department of Agriculture for the Years 1838 to 1893, Inclusive,* USDA, Bulletin no. 1, 1896, which provides a detailed subject index, as well as his *Index to Authors with Titles of Their Publications Appearing in the Documents of the United States Department of Agriculture, 1841–1897,* USDA, Bulletin no. 4, 1898. See also *List by Titles of Publications of the United States Department of Agriculture from 1840 to June 1901, Inclusive,* USDA, Bulletin no. 6, 1902, and *List of Publications of the Agricultural Department, 1862–1902, with Analytical Index* (Washington, D.C.: Government Printing Office, 1904), an extensive index to the department's publications with citations and explanations of earlier indexes and a detailed section on the classification system of the superintendent of documents. This index was updated by Mabel G. Hunt, *List of Publications of the United States Department of Agriculture from January 1901 to December 1925* (Washington, D.C.: Government Printing Office, 1927). In addition, the researcher should see the *Index to Publications of the Department of Agriculture* for the years 1901–1925, 1926–1930, 1931–1935, and 1936–1940 as well as the

Agricultural Index, published from 1916 to 1964, which then became the *Biological & Agricultural Index.*

Agricultural and rural historians also should consult the annual volumes of *Agricultural Statistics.* This useful compendium not only provides the most recent statistical information but often shows trends over a decade or more. This reference work was first published in 1936. Before that date, this kind of information appeared in the *Yearbook of Agriculture,* which has appeared annually since 1894. In addition, consult *List of Bulletins of the Agricultural Experiment Stations in the United States from the Establishment to the End of 1920* and the eleven supplements for 1924–1927 and 1930–1944. See also the cumulative indexes for the *Experiment Station Record* for the period 1889–1946. This publication succeeds the *Digest of the Annual Reports of the Agricultural Experiment Stations for 1888.* Since 1942, the *Bibliography of Agriculture* continues to provide access to experiment-station research literature.

Researchers interested in the Forest Service, which is under the jurisdiction of the USDA, should consult Terry West and Dana E. Supernowicz, *Forest Service Centennial History Bibliography, 1891–1991* (Washington, D.C.: Forest Service, 1993).

INDEX